SPATIAL ABILITIES
Development and
Physiological Foundations

DEVELOPMENTAL PSYCHOLOGY SERIES

SERIES EDITOR
Harry Beilin

Developmental Psychology Program
City University of New York Graduate School
New York, New York

In Preparation

DAVID F. LANCY. *Cross-Cultural Studies in Cognition and Mathematics*

HERBERT P. GINSBURG. (Editor). *The Development of Mathematical Thinking*

Published

MICHAEL POTEGAL. (Editor). *Spatial Abilities: Development and Physiological Foundations*

NANCY EISENBERG. (Editor). *The Development of Prosocial Behavior*

WILLIAM J. FRIEDMAN. (Editor). *The Developmental Psychology of Time*

SIDNEY STRAUSS. (Editor). *U-Shaped Behavioral Growth*

GEORGE E. FORMAN. (Editor). *Action and Thought: From Sensorimotor Schemes to Symbolic Operations*

EUGENE S. GOLLIN. (Editor). *Developmental Plasticity: Behavioral and Biological Aspects of Variations in Development*

W. PATRICK DICKSON. (Editor). *Children's Oral Communication Skills*

LYNN S. LIBEN, ARTHUR H. PATTERSON, and NORA NEWCOMBE. (Editors). *Spatial Representation and Behavior across the Life Span: Theory and Application*

SARAH L. FRIEDMAN and MARIAN SIGMAN. (Editors). *Preterm Birth and Psychological Development*

HARBEN BOUTOURLINE YOUNG and LUCY RAU FERGUSON. *Puberty to Manhood in Italy and America*

RAINER H. KLUWE and HANS SPADA. (Editors). *Developmental Models of Thinking*

The list of titles in this series continues on the last page of this volume.

SPATIAL ABILITIES

Development and Physiological Foundations

Edited by

MICHAEL POTEGAL

Department of Behavioral Physiology
New York State Psychiatric Institute
New York, New York

 1982

ACADEMIC PRESS
A Subsidiary of Harcourt Brace Jovanovich, Publishers
New York   London
Paris   San Diego   San Francisco   São Paulo   Sydney   Tokyo   Toronto

ACADEMIC PRESS, INC.
111 Fifth Avenue, New York, New York 10003

United Kingdom Edition published by
ACADEMIC PRESS, INC. (LONDON) LTD.
24/28 Oval Road, London NW1 7DX

Library of Congress Cataloging in Publication Data
Main entry under title:

Spatial abilities.

(Developmental psychology)
Bibliography: p.
Includes index.
1. Space perception. 2. Space perception in children.
3. Space perception--Physiological aspects.
4. Orientation (Psychology) 5. Perceptual-motor
processes. I. Potegal, Michael. II. Series.
BF469.S67 1982   153.7'52   82-8879
ISBN 0-12-563080-8

PRINTED IN THE UNITED STATES OF AMERICA

82 83 84 85   9 8 7 6 5 4 3 2 1

Contents

〇〇

THE DEVELOPMENT OF SPATIAL ABILITIES

☐☐☐

SPATIAL ABILITIES OF ADULTS:
THE INFLUENCE OF HEREDITY AND GENDER

IV

CORTICAL CONTRIBUTIONS TO SPATIAL ABILITIES

List of Contributors

Numbers in parentheses indicate the pages on which the authors' contributions begin.

Arthur Benton (253), Department of Neurology and Psychology, University of Iowa, Iowa City, Iowa 52242

Maxine Berzok[1] (147), Developmental Psychology Program, Graduate Center, City University of New York, New York, New York 10036

J. Gavin Bremner (79), Department of Psychology, University of Lancaster, Bailrigg, Lancaster LA1 4YF, England

Michael C. Corballis (173), Department of Psychology, The University of Auckland, Private Bag, Auckland, New Zealand

Emerson Foulke (55), Perceptual Alternatives Laboratory, Department of Psychology, University of Louisville, Louisville, Kentucky 40292

Michael E. Goldberg (277), Laboratory of Sensorimotor Research, Building 36, National Institutes of Health, Bethesda, Maryland 20205

Roger Hart (147), Environmental Psychology Program, Graduate Center, City University of New York, New York, New York 10036

Beate Hermelin (35), MRC Developmental Psychology Unit, Drayton House, Gordon Street, London WC1 0AN, England

Mark G. McGee (199), Developmental Psychobiology Research Group, University of Colorado Health Sciences Center, Denver, Colorado 80206, and Brain Sciences Laboratories, National Jewish Hospital and Research Center, 3800 East Colfax Avenue, Denver, Colorado 80206

[1]Present address: The Association for the Advancement of Behavior Therapy, 420 Lexington Avenue, New York, New York 10170.

Nora Newcombe[2] (223), Department of Psychology, The Pennsylvania State University, University Park, Pennsylvania 16802

Niel O'Connor (35), MRC Developmental Psychology Unit, Drayton House, Gordon Street, London WC1 0AN, England

David S. Olton (335), Department of Psychology, The Johns Hopkins University, Baltimore, Maryland 21218

Herbert L. Pick, Jr. (107), Institute for Child Development, University of Minnesota, Minneapolis, Minnesota 55455

Michael Potegal (361), Department of Behavioral Physiology, New York State Psychiatric Institute, 722 West 168 Street, New York, New York 10032

Graham Ratcliff[3] (301), Department of Psychology, University of Reading, Whiteknights, Reading RG6 2A1, England

John J. Rieser (107), Department of Psychology and Human Development, Vanderbilt University, Nashville, Tennessee 37203

Rita G. Rudel (129), Department of Psychiatry, Psychiatric Institute, Colombia University, College of Physicians and Surgeons, New York, New York 10032

H. A. Sedgwick (3), Department of Vision Sciences, State University of New York, State College of Optometry, New York, New York 10010

[2] Present address: Department of Psychology, Temple University, Philadelphia, Pennsylvania 19122.

[3] Present address: Geratric Psychiatry and Behavioral Neurology Module, Western Psychiatric Institute and Clinic, 3811 O'Hara Street, Pittsburgh, Pennsylvania 15261.

Preface

As the cartoon on the next page indicates, verbal communication about spatial orientation is not without hazard, but this sort of hazard is probably inevitable in a field that has been the subject of study by people in a great diversity of disciplines—for example: architecture; city planning; and several branches of psychology, including developmental, environmental, experimental, and physiological psychology. The ability of organizms to organize the space around them and the nature of the brain mechanisms underlying this organization are also of concern to those with more applied educational and clinical concerns: special educators, occupational and physical therapists, orientation and mobility specialists, neurologists, and behavioral optometrists. Because it is difficult to keep up with the developments in all of these areas, investigators have also encountered the hazards of working in relative isolation. As a consequence, similar approaches to related issues have sometimes been developed in parallel without the benefit of full exchange among disciplines. It is the purpose of this book to make available surveys of research in a number of these areas within a single context in the hope that some of the parallels and convergences in research on human and nonhuman subjects, in the field and in experimental and physiological laboratories, will emerge.

The chapters of this book are grouped into sections focused on five different aspects of spatial orientation: sensory bases, development, the role of heredity and gender, cortical contributions, and subcortical contributions. However, a number of common themes span these sections. As Foulke, Hermelin, and

Sedgwick argue, there is little doubt that vision is the "spatial sense" par excellence, contributions from audition (Hermelin) and vestibular input (Potegal) notwithstanding. Is it possible, as Benton suggests, that the "mental maps" invoked by many authors are not amodal or supramodal schemata but are fundamentally visual representations (cf. Corballis) of space onto which other sensory inputs are mapped? If so, serious difficulties may be expected to beset those who must navigate in the absence of vision (cf. Foulke; Pick and Reiser). Is this mapping of other sense modalities onto visual representations one of the functions taking place in the parietal cortex (cf. Goldberg), damage to which in humans can produce profound orientation deficits (Benton, Ratcliff)?

Another common theme is the role of movement in spatial orientation. Movement can be a major factor in adults' acquisition of environmental information (Sedgwick); it may also be crucial in the development of spatial perception in the infant (Bremner). Hart and Berzok suggest that it is the decision aspect of active movement that is particularly important in increasing the comprehension of the large-scale environment in older children. The role of eye movement is of particular interest because of its obvious importance to vision. Possible functions of eye movement in spatial orientation are cited by Bremner, Hermelin, and Rudel; experimental and physiological aspects are discussed in great detail by Sedgwick and Goldberg.

Benton, McGee, Newcombe, and Ratcliff all emphasize the necessity of identifying the various basic spatial abilities. One possible candidate, mental rotation, is discussed in detail by Corballis. An allied theme concerns the

Drawing by Stevenson; © 1976 The New Yorker Magazine, Inc.

spatial strategies or schemata adopted by subjects. These strategies can be empirically evaluated in the laboratory by the use of appropriate transfer tests, as work with human babies (Bremner) and rats (Olton) has shown. This avenue of research, reviewed by Bremner as well as Pick and Reiser, has supported Piaget's arguments for a developmental shift from egocentric to external referent strategies. Egocentric strategies carry with them the computational burden that the internal coordinates of spatial locations must be updated every time the subject moves. For this reason Bremner as well as Hart and Berzok argue that the child abandons egocentric strategies as it matures. However, Potegal and Pick and Reiser consider the alternative possibility that an updating system exists or develops that allows continued use of the egocentric strategy in at least some situations. Conflicting views such as these are the least of the hazards that beset the course of research in spatial orientation; the research efforts that resolve these conflicts will certainly advance our understanding of the spatial abilities.

Acknowledgments

A number of the chapters in this volume were first presented at a conference entitled "The Neural and Developmental Bases of Spatial Orientation," which was held at Teachers College, Columbia University, November 15–18, 1979. The conference, one in a very successful series sponsored by the Neuroscience and Education Program of Teachers College, was made possible only through the kindness of Antoinette M. Gentile, who was then director of the program. I would like to thank Feriha Anwar, Lauren Harris, and John O'Keefe for also participating in the conference. Many other people contributed to the success of the conference, but none more so than Cecily Dell. As administrative assistant she coped heroically with the unrelenting details of conference planning and execution (she also brought to my attention the accompanying cartoon). I would also like to thank Bernard Cohen, Lee Ehrmen, and Celia Fisher, each of whom reviewed selected chapters.

SENSORY BASES OF SPATIAL ABILITIES: ORIENTATION WITH AND WITHOUT VISION

H. A. SEDGWICK

1

Visual Modes of
Spatial Orientation

BASIC CONCEPTS

Visual Spatial Orientation

Every organism has an environment—an ecological niche to which both the evolution of its species and the experiences in its own life have more or less successfully adapted it. This is true of the human species as well, even though the seemingly boundless ingenuity of our technology is continually creating new tests of our ability to adapt to novel environments (outer space, the bottom of the sea) or to interact with our environment in new ways (moving around by automobile or airplane instead of by foot). One essential aspect of our successful adaptation to an environment is our ability to perceive accurately our physical spatial relation to the environment and to monitor accurately the changes in that relation that occur as we move about. This broad adaptive function, which can be subdivided into many, more narrowly defined functions, is what I shall refer to as *spatial orientation*.

All of our senses (even smell and taste) can be used to some degree in spatial orientation, but surely none is more important in this respect than vision (see Hermelin & O'Connor, Chapter 2, this volume). In this chapter, concerned with how vision can contribute to spatial orientation, we shall investigate not

3

SPATIAL ABILITIES
Development and Physiological Foundations

only our visual capacities but also the characteristics of the physical environment in which we live.

The idea that the study of visual perception must consider the environment—
or ecological setting—to which vision is adapted has been energetically advocated
and developed over the past 30 years by J. J. Gibson (1950, 1961, 1966, 1979),
his students, and his associates. Because this chapter builds upon Gibson's
theoretical framework, it is first necessary to consider briefly two key concepts of Gibson's theory: *ecological optics* and the *optic array*.

Ecological Optics

Ecological optics is Gibson's term for the investigation of the information that
is available in the light reaching the eye of an observer in a normal environment
(Gibson, 1961). This investigation is *ecological* because it concentrates on the
normal environment with all of its complexities, regularities, and constraints; it
is a branch of *optics* because it studies the behavior of light in such environments. Information is available in the light reaching the eye because the surfaces of a complex environment systematically structure the light that they
reflect. Thus, the structure of the environment—its spatial layout—can be
mathematically specified by the structure in the light reaching the observer's
eye.

Optic Array

The *optic array* is Gibson's term (1961) for the structured array of light
converging on the observer's eye from the environment. In principle, the optic
array is there even if the observer is not; we can speak of a *point of observation* as
a location in space where an observer's eye could be placed, and we can then
define the optic array at that point of observation. Just as the environment
surrounds the observer, so the optic array is *omnidirectional*—structured light
converges on the point of observation from all directions.

Visual perception, according to Gibson, is the process of picking up the
information that is available in the optic array. This implies that we cannot
understand visual perception unless we understand the structure of the optic
array, and we cannot understand the structure of the optic array without
studying the natural structure of the environment.[1]

---

[1]Critics of Gibson's approach to the study of visual perception have argued that the approach
does not give enough weight or attention to the specifics of the organism's contribution to the
process. For a recent statement of this argument, along with responses from a number of critics and
supporters of Gibson's position, see Ullman(1980); also see Johansson, Hofsten, and Jansson (1980).

Two Visual Modes of Spatial Orientation

For land-based creatures like ourselves, the main features of our physical environment remain fixed and unchanging, for the most part. Because of our need to respond to and interact with our environment, however, we spend much of our time moving around in it. Thus, our own mobility changes our spatial relation to our environment and gives rise to the problem of determining our spatial orientation. The different types of motion of which the human observer is capable can be associated with different visual modes of spatial orientation.

The following analysis centers on the possible motions in space of the basic receptor organ of the human visual system—the eye. Any motion of a rigid physical body, such as the eye, can be analyzed as the sum of no more than six simple motions. Three of these motions are rotational: The eye can rotate to the right or left, up or down, and clockwise or counterclockwise. In every purely rotational motion the center of rotation of the eye remains fixed in space. The other three simple motions are translational. In every purely translational motion the orientation of the eye remains fixed while the eye as a whole, including the center of rotation, moves through space; the center of rotation can translate to the right or left, upward or downward, and forward or backward. Corresponding to the two classes of motion, we can distinguish two visual modes of spatial orientation—a rotational mode and a translational mode. Each mode presents its own problems for spatial orientation and its own sources of visual information.

THE ROTATIONAL MODE: SAMPLING THE OPTIC ARRAY

Characteristics of the Rotational Mode

We shall consider as belonging to the rotational mode any movements of the observer that cause the eye to rotate around its own optical center. As already noted, any given point of observation has an optic array that is omnidirectional. The observer's eye, however, has a field of view, delimited by both the structure of the head and the anatomical extent of the retina, that takes in a roughly elliptical segment of the optic array measuring about 150 degrees horizontally by 120 degrees vertically. This is about one-ninth of the total angular area of the optic array.[2] Even within this delimited field of view the eye's sensitivity varies greatly. Vision is best for that portion of the optic array that is pro-

[2]In binocular vision, the two fields of view only partially overlap, so the total horizontal extent of the combined field of view is about 180 degrees.

jected on the center of the retina—the fovea—and falls off sharply with increasing angular distance from the point of fixation of the retina.

In viewing a stationary environment, the eye normally *samples* the optic array by changing its point of fixation about three or four times every second (Yarbus, 1967, pp. 194–196). Each change in fixation is accomplished by a very rapid rotation of the eye called a *saccade*. These saccadic eye movements do not affect the boundaries of the field of view; they only select which portions of it are to be registered in most detail. A new field of view is obtained by rotation of the observer's head. In normal vision, eye and head rotations often occur together in a complexly coordinated fashion that changes both the point of fixation and the boundaries of the field of view.

Because these movements are primarily rotational, they have little effect on the position in space of the point of observation, which is at the optical center of the eye.[3] Consequently, the structure of the optic array at the point of observation remains essentially unchanged; the purpose of these rotational movements is to sample different portions of that unchanging optic array.

Although these eye rotations, sometimes accompanied by coordinated head and body rotations, serve the important visual function of allowing different portions of the optic array to be sampled with maximum sensitivity, this sampling activity gives rise to a series of related problems concerning spatial orientation: How is the perceived stability of the world maintained during eye and head movements? How is the perception of the constant visual directions of objects maintained during eye and head movements? How is the observer's orientation to the vertical and horizontal maintained during or between these rotations? Several common characteristics of the optic arrays produced by normal environments provide potential visual information relevant to these questions. The potential visual information often seems not only to be used but to be dominant over potential sources of information from other senses (see Hermelin & O'Connor, Chapter 2, this volume). Three of these characteristics of an ecological optic array are discussed in the following sections.

Optic Array Stability

Because the basic structure of the environment is normally stationary and unchanging, the optic array at any point of observation within it provides a

---

[3]A slight approximation is made here because the point in the eye that corresponds to the center of the optic array is the center of the eye's entrance pupil, which is about 1 cm in front of the eye's center of rotation (cf. Nakayama & Loomis, 1974). A somewhat rougher approximation arises because the axes around which the head rotates do not quite coincide with the center of rotation of the eye. Thus, the eye is normally carried to a slightly different point in space when the head turns. This translation is sufficiently slight, however, that for most purposes we can consider the optic array to remain practically unchanged as the head rotates.

stable visual framework for an observer. Evidence for the importance of this optic array stability in visual spatial orientation comes from a variety of experimental situations in which striking losses of spatial orientation are produced by artificially making the observer's optic array unstable.

STABILITY DURING SACCADES

As has already been noted, the eyes scan the stationary environment by means of a series of steady fixations punctuated by saccades. During these saccades, the retina sweeps across the optic array at velocities of hundreds of degrees of visual angle per second, and it has long been regarded as a puzzle why the world is not perceived to jump about as the eyes move. A process of "saccadic compensation" was postulated whereby the perceptual system would be notified by the oculomotor system of what the eyes were doing and so would be able to discount those retinal motions produced by the eyes' movements. Whether, and how much, such "compensation" helps in stabilizing the perception of the world continues to be a controversial question (for some recent discussions of this issue see MacKay, 1973; Matin, E., 1976; Matin, L., 1972, 1976; Shebilske, 1977). One interesting finding that has emerged from this controversy shows how the visual system's presumption of optic array stability enters into the processing of visual information during saccades. By measuring the eyes' movements and using mirrors or electronics to control the position of the optic array it is possible to make the optic array actually rotate in synchrony with the saccadic sweep. Thus, both the eyes and the visual environment rotate at the same time, either in the same or opposite directions. The interesting result is that the visual system's sensitivity to rotations of the optic array is greatly reduced when those rotations occur during saccades (Bridgeman, Hendry, & Stark, 1975; Mack, 1970). The amount of optic array rotation that will go undetected is tied to the size of the saccade; generally, optic array rotations of up to about 30% of the eyes' rotation are not noticed. This indicates that whatever compensation for eye movements occurs during saccades is rather rough. The perceived stability of the world when the eye saccades thus appears to be due in part to the visual system's "assumption" that retinal motions occurring during saccades are due to rotations of the eye rather than to rotations of the optic array.

SACCADIC RECALIBRATION

A second effect of artificially rotating the optic array during a saccade is that the saccade is no longer accurate. The saccade is a *ballistic* movement, that is, it occurs so rapidly that there is no time for the visual system to receive corrective feedback and relay it to the oculomotor system during the course of the saccade's execution. Thus, when the optic array is rotated during the saccade, the eye arrives at where its new fixation target used to be, rather than where it is now. As we have seen in the previous section, the "error" that has occurred appears, within fairly broad limits, to be attributed to an inaccuracy in the eye movement rather than to a rotation of the optic array; the optic array is presumed to be stable. If the same "error" is artificially produced on repeated saccades, the oculomotor system tends to recalibrate itself quickly, adjusting the size of its saccades to compensate for their apparent inaccuracy (Henson, 1978; McLaughlin, 1967; Miller, Anstis, & Templeton, 1981).[4]

INDUCED MOTION

If an observer in a darkened room fixates on a stationary point of light that is surrounded by an illuminated rectangle that is moving slowly to the right, then the observer is likely to perceive the rectangle as being stationary and the spot as moving slowly to the left. This is a very simple example of the phenomenon of *induced motion*. The larger, surrounding figure is taken by the visual system as a stable framework; the relative motion between it and the smaller figure it surrounds is perceived as belonging to the surrounded figure. In this case the presumed stability of the optic array operates in a somewhat more complicated situation than we have yet considered. When there are relative displacements within the optic array the visual system must decide, in effect, which portions of the optic array come from the underlying, stable structure of the environment and which portions come from objects that may be in motion in the environment. As in the simple example of the rectangle and the spot, it is commonly the larger, surrounding figure that is taken as the stable frame of reference. More complicated displays, however, may produce more complicated, less easily understandable perceptions. Induced motion, described and studied by Duncker (1929), has been the subject of a very large number of investigations which have revealed a great variety of related phenomena of motion perception but have not yet succeeded in organizing all of these phenomena into an entirely coherent theoretical account (for some recent studies from varying theoretical perspectives see Day, 1978; Gogel, 1977; McConkie & Farber, 1979; Nakayama & Tyler, 1978; Schulman, 1979).

ADAPTATION OF THE VESTIBULO-OCULAR REFLEX

Rotation of the head is registered by the semicircular canals and the vestibular system generates a counterrotation of the eyes that has the effect of holding them fairly stationary with respect to the environment. This response, known as the *vestibulo-ocular reflex* (VOR), occurs in the dark as well as in the light, although in the light it is somewhat more accurate, presumably because visual

---

[4]In large saccades the eye appears to be deliberately programmed to stop slightly short of its target. Henson (1978) discusses the functional value of this "undershooting."

information is then also present to help stabilize the eyes relative to the optic array.

If an observer wears reversing prisms, which, when the head rotates one way, make the optic array rotate the other, the VOR no longer serves its function as a stabilizing response. Instead, the VOR makes the situation worse by driving the eyes in a direction exactly opposite to the one in which the optic array is rotating. This effect is very disturbing to observers, who sometimes experience discomfort and even nausea. But surprisingly, the VOR can change to adapt to this new situation. Over a number of hours the magnitude of the VOR progressively decreases until it finally reverses direction. That this is a genuine adaptation of the VOR, not just an effect of its being temporarily overridden by the visual system, is shown when the observer is now tested in the dark; the reversed VOR persists (Gonshor & Melvill Jones, 1976). This adaptation phenomenon shows that the presumption of optic array stability is not limited to the visual system but also underlies the perceptual integration of visual information with spatial information from other sensory systems.

CIRCULAR VECTION

If an observer's visual environment, a cylindrical room for instance, rotates around the observer while the observer remains stationary the observer will at first perceive the situation correctly. But after a few seconds, the observer will begin to perceive herself or himself as rotating in the opposite direction while the environment appears to become stationary. This illusion of self-motion, referred to as *circular vection*, strikingly illustrates how the visual system is set to register the optic array from the environment as being stable and motionless. The few seconds' delay in the onset of circular vection is attributed to vestibular input from the observer's semicircular canals, which of course are indicating that there is no motion of the observer's head. The semicircular canals are primarily sensitive to rotary accelerations or decelerations, however, rather than to rotations of constant velocity. Thus, as the rotation of the environment continues, the visual input overrides the vestibular input and the continuing motion in the optic array becomes attributed perceptually to rotation of the observer. The phenomenon of circular vection has been known for many years, but has only recently been the subject of intensive investigation (cf. Dichgans & Brandt, 1978; Lackner, 1978, for recent reviews). One of the most interesting of the recent results (Brandt, Wist, & Dichgans, 1975) shows that circular vection is much more readily induced when the background rather than the foreground of the environment is in motion. This suggests considerable subtlety in the visual system's ability to extract information about a presumably stable background environment from the optic array. Like the VOR, circular vection shows an adaptive reversal to the wearing of reversing prisms (Oman, Bock, & Huang, 1980).

Optic Array Anisotropy

The optic array from a normal environment is characteristically and systematically different in different directions. Such a condition of having unequal properties along different axes is known as *anisotropy*. The anisotropy of the optic array provides very useful visual information about the orientation of the observer to the environment. Information concerning the principal orientations, vertical and horizontal, and their associated directional dichotomies, up–down and left–right, is discussed in the following sections.

VERTICAL

The pervasive effects of gravity in the natural environment produce many indications of verticality (e.g., tree trunks, waterfalls) that are projected as potential visual information in the optic array. Carpentered environments made by humans are, of course, also filled with such visual indications of the vertical. Vertical contours in the environment always project as vertical contours in the optic array (see Figure 1.1). This visual information normally allows observers to orient themselves to the vertical with considerable accuracy. An observer's vestibular system also provides information about the orientation of the observer's head relative to the direction of gravity, so that even in the dark the observer is capable of maintaining a vertical orientation. A great deal of research has been done over the years in which these two sources of information, visual and vestibular, are artificially put in conflict with each other—for instance, by seating the observer in a room that can be tilted independently of the observer. In general, the results have been variable (for reviews, see Howard & Templeton, 1966; Lackner, 1978). Some observers respond more in terms of the visually specified vertical; others respond more in terms of the vestibularly specified vertical. Perhaps the most important point to be made from an ecological point of view is, as Gibson (1952) has suggested, that visual and vestibular information for vertical normally *covary*, rather than conflict. That is, both sources of information normally specify the same vertical and so are mutually reinforcing rather than contradictory.

UP AND DOWN

The vertical orientation can be differentiated into two directions, up and down (e.g., an object moving along a vertical path can be moving in either of two directions: upward or downward). Again because of the effects of gravity

Figure 1.1. Every vertical contour in the environment appears as a vertical contour in the optic array or in a perspective representation such as this, but the projections of contours that are horizontal in the environment may have any orientation, even vertical. (Notice, however, that the nonhorizontal projections of parallel horizontal contours always converge, if extended, to some point on the horizon; this invariant rule helps to specify that the contours are horizontal.) (From Vredeman de Vries, 1604/1968, p. 28.)

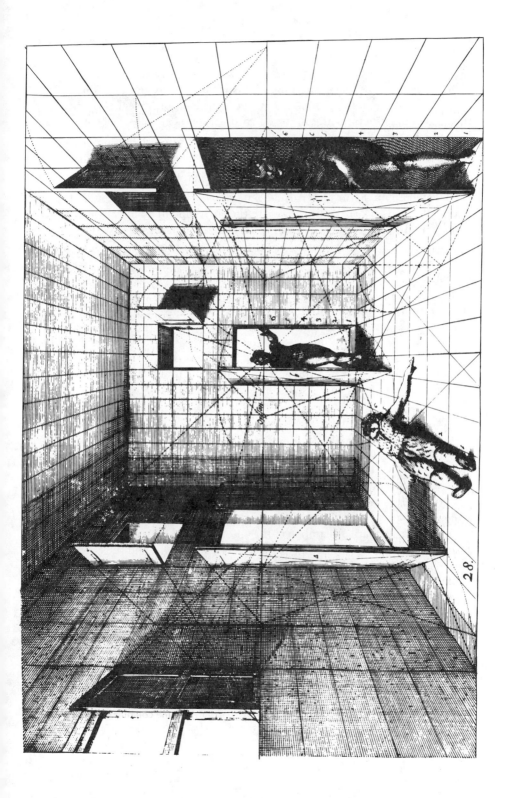

these two directions are clearly differentiated by many visible features of the natural environment (objects rest on the floor, but hang from the ceiling, etc.). Adults and even children normally have little difficulty in selectively orienting themselves or objects in the up or down direction (Corballis & Beale, 1976; Rudel & Teuber, 1963).

HORIZONTAL

Because, again, of the effects of gravity, identifiable horizontal contours are abundant both in natural, terrestial environments and in carpentered ones made by humans. Horizontal contours in the environment do not always project as horizontal contours in the optic array, however. In accordance with the laws of perspective, horizontal contours that are receding from the observer will project as more or less slanted contours in the optic array. The extreme cases are horizontal contours that recede directly away from the observer (like spokes radiating out from the hub of a wheel); these horizontal contours actually project as vertical contours in the optic array (see Figure 1.1). The horizontal orientation is, nevertheless, normally specified visually in the optic array. This is so primarily because these same laws of perspective establish regular mathematical relations between the optic array and the environment, permitting the visual system to determine which environmental contours are horizontal (see page 20 for further discussion related to this point).

RIGHT AND LEFT

The directions right and left are usually thought of as a differentiation of the horizontal orientation into two opposite directions. Right and left, however, are primarily defined in terms of the observer's own body, rather than in terms of the stable environment. Thus, what was on the observer's right becomes on the observer's left when the observer turns around. The environment thus lacks stable designations of right and left except in a few relatively artificial situations where a standard orientation of the observer is assumed so that permanent designations of right and left can be made (e.g., stage right and stage left in the theater). Consequently, except in these special circumstances, there is no available optic array information for distinguishing right and left. Correspondingly, observers, particularly children, have considerable tendencies to confuse right and left in a wide variety of situations (see Corballis & Beale, 1976, for a detailed discussion of this issue and a review of the experimental literature).

Optic Array Density

The optic array is *densely structured.* A normal environment surrounds the observer with textured objects and surfaces so that the optic array is usually filled in every direction with dense, complexly structured projections (the sky on a clear day is an obvious exception to this). A variety of evidence points to the importance of a richly structured optic array in maintaining spatial orientation.

SURFACE PERCEPTION

Much of the visible portion of the environment can be described as a layout of surfaces (Gibson, 1979, pp. 22–32). It is the optical texture of surfaces— that is, their uneven reflection and refraction of light—that makes them visible. The density of their optic array structure provides potential visual information for the existence of continuous, solid surfaces. Experimental work initiated by the Gestalt psychologists has shown that if all visible texture (or microstructure) is removed from the optical array, resulting in a *ganzfeld,* then there is no visual surface perception. Instead, observers variously report that they see a "space-filling fog" or "nothing at all" (Gibson & Waddell, 1952; Koffka, 1935, pp. 110–124). Gibson went on to show that the presence in the optic array of contours belonging to a continuous surface does not in itself result in the perception of a solid surface; the contours must have a certain density before they are perceived as lying on a continuous, solid surface (Gibson, Purdy, & Lawrence, 1955; Gibson, 1979).

AUTOKINETIC EFFECT

The perceived stability of the visual environment also depends upon the optic array having a certain density of structure. If all structure is removed from the optic array then the visual environment becomes unstable perceptually. For example, if a stationary spot of light is displayed in an otherwise completely dark room, the spot does not maintain a stationary appearance. After a little while it appears to begin to drift around, usually appearing to follow a rather random wandering path; the spot may seem to move over fairly large distances and at fairly high velocities. This illusion of movement is referred to as the *autokinetic effect.* It has been very extensively studied and is now known to be affected by many variables, such as position of gaze and suggestion (see Levy, 1972; Royce, Carran, Aftanas, Lehman, & Blumenthal, 1966, for recent reviews). The autokinetic effect clearly demonstrates that maintaining stable visual directions of locations in the environment depends upon having some degree of structure in the optic array. Surrounding the spot by a visible framework tends to eliminate the autokinetic effect (it also creates the possibility of induced motion, however; cf. section on induced motion, page 8).

UNDERCOMPENSATION OF SMOOTH PURSUIT EYE MOVEMENTS

When a spot of light in the dark actually does move, and the eye rotates to pursue it, the extent of the spot's motion may be considerably underestimated (Festinger, Sedgwick, & Holtzman, 1976; Sedgwick & Festinger, 1976; Stoper,

1973). It appears that whatever information the oculomotor system supplies to the perceptual system about the eye's motion during smooth pursuit in the absence of a visual background is rather inadequate, at least for this task. This misperception of motion can lead to misperception of direction as well, as becomes apparent if an observer tries to continually point to a spot while it is moving in the dark (Farber, 1979).[5]

THE TRANSLATIONAL MODE: EXPLORING THE ENVIRONMENT

Characteristics of the Translational Mode

We shall use the term *translational mode* to refer to all movements of the observer that result in a displacement, or motion, through space of the optical center of the observer's eye. Such translational motions occur whenever the observer moves around in the environment. As the observer's point of observation translates through the environment, the optic array available at that point of observation systematically changes, conveying new potential visual information to the observer about the environment (see Figure 1.2). Thus, it is through the translational mode that the observer explores new areas of the environment.

This exploratory activity gives rise to a series of related problems in spatial orientation: How is the stable layout of the environment perceived when the structure of the light reaching the eye is constantly changing? How is the observer's position in relation to the layout of the environment perceived? How are the observer's direction and speed of movement through the environment perceived? How does the observer relate to all those portions of the environment that are hidden from sight at any given time?

Such questions, particularly those concerned with the perception of the spatial layout of the environment, have traditionally belonged to the field of study in experimental psychology referred to as *space perception*. Concern with such questions is older than experimental psychology itself, however, going back at least to Descartes (1637/1965). Many of the researchers who have surveyed and added to this area (e.g., Berkeley, 1709/1910; Brunswik, 1956; Carr, 1935; Helmholtz, 1910/1962; Hochberg, 1971; Ittleson, 1960) have organized their expositions in terms of a list of *cues* (see Harper & Boring, 1948) that can be used by the observer to interpret an essentially two-dimensional retinal image in terms of a three-dimensional world. These cues can be classified as either *binocular* or *monocular,* depending upon whether or not

---

[5]The precise circumstances under which such misperceptions of motion and direction arise remain to be elucidated. Some researchers have found fairly accurate directional responses (Hansen, 1979) and relatively little misperception of motion (Mack & Herman, 1972, 1978) under somewhat similar stimulus situations.

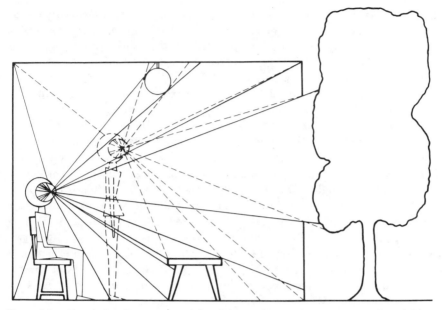

Figure 1.2. Translational movement of the observer changes the optic array. The solid lines indicate the optic array for the seated observer, and the dashed lines indicate the altered optic array after the observer stands up and moves forward. The difference between the two arrays is specific to the difference between the points of observation, that is, to the path of locomotion. (From James J. Gibson: *The Ecological Approach to Visual Perception*. Copyright © 1979 Houghton Mifflin Company. Used by permission. P. 72.)

binocular vision is necessary to obtain them. The monocular cues often are classified further according to whether they can be obtained by a *stationary* observer or are only obtained by *moving* observers (see Gibson, 1950, pp. 19–22, for a critical discussion of this approach).

Our analysis has a somewhat different emphasis, and consequently takes on a somewhat different organization. Because I am proposing that motion of the observer is central to the problems of spatial orientation (cf. Hart & Berzok, Chapter 7, this volume) we shall consider the stationary observer as merely a special case (in mathematical terms, a degenerate case) of the moving observer—the case for which the velocity of motion happens to be zero. All of the information that we shall consider will be information that is available to an observer underoing translational motion. We shall distinguish, however, between information that *only* exists when the observer has a nonzero translational velocity and information that continues to exist in the degenerate case where the velocity of the observer is zero. Also, we shall not consider the information that is available only to binocular observers; this is not to belittle the importance of such information, which has traditionally been regarded as the primary visual resource for space perception. Current treatments of binocular vision do not say much about complexities arising for moving observers in natural environments, and these complexities are the main focus of the present discussion (some recent reviews of binocular space perception are offered by Foley, 1978, 1980; Ono & Comerford, 1977).

Gibson (1958, 1979, pp. 182–184) has proposed that the optic array changes that occur as the observer moves through the environment contain potential visual information about both the stable layout of the environment and the observer's path of motion through that environment. In the remainder of this chapter we shall examine the evidence for this proposal. Our analysis of the potential visual information for spatial orientation in the translational mode will be approached through a *typology of environments*. Advancing step-by-step from the consideration of unrealistically simple environments to the analysis of more naturalistically complex environments will make it easier to consider systematically the variety of potential visual information that is available as an observer explores an environment. As we shall see, what kind of visual information is available to an observer depends upon the specific characteristics of the observer's environment. (For a related, but not identical, nomenclature see Gibson, 1979, pp. 33–35.)

Simple, Uncluttered, Open Environments

Simple, uncluttered environments are defined here as environments in which every surface in the environment is fully projected to the optic array at any point of observation in the environment. These environments are of two types: *open* and *enclosed*. An example of an open environment is a flat plane extending endlessly in every direction; an example of an enclosed environment is an empty room. A great deal of potential visual information for spatial orientation exists even in these two rather minimal environments. Let us consider the open environment first.

TEXTURE OF THE GROUND

The structure of the optic array reflects the structure of the environment. If the environment is nothing more than a flat plane extending endlessly in every direction, it might seem at first to have no features that would project as structure in the optic array. Yet even this environment has what Koffka (1935) refers to as *microstructure,* or texture. The texture of the ground depends, of course, on its composition. A grassy plain, a dirt field, an expanse of sand or pebbles—each has its own distinctive texture and, correspondingly, its own distinctive *optical texture* in the optic array. Nevertheless, there is an important ecological generalization that we can make about most naturally occurring ground textures: The size and distribution of their texture elements tend, statistically, to be homogeneous. This means, for instance, that for an observer standing on a grassy field, the blades of grass will not be systematically smaller and closer together (or larger and farther apart) at increasing distances from the observer. The *projected* size and density of these texture elements in the optic array, of course, does change with increasing distance from the point of observation, but these inhomogeneities in the optical texture are a result of the projective transformation.

As Gibson has pointed out (1950, 1959), the texture of the ground, because of its statistical homogeneity, provides a uniform scale over the entire ground plane. Thus, the distance between any two locations on the ground, or between the observer and some location, can be measured by the number of texture elements, or simply the amount of texture, between them (see Figure 1.3). This distance measure is unaffected by the projective transformation into the optic array because projection does not alter the number of texture elements (see Sedgwick, 1980, for a more detailed discussion). Although the uniform scale provided by the texture of the ground plane is thus an important source of *available* visual information for spatial orientation, adequate experimentation has not yet been done to determine how much use human observers actually make of this available information. Many studies have shown that observers are fairly good at estimating relative distances over the open ground (see Carlson, 1977, for a recent review) but there are typically other forms of visual information present that also specify distance. Experimental work has not yet succeeded in determining which of these redundant forms are actually used by observers in making their judgments.

TEXTURE GRADIENTS

The projection of the homogeneous texture of the ground plane to the optic array produces a *gradient* of optical texture there. The density of optical texture increases for increasingly distant points on the ground (see Figure 1.4). Gibson (1950) pointed out that this texture gradient provides the observer with potential information for a continuous ground surface receding into the distance. Gibson's student Purdy (1960) showed that for an observer of a given height, the texture gradient at any given location on the ground mathematically specifies the distance of that location from the observer. If we let G_L stand for the texture gradient at location L, let D_L stand for the distance of L, and let H stand for the height of the point of observation, then $D_L/H = G_L/3$. (For mathematical derivations of this equation, see Purdy, 1960, or Sedgwick, 1980.) This texture gradient equation for distance thus provides an alternative to counting or estimating the number of texture elements between the observer and the location.

Figure 1.3.   Size and distance are specified by the texture of the ground plane. In this perspective view the relation between each of the two objects and the grid covering the ground plane specifies that the objects both have the same ground plane dimensions (three by two squares) and are separated by a distance of two squares. (From Sedgwick, 1980.)

FLOW FIELDS

The optic array information for spatial orientation discussed in the previous two sections is present in the optic array whether the observer is stationary or moving. Other information from the projected texture of the ground plane becomes available only when the observer is moving. The projection of the ground texture to the moving point of observation creates in the optic array a structured *flow field* (see Figure 1.5) that mathematically specifies the direction and (for an observer of a given height) speed of movement (Gibson, 1950; Gibson, Olum, & Rosenblatt, 1955). Observers are quite accurate in detecting their specified direction of motion when they watch computer-

Figure 1.4. This optic array results from a simple, uncluttered, open environment. The texture of the ground is here depicted by uniform wrinkles or humps. The dashed lines show the angles subtended by these units of texture. There is a gradient of increasing density in the optic array corresponding to increasing distance along the ground. This gradient approaches infinite density at the horizon, which is the optic array boundary between the ground and the sky. (From James J. Gibson: *The Ecological Approach to Visual Perception.* Copyright © 1979 Houghton Mifflin Company. Used by permission. P. 67.)

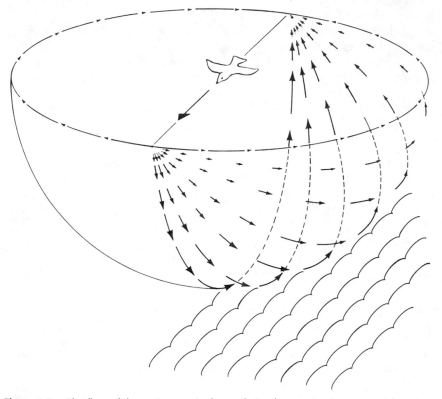

Figure 1.5. The flow of the optic array is shown during locomotion in a terrestrial environment. When a bird moves parallel to the earth, the texture of the lower hemisphere of the optic array flows under its eyes in the manner shown. The flow is centrifugal ahead and centripetal behind (i.e., there are focuses of expansion and contraction at the two poles of the line of locomotion). The greatest velocity of backward flow corresponds to the nearest location on the ground and the other velocities decrease outward from this perpendicularly in all directions, vanishing at the horizon. The vectors in this diagram represent angular velocities. (From James J. Gibson: *The Senses Considered as Perceptual Systems.* Copyright © 1966 Houghton Mifflin Company. Used by permission. P. 161.)

produced movies of such flow fields (Warren, 1976). Gibson refers to any such visually carried information for self-motion as *visual kinaesthesis*. Recent studies have combined sophisticated mathematical analysis of optical flow fields with hypotheses concerning physiological mechanisms that might be capable of registering such flow fields (Clocksin, 1980; Koenderink & van Doorn, 1976).

HORIZON INVARIANTS

Even when the plane of the ground stretches out endlessly in all directions, its projection in the optic array has a definite boundary—the *horizon*. The horizon can be thought of as the limit of the gradient of optical texture that is approached as distance from the observer increases endlessly (see Figure 1.4). Because the horizon is the projection of portions of the ground that are indefinitely far away, its position and orientation in the optic array are unaffected by movements of the point of observation (Sedgwick, 1973, pp. 257–273). For example, if point P on the horizon is directly to the north of the observer, it will stay directly to the north of the observer no matter how far or in what direction the observer moves. (This assumes that the horizon is infinitely far away; the horizon of less idealized open environments can still provide what is effectively the same information when its distance from the observer is large but finite.) Also, the distance from the observer of every location on the ground is inversely related ($D_L/H = 1/\tan V_L$) to the visual angle V_L between the location and the horizon (Sedgwick, 1973, 1980).[6] These horizon invariants are available to both moving observers and, in the degenerate case of zero-velocity, stationary observers.

Simple, Uncluttered, Enclosed Environments

In an *enclosed* environment, such as a room, the observer is surrounded by textured surfaces that are bounded by their intersections with each other. Each surface projects to the optic array a particular optical form and a particular optical texture gradient that depend on the surface's distance and orientation.

TEXTURE AND SLANT

Every surface in an enclosed environment has an orientation, or slant. Gibson and Cornsweet (1952) distinguished between a surface's *geographical slant,* which is its slant relative to some reference plane in the environment, and its *optical slant,* which is the angle that a line of sight from the point of observation makes with the surface. The geographical slant of a surface is constant, but its optical slant varies with movements of the observer and with where the surface is intersected by the line of sight.

The optical slant of a given location on a surface is specified by the texture gradient at that location. The shallower the optical slant R is, the greater is the texture gradient G. The trigonometric relation between these two quantities ($G/3 = 1/\tan R$) is derived elsewhere (see Purdy, 1960, or Sedgwick, 1980). There is a very simple geometrical relation between geographical slant S, the optical slant R at a given location, and the reference plane: If U is the optic array angle between that location and the horizon of the reference plane, then

$$S = R - U.$$

[6] If M is another location on the ground, then D_L/D_M, the distance of L relative to the distance of M, is given by $\tan V_M/\tan V_L$, which for small angles is closely approximated by V_M/V_L.

(See Figure 1.6; also see Purdy, 1960, or Sedgwick, 1980, for a derivation of this expression.)

There is a considerable experimental literature concerning observers' ability to use texture gradients to judge geographical slant (see Carlson, 1977, for a recent review) which suggests that slant is somewhat underestimated when only texture gradient information is available. The conclusions of these experiments, however, tend to be somewhat weakened by a failure to maintain a clear conceptual distinction between optical and geographical slant. In order to judge geographical slant, observers also need information about the orientation of the reference plane, such as the horizontal (see Purdy, 1960), and that information is not always readily available in experimental situations (cf. Perrone, 1980).

OPTICAL FLOW AND SLANT

As the observer moves through the environment, the optic array projection of every surface's texture changes and flows in a way that is mathematically specific to the optical slant of the surface (Koenderink & van Doorn, 1976; Purdy, 1960). If the surface is irregular, having bumps, bends, or depressions, then the local optical slant at every point on the surface is mathematically specified (Clocksin, 1980). It has been shown experimentally that observers can perceive the slant of a surface on the basis of this optical flow information alone (Braunstein, 1976, pp. 127–128; Flock, 1964; Gibson, Gibson, Smith, & Flock, 1959).

EXPANSION PATTERNS

As an observer approaches a surface, its projected texture has a *centrifugal flow pattern* in the optic array; the projected texture streams away in all direc-

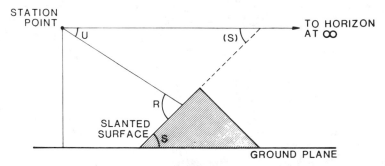

Figure 1.6. A surface slanted up from the ground plane is here seen from the side. If the ground plane is taken as the reference plane, then S is the geographical slant of the surface. The dotted line shows that the surface would make the same angle S if extended to meet the visual ray to the horizon of the ground plane. R is the optical slant of an arbitrarily chosen location on the surface, and U is the projective angle between that location and the horizon. (From Sedgwick, 1980.)

tions from a *center of expansion* that is located at the point on the surface toward which the observer is heading (Gibson, 1950; also see Figure 1.5). In an enclosed environment, the center of expansion in the flow field always provides the observer with potential visual information about the location toward which the observer is heading (Gibson, 1958; Lee, 1974). This information can also be converted into an algorithm for directing locomotion toward a particular goal: Move so as to keep the center of expansion directly over the goal (Gibson, 1979, pp. 227–234). Experiments suggest that observers may not be able to consciously identify the location of the center of expansion with much accuracy (Johnston, White, & Cumming, 1973; Llewellyn, 1971), but that their judgments of their direction of motion are, nevertheless, more accurate when the center of expansion is visible (Warren, 1976). Recent experimental investigations demonstrating visual adaptation to radial flow patterns have led to the suggestion that there may be neural mechanisms or "channels" that would be specifically sensitive to the center of expansion in the flow field (Regan & Beverly, 1978, 1979a, 1979b) and some physiological evidence in support of this hypothesis has also been obtained (Regan & Cynader, 1979).

RATE OF EXPANSION

For an observer moving at a constant speed, the *rate of expansion* of the flow field mathematically specifies the time at which the observer will reach the surface, the *time of collision* (Lee, 1976; Schiff, 1965). Observers' sensitivity to this information has recently been investigated (Schiff & Detwiler, 1979).

LINEAR PERSPECTIVE

When the edges of a room are parallel, they project as converging lines in the optic array. Such systems of projections are part of *linear perspective* and provide powerful potential visual information about the layout of the environment and the observer's relation to it (cf. Hay, 1974; Sedgwick, 1980). Since the Renaissance, artists have made use of this information in creating pictorial representations of spatial layout. Experimental investigations have also demonstrated that observers make use of linear perspective information, but they suggest that a modified linear perspective may be more effective, at least in some contexts, than the correct linear perspective (Hagen & Elliott, 1976; Hagen & Jones, 1978).

PERSPECTIVE TRANSFORMATIONS

When an observer moves around in an enclosed environment, the projected forms of all the surfaces are transformed in accordance with the laws of perspective. Such *perspective transformations* leave certain complex mathematical properties of the projected forms unchanged, however. These invariant properties mathematically specify the true shapes and orientations of the surfaces (Gibson, 1957; Hay, 1966; Lappin, Doner, & Kottas, 1980). The transformation also mathematically specifies the change in spatial orientation of the observer relative to each surface (Gibson, 1979).

Perspective transformations provide potential information in a much wider range of natural environments than do the rules of linear perspective. This is because linear perspective only applies to sets of *parallel* edges, such as exist in our synthetic, carpentered environments, whereas the invariants of perspective transformations are present for any flat surfaces, no matter how irregularly shaped they may be.

LINEAR VECTION AND POSTURAL CONTROL

One of the most general of ecological principles is that the terrestial environment is stable and unmoving. Thus, it is ecologically appropriate for the visual system to interpret relative motion between observer and environment as arising from motion of the observer rather than motion of the environment (see Turvey, 1977). This perceived stability of the environment, which has already been discussed with respect to the rotational mode, also is found in the translational mode. Thus, if an artificial situation is experimentally produced in which observers remain stationary while the environment translates past them, there will be a strong tendency for the observers to perceive the environment as stationary and themselves as moving. This effect, which is referred to either as *linear vection* or as a form of *visual kinaesthesis*, has been demonstrated repeatedly (Butterworth & Hicks, 1977; Dichgans & Brandt, 1978; Johansson, 1977; Lishman & Lee, 1973). It is so strong that it tends to override vestibular and other information that would specify the true movement of the observers. Similarly, observers standing with their eyes closed on a platform that is moving back and forth have no difficulty in correctly perceiving their motion; yet, if they have their eyes open and if the visual surroundings are made to move back and forth with them, then the observers will generally perceive themselves as being stationary (Lishman & Lee, 1973). Such visual information has also been shown to be used in controlling the postural stability of observers (Lee & Aronson, 1974; Lee & Lishman, 1975; Lestienne, Soechting, & Berthoz, 1977).[7]

Cluttered Environments

Cluttered environments are defined here as environments, either open or enclosed, that have objects in them other than the basic structural surfaces of the environment. For example, an open plain might have trees growing on it or boulders resting on it; a room might contain tables, chairs, etc. In a cluttered

[7]In addition to visual and vestibular information, somatosensory information from the limbs has been shown to make an important contribution to postural control and to have visual information closely integrated with it (Nashner & Berthoz, 1978; Nashner & Wollacott, 1979).

environment, some surfaces may be wholly or partially hidden from view (*occluded*) by other surfaces; some surfaces may be located on the hidden sides of objects. Cluttered environments raise a significant new problem for spatial orientation. How does an observer maintain orientation to portions of the environment that are not in view?

OCCLUSION AND DISOCCLUSION

As the observer moves, portions of the environment become hidden or revealed in a variety of complicated ways (cf. Gibson, 1979). When one surface is behind another, its projected texture in the optic array is progressively deleted (or accreted) by the parallax between the two surfaces as the observer moves. Computer-generated displays of texture patterns have been used to show that progressive deletion–accretion of texture produces powerful perceptions of one surface being behind another (Kaplan, 1969; Mace, 1974).

EDGE DETECTION AND LABELING

Edges in the environment produce discontinuities in the pattern of optical flow in the optic array. As the observer moves through the environment, every unhidden surface projects a gradient of optical flow—the projection of that part of the surface which is nearer flows relatively faster than the projection of the farther part. When two surfaces meet to form a corner, there will be an abrupt change in the gradient of optical flow along the projection of the edge. Likewise, if one surface is in front of another there will be an abrupt change in the rate of optical flow along the projection of the occluding edge. These discontinuities in the optic flow thus can provide the observer with information about the existence of edges in the environment. Nakayama and Loomis (1974) have suggested a very simple physiological model in which velocity sensitive cells with center–surround organizations could be combined into *convexity detectors* that would respond to these discontinuities in the optical flow. Nakayama and Loomis observe that if the visual system were to pass the optical flow information through a layer of convexity cells, they would automatically extract a map of all the edges in the environment (see Figure 1.7). In a somewhat similar physiological model by Clocksin (1980) edges are also labeled, indicating whether they are from corners that face toward or away from the observer, or are occluding, disoccluding, or contour edges (contour edges have no surface on one side, as when a building is seen against the sky). These models do not have direct support from physiological evidence, but appear at least to be consistent with known physiology.

SEEING WHAT IS NOT IN VIEW

The progressive occlusion of an object is potential visual information that it continues to exist even after the object is out of sight. For instance, the

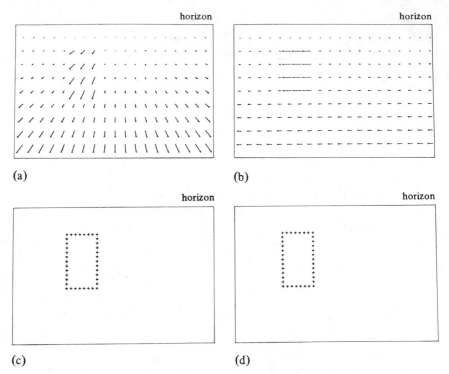

Figure 1.7. This shows planar representations of the velocity field of an observer who is (a) approaching a vertical rectangular screen that is off to the side and above the surface plane, and (b) passing a vertical rectangular screen in a direction parallel to the screen; and spatial response profiles (c) and (d) of a layer of the hypothesized "convexity" cells for the motion in (a) and (b) respectively. (From Nakayama & Loomis, 1974.)

progressive occlusion of an object passing at a constant speed through a short tunnel can produce a powerful *perception* of the object's continuing presence and motion which persists even while the object is totally hidden from view within the tunnel (Michotte, Thinès, & Crabbé, 1964). The optical transitions by which an object *goes out of sight* or *goes out of existence* are different and produce different perceptions (Gibson, 1979; Gibson, Kaplan, Reynolds, & Wheeler, 1969).

Complex Environments

Environments in which not all of the basić structural surfaces of the environment are projected to every point of observation are defined here as *complex environments*. Environments may be made complex by *local convexities* and *concavities* (such as hills and dales in open environments, or alcoves in enclosed environments). Environments may also be made complex by *linking* together environments; for example, the successive *vistas* of a hilly terrain in an open environment, the successive enclosures of adjoining rooms, or mixed environments—enclosures, such as houses, contained with an open environment. The *urban* environment with its complicated mix of interior and exterior spaces is an important example of a complex environment. A special class of complex environments comprises those environments that contain representations (e.g., pictures) of other environments (see Hagen, 1980).

It is primarily with complex environments that the study of "cognitive maps" is concerned (see Pick & Reiser, Hart & Berzok, Chapters 5 and 7, this volume). The problems of how cognitive maps are put together over time from information available in the optic array has hardly begun to be approached (see Gibson, 1979, for an introductory discussion). All that we can do here is consider a few rather broad, admittedly speculative, suggestions.

THE TYPOLOGY OF ENVIROMENTS AND VISUAL EXPLORATION

As has already been shown, the observer explores the surroundings by moving through them. Movement through the environment makes available new visual information in two ways. First, movement of the point of observation causes changes in the optic array, such as perspective transformations and optical flow patterns, that are powerful sources of information about the spatial layout of the environment. Second, movement of the point of observation may reveal portions of the environment that were previously hidden (see Figure 1.8). The particular way in which visual exploration serves to reveal the environment depends on the type of environment in question. It is in complex environments that movement becomes essential. A complex environment may require extensive visual exploration before complete information has been obtained about the layout and the observer's orientation to it.

VISUAL EXPLORATION OF COMPLEX ENVIRONMENTS

It seems clear that in considering the visual exploration of complex environments a detailed typology of environments becomes necessary, because the type of complex environment being explored will determine how visual exploration can be made most effectively. The analysis and study of this area, however, has scarcely begun, so we shall limit outselves to a few rather general observations. First, we conjecture that *complete* information about the layout of the environment can normally be made available to the observer through a *limited* path of visual exploration. Second, we conjecture that there are normally an infinite number of different paths of visual exploration that can make this same visual information available to the observer, although some of these paths may be more efficient than others. Third, we conjecture that the optic arrays at points of observation within a complex environment normally contain potential visual

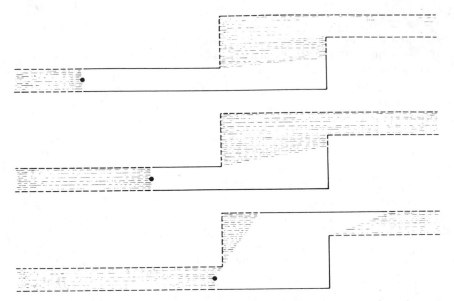

Figure 1.8.   This shows the opening up of a vista at an occluding edge, as seen from above. As an observer moves along the corridor, some surfaces progressively go out of sight behind the observer's head and surfaces in front progressively come into sight at one occluding edge and then at the other. The hidden portions of the ground are indicated by hatching. The hidden portions of the wall are indicated by dashed lines. The position of the observer is indicated by a black dot. (From James J. Gibson: *The Ecological Approach to Visual Perception.* Copyright © 1979 Houghton Mifflin Company. Used by permission. P. 199.)

information that *specifies* fairly efficient paths of visual exploration for the observer to follow and that also specifies when the visual exploration of the environment has been *completed.* If this last conjecture is correct, then there is normally information available within the optic array to help guide visual exploration.

These conjectures are obviously rather rudimentary. This is a reflection of how little is yet known about the visual exploration of complex environments. An effort to develop, explore, and test these conjectures would be one way of beginning to expand our knowledge in this area.

Summary

In this section we have considered a basic typology of natural environments that are open or enclosed, uncluttered or cluttered, simple or complex, and in doing so we have seen that each type of environment poses somewhat different problems for spatial orientation. We have also seen, however, that each type of environment makes available distinctive kinds of visual information by means of which spatial orientation can potentially be maintained during translational movement. Some empirical evidence concerning observers' use of such information has been noted, but it is clear that much empirical work remains to be done, particularly concerning the more complicated types of environments and visual information.

ACKNOWLEDGMENTS

I am grateful to Michael Potegal and Eve Sedgwick for their thorough critical readings of this chapter and to James Farber, Barbara Gillam, and the late James Gibson for their helpful comments on an earlier version of it.

REFERENCES

Berkeley, G. *An essay towards a new theory of vision*. New York: E. P. Dutton, 1910. (Originally published in 1709.)

Brandt, T., Wist, E. R., & Dichgans, J. Foreground and background in dynamic spatial orientation. *Perception and Psychophysics*, 1975, *17*, 497–503.

Braunstein, M. L. *Depth perception through motion*. New York: Academic Press, 1976.

Bridgeman, B., Hendry, D., & Stark, L. Failure to detect displacement of the visual world during saccadic eye movements. *Vision Research*, 1975, *15*, 719–722.

Brunswik, E. *Perception and the representative design of psychological experiments*. Berkeley: University of California Press, 1956.

Butterworth, G., & Hicks, L. Visual proprioception and posture stability in infancy. A developmental study. *Perception*, 1977, *6*, 255–262.

Carlson, V. R. Instructions and perceptual constancy judgments. In W. Epstein (Ed.), *Stability and constancy in visual perception*. New York: Wiley, 1977. Pp. 217–254.

Carr, H. A. *An introduction to space perception*. New York: Hafner, 1935.

Clocksin, W. F. Perception of surface slant and edge labels from optical flow: A computational approach. *Perception*, 1980, *9*, 253–269.

Corballis, M. C., & Beale, L. *The psychology of left and right*. Hillsdale, New Jersey: Lawrence Erlbaum Associates, 1976.

Day, R. H. Induced visual movement as nonveridical resolution of displacement ambiguity. *Perception and Psychophysics*, 1978, *23*, 205–209.

Descartes, R. [*Discourse on method, optics, geometry,* and *meteorology*] (P. Olscamp, trans.). Indianapolis: Bobbs–Merrill, 1965. (Originally published in French, 1637.)

Dichgans, J., & Brandt, T. Visual–vestibular interaction: Effects on self-motion perception and posture control. In R. Held, H. W. Leibowitz, H.-L. Teuber (Eds.), *Handbook of sensory physiology* (Vol. 8: *Perception*). New York: Springer–Verlag, 1978. Pp. 655–673.

Duncker, K. Über induzierte Bewegung (Ein Beitrag zur Theorie optisch wahrgenommener Bewegung). *Psychologische Forschung*, 1929, *12*, 180–259. (Excerpted in W. D. Ellis, *A source book of gestalt psychology*. London: Routledge & Kegan Paul, 1938. Pp. 161–172.)

Farber, J. *Manual tracking of induced object motion*. Paper presented at the annual meeting of the Association for Research in Vision and Ophthalmology, Sarasota, Florida, April 30–May 4, 1979.

Festinger, L., Sedgwick, H. A., & Holtzman, J. D. Visual perception during smooth pursuit eye movements. *Vision Research*, 1976, *16*, 1377–1386.

Flock, H. R. Some conditions sufficient for accurate monocular perceptions of moving surface slants. *Journal of Experimental Psychology*, 1964, *67*, 560–572.

Foley, J. M. Primary distance perception. In R. Held, H. W. Leibowitz, & H.-L. Teuber (Eds.), *Handbook of sensory physiology* (Vol. 8: *Perception*). New York: Springer–Verlag, 1978. Pp. 181–213.

Foley, J. M. Binocular distance perception. *Psychological Review*, 1980, *87*, 411–434.

Gibson, E. J., Gibson, J. J., Smith, O. W., & Flock, H. Motion parallax as a determinant of perceived depth. *Journal of Experimental Psychology*, 1959, *58*, 40–51.

Gibson, J. J. *The perception of the visual world*. Boston: Houghton Mifflin, 1950.

Gibson, J. J. The relation between visual and postural determinants of the phenomenal vertical. *Psychological Review*, 1952, *59*, 370–375.

Gibson, J. J. Optical motions and transformations as stimuli for visual perception. *Psychological Review*, 1957, *64*, 288–295.

Gibson, J. J. Visually controlled locomotion and visual orientation in animals. *British Journal of Psychology*, 1958, *49*, 182–194.

Gibson, J. J. Perception as a function of stimulation. In S. Koch (Ed.), *Psychology: A study of a science* (Vol. 1). New York: McGraw–Hill, 1959. Pp. 456–501.

Gibson, J. J. Ecological optics. *Vision Research*, 1961, *1*, 253–262.

Gibson, J. J. *The senses considered as perceptual systems*. Boston: Houghton Mifflin, 1966.

Gibson, J. J. *The ecological approach to visual perception*. Boston: Houghton Mifflin, 1979.

Gibson, J. J., & Cornsweet, J. The perceived slant of visual surfaces—optical and geographical. *Journal of Experimental Psychology*, 1952, *44*, 11–14.

Gibson, J. J., Kaplan, G. A., Reynolds, H. N., & Wheeler, K. The change from visible to invisible: A study of optical transitions. *Perception and Psychophysics*, 1969, *5*, 113–116.

Gibson, J. J., Olum, P., & Rosenblatt, F. Parallax and perspective during aircraft landings. *American Journal of Psychology*, 1955, *68*, 373–385.

Gibson, J. J., Purdy, J., & Lawrence, L. A method of controlling stimulation for the study of space perception: The optical tunnel. *Journal of Experimental Psychology*, 1955, *50*, 1–14.

Gibson, J. J., & Waddell, D. Homogeneous retinal stimulation and visual perception. *American Journal of Psychology*, 1952, *65*, 263–270.

Gogel, W. C. The metric of visual space. In W. Epstein (Ed.), *Stability and constancy in visual perception*. New York: Wiley, 1977. Pp. 129–181.

Gonshor, A., & Melvill Jones, G. Extreme vestibulo–ocular adaptation induced by prolonged optical reversal of vision. *Journal of Physiology*, 1976, *256*, 381–414.

Hagen, M. (Ed.) *The perception of pictures* (Vols. 1 and 2). New York: Academic Press, 1980.

Hagen, M. A., & Elliott, H. B. An investigation of the relationship between viewing condition and preference for true and modified linear perspective with adults. *Journal of Experimental Psychology: Human Perception and Performance*, 1976, *2*, 479–490.

Hagen, M. A., & Jones, R. K. Differential patterns of preference for modified linear perspective in children and adults. *Journal of Experimental Child Psychology*, 1978, *26*, 205–215.

Hansen, R. M. Spatial localization during pursuit eye movements. *Vision Research*, 1979, *19*, 1213–1221.

Harper, R. S., & Boring, E. G. Cues. *American Journal of Psychology*, 1948, *61*, 119–123.

Hay, J. C. Optical motions and space perception: An extension of Gibson's analysis. *Psychological Review*, 1966, *73*, 550–565.

Hay, J. C. The ghost image: A tool for the analysis of the visual stimulus. In R. B. MacLeod & H. L. Pick, Jr. (Eds.), *Perception: Essays in honor of James J. Gibson*. Ithaca: Cornell University Press, 1974. Pp. 268–275.

Helmholtz, H. von [*Treatise on physiological optics* (Vol. 3)] (J. P. C. Southal, Ed. and trans.). New York: Dover, 1962. (From the third German edition, 1910.)

Henson, D. B. Corrective saccades: Effects of altering visual feedback. *Vision Research*, 1978, *18*, 63–67.

Hochberg, J. Perception II. Space and movement. In J. W. Kling & L. A. Riggs (Eds.), *Experimental psychology* (3rd ed.). New York: Holt, Rinehart and Winston, 1971. Pp. 475–550.

Howard, I. P., & Templeton, W. B. *Human spatial orientation.* New York: Wiley, 1966.

Ittelson, W. H. *Visual space perception.* New York: Springer, 1960.

Johansson, G. Studies on visual perception of locomotion. *Perception*, 1977, *6*, 365–376.

Johansson, G., Hofstein, C. von, & Jansson, G. Event perception. *Annual Review of Psychology*, 1980, *31*, 27–63.

Johnston, I. R., White, G. R., & Cumming, R. W. The role of optical expansion patterns in locomotor control. *American Journal of Pyschology*, 1973, *86*, 311–324.

Kaplan, G. A. Kinetic disruption of optical texture: The perception of depth at an edge. *Perception and Psychophysics*, 1969, *6*, 193–198.

Koenderink, J. J., & van Doorn, A. J. Local structure of movement parallax of the plane. *Jourhal of the Optical Society of America*, 1976, *66*, 717–723.

Koffka, K. *Principles of Gestalt Psychology.* New York: Harcourt, Brace and World, 1935.

Lackner, J. R. Some mechanisms underlying sensory and postural stability in man. In R. Held, H. W. Leibowitz, & H.-L. Teuber (Eds.), *Handbook of sensory physiology* (Vol. 8: *Perception*). New York: Springer–Verlag, 1978. Pp. 805–845.

Lappin, J. S., Doner, J. F., & Kottas, B. L. Minimal conditions for the visual detection of structure and motion in three dimensions. *Science*, 1980, *209*, 717–719.

Lee, D. N. Visual information during locomotion. In R. B. MacLeod & H. L. Pick, Jr. (Eds.), *Perception: Essays in honor of James J. Gibson.* Ithaca: Cornell University Press, 1974.

Lee, D. N. A theory of visual control of braking based on information about time-to-collision. *Perception*, 1976, *5*, 437–459.

Lee, D. N., & Aronson, E. Visual proprioceptive control of standing in human infants. *Perception and Psychophysics*, 1974, *15*, 529–532.

Lee, D. N., & Lishman, J. R. Visual proprioceptive control of stance. *Journal of Human Movement Studies*, 1975, *1*, 87–95.

Lestienne, F., Soechting, J., & Berthoz, A. Postural readjustments induced by linear motion of visual scenes. *Experimental Brain Research*, 1977, *28*, 363–384.

Levy, J. Autokinetic illusion: A systematic review of theories, measures, and independent variables. *Psychological Bulletin*, 1972, *6*, 457–474.

Lishman, J. R., & Lee, D. N. The autonomy of visual kinaesthesis. *Perception*, 1973, *2*, 287–294.

Llewellyn, K. R. Visual guidance of locomotion. *Journal of Experimental Psychology*, 1971, *91*, 245–261.

Mace, W. M. Ecologically stimulating cognitive psychology: Gibsonian perspectives. In W. B. Weimer & D. S. Palermo (Eds.), *Cognition and the symbolic processes.* Hillsdale, New Jersey: Lawrence Erlbaum Associates, 1974. Pp. 137–164.

Mack, A. An investigation of the relationship between eye and retinal image movement in the perception of movement. *Perception and Psychophysics*, 1970, *8*, 291–297.

Mack, A., & Herman, E. A new illusion: The underestimation of distance during pursuit eye movements. *Perception and Psychophysics*, 1972, *12*, 471–473.

Mack, A., & Herman, E. The loss of position constancy during pursuit eye movements. *Vision Research*, 1978, *18*, 55–62.

MacKay, D. Visual stability and voluntary eye movements. In R. Jung (Ed.), *Handbook of sensory physiology* (Vol. 7, Pt. 3A: *Central visual information*). New York: Springer–Verlag, 1973. Pp. 307–331.

Matin, E. Saccadic suppression and the stable world. In R. A. Monty & J. W. Senders (Eds.), *Eye movements and psychological processes.* Hillsdale, New Jersey: Lawrence Erlbaum Associates, 1976. Pp. 113–119.

Matin, L. Eye movements and perceived visual direction. In D. Jameson & L. Hurvitch (Eds.), *Handbook of sensory physiology* (Vol. 7, Pt. 4: *Visual psychophysics*). New York: Springer–Verlag, 1972. Pp. 331–380.

Matin, L. A possible hybrid mechanism for modification of visual direction associated with eye movements: The paralyzed-eye experiment reconsidered. *Perception,* 1976, 5, 233–239.

McConkie, A. B., & Farber, J. Relation between perceived depth and perceived motion in uniform flow fields. *Journal of Experimental Psychology: Human Perception and Performance,* 1979, 5, 501–508.

McLaughlin, S. C. Parametric adjustments in saccadic eye movement. *Perception and Psychophysics,* 1967, 2, 359–362.

Michotte, A., Thinès, G., & Crabbé, G. Les compléments amodaux des structures perceptives. In A. Michotte & J. Nuttin (Eds.), *Studia Psychologia.* Louvain: Publications Université de Louvain, 1964.

Miller, J. M., Anstis, T., & Templeton, W. B. Saccadic plasticity: Parametric adaptive control by retinal feedback. *Journal of Experimental Psychology: Human Perception and Performance,* 1981, 7, 356–366.

Nakayama, K., & Loomis, J. M. Optical velocity patterns, velocity-sensitive neurons and space perception: A hypothesis. *Perception,* 1974, 3, 63–80.

Nakayama, K., & Tyler, C. W. Relative motion induced between stationary lines. *Vision Research,* 1978, 18, 1663–1668.

Nashner, L., & Berthoz, A. Visual contribution to rapid motor responses during postural control. *Brain Research,* 1978, 150, 403–407.

Nashner, L. M., & Woolacott, M. The organization of rapid postural adjustments of standing humans: An experimental–conceptual model. In R. E. Talbott & D. R. Humphrey (Eds.), *Posture and movement.* New York: Raven Press, 1979. Pp. 243–257.

Oman, C. M., Bock, O. L., & Huang, J. K. Visually induced self-motion sensation adapts rapidly to left–right visual reversal. *Science,* 1980, 209, 706–708.

Ono, H., & Comerford, J. Stereoscopic depth constancy. In W. Epstein (Ed.), *Stability and constancy in visual perception.* New York: Wiley, 1977. Pp. 91–128.

Perrone, J. A. Slant underestimation: A model based on the size of the viewing aperture. *Perception,* 1980, 9, 285–302.

Purdy, W. C. *The hypothesis of psychophysical correspondence in space perception.* General Electric, Technical Information Series, No. R60ELC56. Ithaca, New York: General Electric, 1960.

Regan, D., & Beverly, K. I. Looming detectors in the human visual pathway. *Vision Research,* 1978, 18, 415–421.

Regan, D., & Beverly, K. I. Binocular and monocular stimuli for motion in depth: Changing-disparity and changing-size feed the same motion-in-depth stage. *Vision Research,* 1979, 19, 1331–1342. (a)

Regan, D., & Beverly, K. I. Visually guided locomotion: Psychophysical evidence for a neural mechanism sensitive to flow patterns. *Science,* 1979, 205, 311–313. (b)

Regan, D., & Cynader, M. Neurons in area 18 of cat visual cortex selectively sensitive to changing size: Nonlinear interactions between responses to two edges. *Vision Research,* 1979, 19, 699–711.

Royce, J. R., Carran, A. B., Aftanas, M., Lehman, R. S., & Blumenthal, A. The autokinetic phenomenon: A critical review. *Psychological Bulletin,* 1966, 65, 243–260.

Rudel, R. G., & Teuber, H.-L. Discrimination of direction of line in children. *Journal of Comparative and Physiological Psychology,* 1963, 56, 892–898.

Schiff, W. Perception of impending collision: A study of visually directed avoidant behavior. *Psychological Monographs*, 1965, *79*, whole number 604.

Schiff, W., & Detwiler, M. L. Information used in judging impending collision. *Perception*, 1979, *8*, 647–658.

Schulman, P. H. Eye movements do not cause induced motion. *Perception and Psychophysics*, 1979, *26*, 361–383.

Sedgwick, H. A. The visible horizon: A potential source of visual information for the perception of size and distance (Doctoral dissertation, Cornell University, 1973). *Dissertation Abstracts International*, 1973, *34*, 1301B–1302B. (University Microfilms No. 73-22,530)

Sedgwick, H. A. The geometry of spatial layout in pictorial representations. In M. Hagen (Ed.), *The perception of pictures* (Vol. 1). New York: Academic Press, 1980. Pp. 33–90.

Sedgwick, H. A., & Festinger, L. Eye movements, efference, and visual perception. In R. A. Monty & J. W. Senders (Eds.), *Eye movements and psychological processes*. Hillsdale, New Jersey: Lawrence Erlbaum Associates, 1976. Pp. 221–230.

Shebilske, W. Visuomotor coordination in visual direction and position constancies. In W. Epstein (Ed.), *Stability and constancy in visual perception*. New York: Wiley, 1977. Pp. 23–69.

Stoper, A. E. Apparent motion of stimuli presented stroboscopically during pursuit movement of the eye. *Perception and Psychophysics*, 1973, *13*, 201–211.

Turvey, M. T. Preliminaries to a theory of action with reference to vision. In R. Shaw & J. Bransford (Eds.), *Perceiving, acting, and knowing*. Hillsdale, New Jersey: Lawrence Erlbaum Associates, 1977. Pp. 211–265.

Ullman, S. Against direct perception. *The Behavioral and Brain Sciences*, 1980, *3*, 373–415.

Vredeman de Vries, J. [*Perspective*]. Illustrations reprinted. New York: Dover, 1968. (Originally published Leiden: Henricus Hondius, 1604 and, *pars altera*, 1605.)

Warren, R. The perception of ego motion. *Journal of Experimental Psychology: Human Perception and Performance*, 1976, *2*, 448–456.

Yarbus, A. L. [*Eye Movements and Vision*]. (B. Haigh, trans.) New York: Plenum Press, 1967.

BEATE HERMELIN
NEIL O'CONNOR

2

Spatial Modality Coding in Children with and without Impairments

If one asks how either sensory or cognitive impairments might affect the perception and conception of the spatial framework, the research strategy one adopts will at least partially be determined by one's own view of the nature of space. There are essentially three such possible views. The first was put forward most clearly by the philosopher Emanual Kant in 1781, who held that humans conceived of space intuitively and that this intuition, together with those of time and causality, rested neither on experience nor on inferences from experience. This view was reflected by Schopenhauer, who wrote in 1819:

> A man must be forsaken by all the gods to dream that the world we see outside of us, filling space with its three dimensions, moving down the inexorable stream of time, governed at each step by Causality's invariable law—but in all this only following rules which we may prescribe for it in advance of all experience—to dream, I say, that such a world should stand there outside of us, quite objectively real with no complicity of ours, and thereupon by a subsequent act, through the instrumentality of mere sensation, that it should enter our head and reconstruct a duplicate of itself as it was outside [p. 316].

SOME VIEWS ON SPACE PERCEPTION

In 1890, in the early history of scientific psychology, James (1890/1980) put forward different conclusions, thinking that spatial qualities were contained in

35

SPATIAL ABILITIES
Development and Physiological Foundations

sensations and that the senses were responsive to aspects of space such as, for instance, volume or extent. Although he believed that those characteristics were extracted by all the spatial senses, he conceded that although the whole of a scene could be immediately and simultaneously present to the visual system, tactual exploration of the environment was a successively organized process. He concluded from this that although sighted infants would have to learn to isolate separate stimuli in space from a total scene by a process of analysis, those born blind would have to synthesize serially experienced spatial components into a total and simultaneously present space.

According to Wundt (1894), there was no spatial quality as such in sensation; spatial attributes, such as distance, were inferences based on experience and learning. In our current psychological terminology, we might say that Kant held that the experience and the concept of space were predetermined by, and programmed into, the structure of the nervous system. Aspects of James's theory are reflected in contemporary discussions about the dependence of coding processes on modality specific characteristics which, together with other aspects of stimulation, may be preserved in memory. Finally, Wundt puts the emphasis on learning and on advancing hypotheses about space derived from various forms of experience. His view is taken by much of the current workers in cognitive psychology.

Whatever view on the origins of space perception is held, each must take account of the various aspects of spatial experience. The perception of spatial relationships by a stationary observer may be determined by factors other than those that operate when either the observer or the surroundings move. The perception of one's own body in space may be distorted in different ways when either body position or the environment are altered. In addition, though there is extensive integration of data from different sensory modalities, different perceptual systems may be biased toward extracting different aspects of space from the environment. Though we normally perceive objects and events through several senses, memory will preserve not only the general meaning, but also some of the modality specific features of such stimulation.

Words

Other than the sensory component in brief iconic and echoic storage (Neisser, 1967; Sperling, 1960), there is evidence that specific perceptual properties of stimuli are recorded in more persistent codes. For instance, Nilsson (1974) found that when lists of words were presented in mixed auditory and visual modes to subjects, they remembered the modality of presentation of the separate items in free recall. Paivio (1969) demonstrated that concrete and imagable words were represented in a dual, verbal and pictorial code, and later concluded (1971) that pictorial memory codes were specialized for spatial structures whereas verbal memory codes favored temporal structuring. Snodgrass, McClure, and Pirone (1978) have confirmed this interaction of pictures and words with space and time and have demonstrated that image codes were more easily associated with spatial locations whereas verbal codes were more easily associated with temporal locations.

Thus, for some words at least, verbal and visual codes seem to overlap. Posner (1967) has reported that an interpolated verbal task affected not only verbal but also visual retention, whereas such verbal interference had no effect on kinesthetic memory. However, there seems to be a close association between spatial representations derived from movement and vision. Stelmach and Wilson (1970) as well as Posner and his colleagues, have argued that movement information is at least partly visually recoded (see also Abramsky, Carmon, & Benton, 1971; Coquery & Amblard, 1973). To account for such recoding strategies, Attneave and Benson (1969) have suggested that different modalities may have different facilities for data processing. They suggest that space is represented primarily in visual terms. We can, for example, inspect a route on a map, and actively follow it either forward or backward with equal ease. Even if the route was initially learned in a sequential fashion, an internal "map" would eventually be built up and, at this stage of learning, following the route will have become direction independent. In addition, the visual system leads to a code that is biased to extract the place rather than the time of visually presented components. If one wants to revisit a painting in a gallery, one probably remembers the place on a particular wall where it was hanging, rather than where it occurred in the sequence of paintings that had been viewed.

Audition

In contrast to input to the visual system, which tends to favor a spatially determined organization of stimuli, the temporal–sequential order of auditory stimulation tends to be retained. Touch, like audition, also seems to be better retained when the stimuli are successively rather than simultaneously presented (Geldard, 1966; Geldard & Sherrick, 1965). The same point can be made by referring to everyday experience. We realize the temporal–sequential dependence of auditory material if we try to recite a poem backward, or attempt to recognize a piece of music played in the reverse direction from the last note to the first. Hearing and remembering what we have heard rely on the perception and retention of the temporal order of the stimuli.

Movement

Where movement is concerned the situation is more complex. Because it occurs in space but also often consists of a temporally ordered sequence of motor components, spatial as well as temporal aspects are important. This association between particular perceptual systems and the formal properties of the codes extracted through them must be regarded as representing certain biases rather than limits. We can, of course, say where we heard or when we saw something. Nevertheless, it seems true that vision is a predominantly spatial sense and that hearing is temporally organized, whereas the motor system generates kinesthetic feedback about spatial as well as temporal components from a movement sequence.

Intersensory Recoding

However, as pointed out earlier, it would be an oversimplification to state that coding processes invariably tend to be determined by the sensory modality of the stimulation. Quite frequently information is not treated in terms of the sensory system through which it was perceived but by a process of intersensory recoding. Gibson (1966) has proposed that during the course of development, children become increasingly able to abstract supramodal stimulus features, which enable them to appreciate those features in any sensory modality. This ability allows information to be processed in terms of mental schemas and representations and enables subjects to restructure it into its most appropriate format. Thus Conrad (1959) has shown in his classical experiment that if subjects were visually presented with a series of letters that they were asked to recall, confusion did not occur between letters that looked alike, but did occur between those that sounded alike. Though the letters were seen, they were treated as if they had been heard. Similarly, Paivio (1969) has demonstrated that words that are readily imaged are stored not only verbally but also in pictorial form. Such words, whether heard or read, are also remembered in terms of the pictorial images of the objects they signify (but see Light, 1975).

Simpson (1973) has performed experiments on intermodal spatial localization and has concluded that adults tend to use a visual reference system for spatial localization of stimuli in all modalities. Platt and Warren (1972) and Jones and Kabanoff (1975) have shown that auditory space perception is facilitated by appropriate eye movements and that performance deteriorates when the subject's eye movements are cued away from a tone to be localized. This indicates that auditory–spatial information may be visually (or visual–motorically) coded. Jones (1975) also showed that eye movements in the light were more effective than those in the dark, thus indicating the role of visual information processing. Warren (1970) failed to find such visual facilitation effects on auditory localization in children, whereas Fraiberg, Siegel, and Gibson (1966) reported that blind babies of 28 weeks did not reach for a sound-producing object. This might indicate that experience and learning may be necessary to map auditory spatial information onto a visual representation of space. However, Bower and Wishart (1973) found that sighted babies, too, did not look at or reach for a sound producing object in the dark until they were 44 weeks old.

Furthermore, Jones (1975) subsequently did obtain results from children which indicated that they were more sensitive to auditory position with eyes open than eyes closed. It may be that the construction of a common visual–auditory space (Auerbach & Sperling, 1974; see Marks, 1978 for review) is dependent upon experience. In the case of visual–kinesthetic/motor linkage, experience may be necessary to maintain, rather than organize, the linkage. Freedman (1964) observed that infants who are born blind nevertheless tend to track their hands with their unseeing eyes, though this behavior disappears after about 16 weeks. Bower and Wishart (1972) found that if infants of around 26 weeks are shown an object and the lights are then extinguished, the infant can reach out in total darkness and grasp the object with very high accuracy. Bruner (1970) demonstrated that babies even tended to shut their eyes while reaching for an object. Thus a visually remembered location seems to be used, rather than continuous visual guidance toward the object.

Sensory Impairment

Everday experience indicates that people are not solely dependent on the modality information that is available at any given moment. When we read a poem or a piece of music we tend to "hear it in our heads" and when we enter a dark familiar room, we do not rely solely on our sense of touch but tend to "see" the arrangement of the furniture in our "mind's eye." The hypothesis tested in the following experiments states that cognitive or perceptual maldevelopment may impede children's evocation of such relevant mental representations. The intellectually handicapped child may be unable to perform the abstractions and transformations necessary to derive codes other than those inherent in the modality of stimulus presentation. However, the child with a restricted reper- toire of perceptually based codes may have the necessary mental ability for recoding but may lack schemas based on alternative input modalities. More concretely, this hypothesis implies that, for instance, mentally handicapped children may not use a strategy of verbal rehearsal to retain visually presented letters. Similarly, congenitally deaf children cannot imagine the sound of printed words, and those born blind cannot evoke visual images. This leads to processing and memory codes in children with cognitive or perceptual impair- ment different from those of normal children.

It must be kept in mind that we have outlined two possible approaches to the selection of a code; they are not mutually exclusive. Thus if task requirements coincide with the presentation of stimuli so that the perceptual system ad- dressed is an efficient one for handling the data, coding will make use of the modality characteristics of the stimulus. Under these conditions we predict no differences between groups of children with or without cognitive impairment. However, when the presentation modality does not provide sufficient or appropriate information for an efficient coding strategy, normal children may restructure and map material perceived through one sense onto another modality schema that has been derived from previous experience. Perceptually impaired children will in some instances lack the appropriate experience and will thus be unable to adopt such a strategy, whereas cognitively impaired children may not be able to make the abstractions necessary for such transformations. Thus one might expect that different groups of children will use different coding strategies in such circumstances. The experiments to be reported here tested these predictions by comparing coding strategies of groups of normal children and perceptually or cognitively impaired children.

EXPERIMENTS ON SENSORY SPECIFIC CODING

Seeing and Hearing

In the first experiment to be reported we studied blind, deaf, and normal children about 10 years old and cognitively subnormal children of about 15. We tested the effects on a successive discrimination task of either a spatially or temporally organized series of stimuli presented in either a visual or auditory mode. The cognitive demands of the auditory and visual tasks were identical, so that any differential coding strategies could be attributed to the modality differences of the presented material.

The subject was placed at a distance of 3 m from the stimulus sources, which consisted of 2½-inch loudspeakers and florescent lights of 1⅛-inch diameter. The required signals were prerecorded onto standard cassettes. For the auditory output the amplified signals were switched to the loudspeakers, whereas for the visual stimuli an amplified 1 KHz tone was switched to the secondaries of the output transformer and the primaries were connected to the lights. The task was to decide whether two successive series of stimuli were the same or were different. A stimulus series consisted either of a number of light flashes or of pure tones of a single frequency. Light flashes or tones followed each other with an interstimulus interval of 100 msec within each series, and an interval of 1 sec between the two comparison sequences. How many items a series contained for each child was determined by first establishing the immediate memory span for digits for each subject, and then by increasing the length of the series to be presented by one-third of this span. By using this procedure effects due to differential short-term memory capacity were avoided.

For either visually or auditorily presented stimulus series there were two conditions. First, the stimulus source for either light or sound was directly in front of the subject. In each series the items varied in duration, rather like in a Morse code: Long stimuli lasted for 500 msec and short for 150 msec, so that they were easily discriminable. Thus a series of stimuli might consist of items arranged in order of "short–long–long–short–short–long." This would be followed by another series of items in the same modality, having either the same or a different temporal structure. The subject had to decide whether two such sequences were exactly the same or different. The two sequences thus to be compared were always in the same modality, and light and sound sequences were presented in blocks.

In the second condition, light as well as sound sources were placed to the right and left of the subject, at an angle of 45 degrees from midline. All stimulus items were of the same duration (300 msec) but came from either the right or left. The spatial distribution of the items in two successive sequences might, for instance, be "left–left–right–right–right–left" followed by "left right–left–left–right–right." In this instance the series would, of course, have to be judged as being different from each other. Deaf children were presented with visual and blind children with auditory stimulation, whereas all normal and subnormal children were presented with light and sound. The sighted groups were blindfolded for the latter condition.

With visual presentations, subjects recognized the light sequences significantly better when the stimuli came from the left and right than when they were of short and long duration ($p = .005$). Thus, two sequences of spatially distributed visual stimuli were easier to compare than temporally structured series of lights. The reverse results were found when two sequences of tones had to be compared. Such tone sequences were significantly more often correctly compared when the items in them varied in duration than when they came from different directions ($p = .01$). Thus with tones temporally rather than spatially structured sequences led to better recognition and this was true for all subject groups tested. There were no significant group differences or groups-by-conditions interactions.

These findings confirm the first of the stated hypotheses, that is, that different features tend to be abstracted or stored from different modalities of stimulation. The present study does not allow us to decide whether it was the initial extraction of certain modality features, or the memory for specific stimulus characteristics that were primarily responsible for the results.

In this study the stimuli were nonverbal and meaningless and presentation rates were too fast to allow storage of a verbally mediated sequence, as, for example, in sets of words such as *long–short–short–long–short*. Therefore one would not have expected any transformations to have occurred and the lights should have been processed and stored in the visual mode and the tones in the auditory mode. Consequently, one would not have predicted that perceptually or cognitively impaired children would differ from normal children in their coding strategies. Group differences would be predicted only in those situations in which recoding into a system other than the stimulated modality system would seem advantageous. In fact, the expectation that perceptually and cognitively impaired children would use modality specific codes similar to those of normal children was confirmed. All children—normal, blind, and subnormal—responded more correctly to sounds when they were temporally structured than when they were spatially structured, whereas the deaf as well as the normal and subnormal subjects were more often correct when the stimuli in two series of lights were spatially rather than temporally distinguishable.

Coding of Movement

We will now turn from auditory–temporal and visual–spatial coding, to processes and mental representations that are characteristically associated with movement information. This area is of particular interest because movements have both spatial and temporal aspects. They occur within a well-defined spatial framework but are also temporally organized, as any movement sequence takes place in a predetermined temporal order. This dual aspect of movement reflects the close intersensory integration between the various spatial senses (i.e., proprioception, kinesthesis, and vision). This sensory interdependence between movement and vision for orientation in space has been shown in many experiments (e.g., in the classical studies of Held and his colleagues, 1961). They found that kittens reared in the dark remained functionally blind when they were not allowed to experience active movement together with visual information. Held (1961) assumed that a self-instruction to make a movement was accompanied by a "copy" of previous results of such movements in visual terms. Coquery and Amblard (1973) have also suggested an overlap between verbal and tactile retention and particularly between visual and tactile retention.

If input from vision and from movement is not concurrent, vision will have had to precede such movement relatively recently in order to affect it. Worchel (1951) compared the motor development of children blinded at different times. He concluded that for movement in space, visual images seem to be relevant only when the visual experience had been recent. But even with such recent experience, the exploration of objects through touch and movement may lead to different results than when vision is directly involved. Pick, Klein, and Pick (1966) asked blind children and sighted children with or without blindfolds to feel a pair of shapes one of which was rotated by 180 degrees relative to the other. The subjects had to decide which shape was "upside down." The choices of the children who could see the shapes were consistent across subjects. No such consistencies were found in the blind and blindfolded groups who explored the shapes tactually. One possible explanation for the lack of consistency in identifying the upper part of a form by touch may be accounted for not only by the frequent experience of handling objects independently of their orientation, but also by the turning of the hand, so that no part of it consistently felt the same part of an object. This would correspond not only to seeing objects at different orientations, but also to standing on one's head and viewing the world while upside down. Thus in this particular experiment touch and vision seem to have resulted in the extraction of different shape features which were regarded as crucial for judgments of orientation.

In our experiment we compared spatial judgments made by sighted normal, autistic, and cogenitally blind children when they were either given visual or tactual cues (O'Conner & Hermelin, 1978). All children were about 10'years old, but the autistic children had a mental age of about 6 years. The task was adapted from one used by Attneave and Benson (1969). Half the normal and autistic children were blindfolded. The children were asked to place their second and third fingers of each hand before them, with alternate children putting either the left or right hand in front of the other. In a training period, each of these four fingers was touched in turn, while the experimenter simultaneously uttered a word that was to serve as the paired associate response for this particular stimulation. The child had to learn to respond with the appropriate word, which was either *run, sit, walk,* or *stand,* when the appropriate finger was touched. The training proceeded from the finger closest to the child's body outward until he had learned to respond with each word corresponding to the finger being touched. The fingers were then touched in a predetermined random order, until 19 correct verbal responses were obtained in 20 trials. Immediately following such criterion performance, the experimenter reversed the relative position of the child's hands.

Without commenting on this change of hand position, stimulation was continued. The children now had two possible options for their responses. They could either retain an association between the touch of a particular finger and the previously learned verbal response to it, disregarding the fact that this finger now occupied a different spatial position, or they could give a response to a particular spatial position, though a different finger now occupied this position. No corrections of the responses were given.

The results showed that children who performed the task visually tended to retain their responses to fixed locations in space after the change in the position of their hands. This was so even though the fingers occupying these positions were now different ones. This response tendency in terms of stable spatial locations was observed in normal and autistic children in the visual condition. However, when the task was performed without vision, blindfolded sighted and blind children alike gave verbal responses that were appropriate to the finger initially stimulated before hand reversal, though this finger was now in a different spatial position.

The results of this experiment are similar to those found by Attneave and

Benson (1969) and confirm the conclusions from our previous study using light and sound. In both of these experiments the sensory system involved determined the organizational properties of the code used to structure the items. Thus in the first experiment lights were more easily remembered as spatially and sounds as temporally structured sequences, whereas in the present experiment, tactual stimulation led to a relative and "egocentric" organization of space and visual stimulation led to a stable and objective organization of space. Piaget and Inhelder (1956) have shown a developmental trend leading from a strong tendency to locate objects in a display from the observer's point of view to a coordination of such objects within a stable invariant spatial framework. In our experiment the reliance on either projectively or metrically conceived space was not related to the developmental stage of the child, but to the modality system that was addressed. Pick, Acredo, and Gronseth (1973) have reported similar results from children temporarily deprived of vision.

EXPERIMENTS ON REPRESENTATIONAL CODING

We will now turn to tests of the second part of our hypothesis. We had suggested that whenever information was addressed to a perceptual system that had no special facility to process such data in accordance with the task requirements, the information would additionally, or instead, be coded in terms of schemas derived from another, more appropriate modality. We also proposed that the ability for such intersensory recoding may be limited in children with perceptual or cognitive impairments, because the former would have a restricted repertoire of perceptually derived schemas, whereas the latter might suffer from an inability to evoke the appropriate mental representations.

Temporal–Spatial Incongruity

The first experiment to be reported in this context (O'Connor & Hermelin, 1978) used deaf, normal, and autistic children who were between 8 and 9 years old. We asked whether task requirements or the nature of the presentation modality would determine coding strategies. The material consisted of either Arabic or Roman letters. Three such letters were visually presented for 300 msec each and the child had to remember the order of the items. However, the temporal (i.e., the first to last) and the spatial (i.e., the left to right) order in which the stimuli appeared did not correspond. On a given trial, for example, the first stimulus might appear in the middle one of three horizontally arranged windows, the second one might then appear in the rightmost window, and the third would follow in the leftmost window.

The interstimulus interval (ISI) between each letter presentation in any one trial was 200 msec, which allowed for implicit verbalization of the preceding letter, but not for rehearsal of more than one item. A previous pilot experiment had shown that with very brief ISI, where no verbalization was possible, responses were biased toward a spatial order, whereas long ISI, allowing for rehearsal, resulted in a temporally ordered sequence. The ISI selected may thus have been crucial for the pattern of results that was obtained.

In the first experimental condition, three Arabic letters were presented in the described temporal–spatial incongruent order. Since these letters could not be read by the children, these complex, briefly presented shapes were unlikely to be verbally coded. The children were told that they would see letters from a foreign script and would have to remember them later on. There were 20 trials for each child; for 10 of them recall memory was required whereas for the other 10 trials recognition memory was asked for. For recall, the child was given three separate cards, each showing one of the Arabic letters, and was asked to put those on a table in front of him or her in the order remembered. For recognition, the child had to select one of three vertically arranged columns of the three letters. In the first column the letters were printed from the first to the last as they had appeared, but were arranged from top to bottom. The second column reflected the left to right order read from top to bottom, and the third was a random order arrangement that had not been shown. No significant group differences and no differences between recall and recognition were found. With both methods the responses were significantly more frequently spatially ordered than temporally ordered (i.e., the left to right rather than the first to last order of presentation seems to have been stored in memory). As has already been pointed out, such spatial coding seems to reflect the visual, nonverbal, nature of the presented material. Snodgrass *et al.* (1978) had confirmed Paivio's (1971) conclusions that image codes were associated with a spatial code. This association was also found in the present study though the stimuli were not pictures but were nonverbalizable complex shapes. However, a display might in appropriate conditions not be derived from its sensory modality characteristics, but could be evoked from memory in another form, which would be more appropriate for the material. In the alternative condition of this experiment, the visually presented items were Roman rather than Arabic letters. The same incongruent presentations with regard to the temporal and spatial order of the items was used. The first letter was never presented on the extreme left, the second was never the middle one, and the third never appeared on the right. The children were again asked to either arrange separate letter cards or to select an arrangement displayed on a response card to match the series just presented.

For the normal children, a change of response strategy was clearly observable in this condition. Whereas there had been a significant bias toward a left to right arrangement with Arabic letters, there was now a clear tendency toward remembering the first to last order of presentation. However, congenitally deaf and hearing-autistic children, who had also been presented with this material, tended to remember the Roman letters, like the Arabic letters, according to their left to right spatial rather than their first to last, temporal, presentation order.

The conclusion from these results must be that although nonverbalizable Arabic letters were stored as a visual image that was scanned from left to right, the Roman letters seem to have been verbalized in the same temporal order in which they occurred, they were then remembered in this same order. Thus, though the letters were visually presented, these normal children coded them as if they had been heard. Deaf children, as has already been pointed out, are unable to transfer verbal material from visual into auditory form. They are also less inclined toward implicit verbalization of visual displays, and may therefore rely more on features characteristically associated with visual stimulation. Space, not time, thus provided the framework within which these deaf children coded.

A similar explanation to that accounting for these results for the deaf could also be used to interpret the results from the autistic children, although for different reasons. One could argue with Piaget that the representations of a visually presented letter by an auditory symbol requires the manipulation of symbols and that this is the area of cognitive functioning where autistic children show the most severe deficit. They remained stimulus bound and used a visually based, rather than auditory–verbally derived memory code in the above experiment. To investigate whether the response mode (i.e., ordering printed letters) rather than the visual presentation modality determined the results, the second part of the experiment was repeated using verbal recall (O'Connor & Hermelin, 1978). Normal, deaf, and autistic children were individually presented with three digits that were shown such that the first to last and the left to right order of presentation did not correspond. Each child was given 16 trials; the mean results are shown in Table 2.1.

The pattern of these results corresponds to that previously obtained. Thus it seems that this pattern is determined by the memory code used to store the items rather than by the particular recall mode.

Kinesthetic Processing

We will now turn from the study of the relationship between auditory–temporal and visual–spatial coding strategies to aspects of coding that are characteristic of kinesthetic information processing. This is of particular interest to our discussion because movements have both spatial and temporal aspects.

TABLE 2.1
Recall Order

Subjects	Temporal	Spatial	Errors
Normal	12	3	1
Deaf	2	13	1
Autistic	4	10	1

Forward and Backward Tracking

In the first of these experiments we compared sighted and congenitally blind 11- to 12-year-old children on a tracking task (O'Connor & Hermelin, 1978). The sighted children were blindfolded for the experiment. The tracks varied in length from three to nine lines and they were continuous or set at right angles to each other. For example, the lines of a track might run "up–right–down–down–left–up–right–down–right." After a training trial in which the child ran the right index finger along such a track he or she was presented, on a recognition trial, with a second track of equal length. This track would be either identical to the first or its component lines might be differently arranged (e.g., "up–left–left–down–right–down–down–up." The task was to decide after going along the second track with the same finger as before whether or not it was identical to the first. During tracking the child was asked to count to prevent implicit verbalization of structural features.

There were two conditions for the recognition trials. In one the child had to begin tracking the second track of a pair from the same starting point and in the same direction as in the initial trial. In the second condition, tracking for recognition proceeded in the direction opposite that in the first of the pair of tracks. Thus a subject who had tracked left to right on the training trial had to track right to left on the recognition trial. For these reversed recognition trials the children's hands were placed on the new starting point and it was explained to them that they were now going to go backward, but still had to decide whether the finger was traveling along a path that was the same or different from the first one.

It was found that blind and blindfolded children were equally efficient when a pair of tracks about which they had to make a decision had both been traced in the same direction. There was also no significant difference in the degree of accuracy with which longer and shorter pairs of tracks were compared. It thus seems that two successive series of arm movements could be adequately remembered for comparison, which implies that at least for the blind, kinesthetic memory was sufficiently distinct.

The pattern of results obtained for reverse tracking was more complex. For short tracks consisting of three or four lines there was no significant difference between the groups in the number of accurate judgments. However, with longer tracks blind children had increasingly more difficulty than the blindfolded children. When the tracks consisted of six or more lines this trend became statistically highly significant.

We mentioned before that although visual information can presumably be reviewed independently of direction, a series of auditory stimuli tends to be stored and reviewed in the same order in which it initially occurred. From the present study it seems that movement components, like sounds, are also stored in a temporal–sequential order. As with sounds, reversal of such a temporally ordered series is relatively easy when the series is short and contains only a few elements. However, such mental reversal seems to become increasingly difficult as the series gets longer, and from our results one could assume that the sighted children might have used visual memory images of the tracks to provide them with a direction independent comparison. Such a strategy was, of course, not possible for the blind.

Location and Distance

The crucial importance of vision for the development of a spatial framework had been emphasized by von Senden (1960) who even thought that an adequate conceptual grasp of space could not occur without at least some experience of vision. Other investigators, such as Adams and Dijkstra (1960), Posner and Konic (1966), Posner (1967), and Stelmach and Wilson (1970) have suggested that movement information is at least partly coded in visual form. A paradigm often used in this context has been the recall ability of either the terminal location or the distance traveled by an arm movement, when the subject was prevented from seeing the execution of this movement. Martiniuk and Roy (1972) concluded that distance information was not easily codable when the only data available was derived from limb movement. They, as well as Stelmach and Wilson (1970) and Roy (1977) found recall of terminal location to be more accurate than recall of distance. Posner (1967) and Laabs (1973) found distance information to be nonrehearsable, whereas Diewert (1975) reported that provided explicit instructions were given distance information could be rehearsed. Laabs (1973) has proposed that distance cues may be coded by speed of movement derived from timing or counting and Roy and Diewert (1978) thought that timing or extrapolation from the changing location information provided the data for distance estimates. However, Stelmach and Wilson (1970) as well as Doody (1977) have reported that coding of distance was not enhanced through the provisions of more location information related to the change in starting position. Recently, Walsh, Russell, Imanaka, and James (1979) found that subjects were unable to utilize the change of starting position to recall distance more accurately. They found that the greater the change in starting location, the larger the errors in distance as well as location estimate became, with distance tending to be under- and terminal location to over-estimated. They concluded that memory for movement was based on an interaction between location and distance cues. The nature of location and distance coding is still undetermined and under discussion.

We carried out an experiment with normal, autistic, and congenitally blind 12-year-old children on their memory for location or distance information (Hermelin & O'Connor, 1975). The autistic and normal children were blindfolded during the experiment. The apparatus consisted of a 50-cm rod that was mounted vertically on a platform placed on a table. A centimeter scale ran along the rod, which had a movable pointer attached to it, as well as a movable stop which could be put at different points. For the initial trials this end stop was a set at a height of either 10, 15, 20, 25, 30, 35, or 40 cm in predetermined random order, so that the required movement had to cover these distances. The subject was asked to move the lever from the base up to the stop three times with the right hand. The stop was then removed and the starting position of lever either remained as before or was set at a new point (e.g., 10 cm above the base).

For the first condition, the subject's task was to stop at the same endpoint as in the three preceding training trials, either from the same or from an altered starting position. Thus what had to be reproduced was the terminal location of a previously experienced arm movement when the starting point was either the same or different. The second condition tested retention of the extent of a previously experienced movement. In this condition the subject was instructed to push the lever along the rod until it had covered the same distance as it had traveled in the three preceding training trials. Thus if in these trials the lever had been set at a height of 15 cm above base and the stop at 25 cm, it would have traveled 10 cm. From a new starting position of 20 cm above base the pointer would now have to be pushed up to a height of 30 cm to match the distance covered in the training trials.

An error analysis showed that there was no significant difference in the size of error of location estimates between the normal sighted and the blind children. The errors of the autistic children, however, were significantly greater than those of the other two groups whether or not the starting position was changed. All three groups tended to overshoot the previously experienced terminal position. In distance estimates the blind children were significantly worse than the blindfolded normal children, particularly in the estimation of longer distances. The blind and the autistic children did not differ from each other. All children tended to overestimate the shorter distances and to underestimate the longer distances.

The first point to be made about the results is that the autistic children seem to have found the tasks very difficult. However, their errors were not large enough to indicate that they had not understood the task requirement. The reasons for their particular difficulties will need further investigation. As for the blind children, it would seem that lack of sight does not specifically impair recall for a terminal arm position. Thus even if location should normally be stored in a visual–kinesthetic code, as Diewert (1975) postulates, this seems not to be the only possible efficient coding mechanism. As Martiniuk and Roy (1972) have pointed out, the position of a limb in space is signaled very precisely through feedback mechanisms by cells in the muscles and joints; apparently this information can be recalled adequately without any visually derived coding. However, differences in distance estimates were found between blind and sighted children. In view of this difference, it seems unlikely that sophisticated time-based codes were used by the children to recall distance, since such a strategy would have been equally available to all subjects. One may thus conclude from the results on distance estimates that for the sighted children this task was facilitated by the use of a visually derived image code. This interpretation is in conflict with that of Diewert (1976), who holds that evidence indicates that visual imagery plays no decisive part in distance coding. Dickinson (1977), who found that a blind subject could retain location information better than distance information, suggests that there may be two different kinds of proprioceptive stores, one more efficient than the other.

CONCLUSIONS

These few experiments are examples from a larger number published elsewhere (O'Connor & Hermelin, 1978), all of which illustrate the themes outlined at the beginning of this chapter. The first of these is that stimulation in different sensory modalities results in different coding strategies. The codes that are extracted not only contain the essential meaning of the stimulation, but also preserve some of the sensory specific characteristics of the material. Two of the described experiments illustrated this. In the first of these, series of sounds were more easily recognized when they were temporally than when they were spatially structured; the reverse was found for sequences of lights. In the second study, visual processing led to a stable and invariant organization of points in space, whereas when touch was applied without vision, the spatial framework was structured in subjective and relative terms. The findings from these two studies applied to normal children as well as to those subjects with perceptual impairments or cognitive deficits.

The results from both experiments confirm conclusions of Simpson (1973), Jones and Kabanoff (1975), Warren (1970) and others about the importance of the visual system for coding spatial aspects of stimulation. In both experiments the availability of vision contributed to the structuring and stabilization of the spatial framework within which the stimuli were perceived. However, neither auditory nor kinesthetic information tended to be processed primarily in spatial terms. Since the cognitive requirements of the tasks were the same for all modalities, this must indicate a particularly close connection between space and vision. Our experiments were designed so that qualitatively different coding processes rather than differential efficiency for different modalities could be demonstrated. These coding differences reflect strategy biases and preferences rather than absolute limitations in the ability to extract appropriate codes from different sensory information. Nevertheless, such biases may reflect real processing and storage differences and indicate limitations of the concept of a modality independent spatial reference schema.

The second set of studies was concerned with the suggestion that in many instances codes may be based on mental images and representations rather than on the characteristic features of the perceptual system that has been addressed. In one experiment that demonstrated this, normal children responded to visually presented, nonverbal material by remembering its spatial presentation order. However, when the visually presented items were verbal, they were remembered in a first to last temporal order as if the children had heard rather than seen them. However, deaf as well as autistic children, who do not tend to verbalize spontaneously, remembered both verbal and nonverbal items in terms of a visual–spatial modality code. Similarly, when asked to reproduce a distance covered by a previously experienced arm movement, congenitally blind children were found to be less accurate than blindfolded sighted subjects. As both groups were found to be equally able to estimate a previous terminal position, it is difficult to escape the conclusion that the possibility of evoking visually derived images helped the sighted children to recall distance. However, this seems not to have given them an advantage in location recall, for which the directly provided proprioceptive information seems to have been sufficient.

One other experiment was concerned with the visual representation of kinesthetically experienced space. It was concluded that normal children when blindfolded seem to have evoked visual images to reverse a sequence of movements. Blind and autistic children had difficulty with such reversals suggesting that the kinesthetic information on which they were presumably relying was less efficient.

As far as the perceptually impaired children are concerned, these findings are clearly interpretable. Deaf children cannot evoke the sound of words and the blind cannot form visual images. Thus the available coding repertoire of the congenitally blind and deaf children is restricted. The deaf, even though their vision is intact, may nonetheless extract different features from visually presented material than would the hearing. Similarly, the blind may respond differently to kinesthetic stimuli than the sighted and may be less efficient in processing such information.

It was also found in these studies that children with general cognitive deficits responded in a manner that was similar to that observed in the deaf and blind. Thus a general deficit resulted in apparently specific perceptual abnormalities. This was particularly marked in children with infantile autism. These findings will have to be interpreted in conjunction with the relevant clinical features of the syndrome. One possible explanation lies in the characteristic memory processes found in infantile autism. These are best illustrated by comparing the serial position effects in immediate memory tests. With normal children, immediate recall of a series of presented items typically results in strong primacy as well as recency effects. The primacy effect (i.e., the efficient recall of the first few items in a list) has been attributed to the possibility that these early items might already have been rehearsed and transferred into a more permanent longer term memory store, where they would be better retained. The recency effect (i.e., the better retention of the last few items) may be caused by additional sensory after effect, which persist briefly (Sperling, 1960; Neisser, 1967).

Autistic children, who have very good short-term memories, show very strong recency effects. These recency effects frequently cover more items from the end of the list than one would normally expect; the same tendency can be observed in echolalia which is typical for autistic children. Indeed, one can also observe long-term echolalia in autistic children which makes one suspect that they use the same processing strategies that are typical of normal short-term memory processes also for long-term retention. This may explain their overdependence on perceptual stimulus features and their reluctance to evoke mental representations, which was deduced from our experiments.

There is one further point to be made in conclusion. The position taken throughout this chapter has been that vision was particularly suited to provide data for spatial coding. This conclusion should be extended to those instances in which information is directly available but has to be evoked from previously established visually derived spatial schemas. Hearing and proprioception may at times provide sufficient information for effective spatial coding. However, the relative restrictions for structuring information within a stable spatial framework which we found in the blind indicates the crucial role of vision and visual images for the development of a coordinated schema of space.

REFERENCES

Abramsky, O., Carmon, A., & Benton, A. Masking of and by tactile pressure stimuli. *Perception and Psychophysics*, 1971, *10*, 353–355.

Adams, J., & Dijkstra, A. Short-term memory for motor responses. *Journal of Experimental Psychology*, 1960, *71*, 314–318.

Attneave, F., & Benson, L. Spatial coding and tactual stimulation. *Journal of Experimental Psychology*, 1969, *81*, 216–222.

Auerbach, C., & Sperling, P. A common auditory–visual space: Evidence for its reality. *Perception and Psychophysics*, 1974, *16*, 129–135.

Bower, T. G. R., & Wishart, J. G. The effects of motor skill on object performance. *Cognition*, 1972, *1*(2), 9–27.

Bower, T. G. R., & Wishart, J. G. Development of auditory–manual co-ordination. *Cognition*, 1973, *4*(2), 56–74.

Bruner, J. S. The growth and structure of skill. In K. Conolly (Ed.), *Mechanisms of motor skill development*. New York: Academic Press, 1970.

Conrad, R. Errors of immediate memory. *British Journal of Psychology*, 1959, *50*, 349–359.

Conquery, J., & Amblard, B. Backward and forward masking in the perception of cutaneous stimuli. *Perception and Psychophysics*, 1973, *13*, 161–163.

Diewert, G. L. Retention and coding in motor short term memory: A comparison of storage for distance and location information. *Journal of Motor Behaviour*, 1975, *7*, 183–190.

Diewert, G. L. The role of vision and kinesthetics in coding of two dimensional movement information. *Journal of Human Movement Studies*, 1976, *3*, 191–198.

Dickinson, J. Distance and location in retention of movements by a congenitally blind subject. *Journal of Psychology*, 1977, *97*, 215–219.

Doody, S. G. Subject strategies in motor short term memory. In D. M. Landers & R. Christina (Eds.), *Psychology of motor behaviour and sport* (Vol. 1). Champaign, Illinois: Human Kinetics, 1977.

Fraiberg, S., Siegel, B. L., & Gibson, R. The role of sound in the search behaviour of a blind infant. *Psychoanalytic Study of the Child*, 1966, *21*, 327–357.

Freedman, D. G. Smiling in blind infants and the issue of innate versus acquired. *Journal of Child Psychology, Psychiatry and Allied Disciplines*, 1964, *5*, 171–184.

Geldard, F. A. Cutaneous coding of optical signals. *Perception and Psychophysics*, 1966, 377–381.

Geldard, F. A., & Sherrick, C. E. Cutaneous stimulation: The discrimination of vibratory patterns. *Journal of the Acoustic Society of America*, 1965, *37*, 797–801.

Gibson, E. J. *The senses considered as perceptual systems*. Boston: Houghton–Mifflin, 1966.

Held, R. Exposure history as a factor in maintaining stability of perception and co-ordination. *Journal of Nervous and Mental Diseases*, 1961, *132*, 26–32.

Hermelin, B., & O'Connor, N. Location and distance estimates by blind and sighted children. *International Journal of Experimental Psychology*, 1975, *27*, 295–301.

James, W. The principles of psychology. In *The perception of space* (Vol. 2, Chap. 20). New York: Dover Publications, 1980. (Originally published, 1890.)

Jones, B. Visual facilitation of auditory localization in school children. *Perception and psychophysics*, 1975, *17*, 217–220.

Jones, B., & Kabanoff, B. Eye movements in auditory space perception. *Perception and Psychophysics*, 1975, *17*, 241–245.

Laabs, G. J. Retention characteristics of motor short term memory cues. *Journal of Motor Behaviour*, 1973, *5*, 249–259.

Light, L. L. Trade off between memory for verbal items and their visual attributes. *Journal of Experimental Psychology*, 1975, *104*, 188–193.

Marks, L. E. Multimold perception. In E. C. Carterette & M. P. Friedman (Eds.), *Handbook of perception* (Vol. 8). New York: Academic Press, 1978.

Martiniuk, R. G., & Roy, E. A. The codability of kinesthetic location and distance information. *Acta Psychologica*, 1972, *36*, 471–479.

Neisser, U. *Cognitive psychology.* New York: Appleton–Century–Crofts, 1967.

Nilsson, L. O. Further evidence for organization of modality in immediate memory. *Journal of Experimental Psychology,* 1974, *103*(5), 948–957.

O'Connor, N., & Hermelin, B. *Seeing and hearing and space and time.* New York: Academic Press, 1978.

Paivio, A. Mental imagery in associative learning. *Psychological Review*, 1969, 76, 241–264.

Paivio, A. *Imagery and verbal processes.* New York: Holt, Rinehart & Winston, 1971.

Piaget, J., & Inhelder, B. *The child's conception of space.* London: Routledge & Kegan Paul, 1956.

Pick, H. L., Acredolo, L. P., & Gronseth, M. *Children's knowledge of the spatial layout of their homes.* Paper presented at the Society for Research in Child Development, Philadelphia, 1973.

Pick, H. L., Klein, R. E., & Pick, A. D. Visual and tactual identification of form orientation. *Journal of Experimental Child Psychology,* 1966, *4,* 391–397.

Platt, B. B., & Warren, D. H. Auditory localization: The importance of eye movements and a textured visual environment. *Perception and Psychophysics,* 1972, *12,* 245–248.

Posner, M. I. Characteristics of visual and kinesthetic memory codes. *Journal of Experimental Psychology,* 1967, *75,* 103–107.

Posner, M. I., & Konick, A. F. Short term retention of visual and kinesthetic information *Organizational Behaviour and Human Performance,* 1966, *1,* 71–86.

Roy, E. A., & Diewert, G. L. The coding of movement extent information. *Journal of Human Movement Studies,* 1978, *4,* 94–101.

Schopenhauer, A. Vierfache wurzel des satzes vom zureichenden grunde. In Die welt als wille und vorstellung. Leipzig: Cotter Verlag, 1819.

Senden, M. von. *Space and Sight.* London: Methuen, 1960.

Simpson, W. E. Latencies in intermodal spatial localization. *Journal of Experimental Psychology,* 1973, *99,* 148–150.

Snodgrass, J. G., McClure, B. P., & Pirone, G. V. Pictures and words and space and time: In search of the elusive interaction. *Journal of Experimental Psychology,* 1978, *107,* 206–230.

Sperling, G. The information available in brief visual presentations. *Psychological Monographs,* 1960, *74*(11).

Stelmach, G. E., & Wilson, M. Kinesthetic retention, movement extent and information processing. *Journal of Experimental Psychology,* 1970, *85,* 425–430.

Walsh, W. D., Russell, D. G., Imanaka, K., & James, B. Memory for constrained and preselected movement: Location and distance, effects of starting position and length. *Journal of Motor Behaviour,* 1979, *11,* 210–214.

Warren, M. Intermodality interactions in spatial localizations. *Cognitive Psychology,* 1970, *1,* 114–133.

Worchel, P. Space perception and orientation in the blind. *Psychological Monographs,* 1951, 65(332).

Wundt, W. *Lectures on human and animal psychology.* London: Sonnenschein, 1894.

EMERSON FOULKE 3

Perception, Cognition, and the Mobility of Blind Pedestrians

Many handicaps can be overcome, or at least reduced, by acquiring new skills and by using special instruments. However, the limited ability of blind pedestrians to travel independently (i.e., to achieve mobility) is their most serious handicap, because the efficient performance of so many tasks demands this ability. Blind persons who can travel independently are often able to carry out plans of their own devising whereas blind persons with little ability to travel independently are often compelled to lead lives characterized by passive acquiesence to plans made for them by others and they are deprived of opportunities to employ whatever skills they may have acquired.

There have always been some blind persons who have been able to achieve mobility; however, the development of a systematic method for teaching mobility to blind pedestrians is a recent accomplishment, made possible by the insights of Richard Hoover (1968) who, while stationed at Valley Forge Army Hospital, was assigned the task of seeking ways to rehabilitate blinded soldiers. Hoover had observed that many blind persons seemed to use canes to some advantage and realized that the usefulness of a cane could be greatly increased by making it long enough so that when its tip made contact with an obstacle in the path ahead there would be time for corrective action, and by teaching its users a scanning procedure that would ensure that with each new step the foot would fall on a portion of the surface already examined by the tip of the cane (Hill & Ponder, 1976; Suterko, 1973). Thus, the long cane was born and a new

55

SPATIAL ABILITIES
Development and Physiological Foundations

mobility (O&M) *specialists*. They are now regarded as essential staff members of the schools where blind children are educated and the rehabilitation agencies where blinded adults are helped to regain their independence (Wiener, 1980; Wiener & Welsh, 1980). Although methods and procedures appear to vary considerably among O&M specialists (Warren & Kocon, 1974), the usefulness of the training they provide is beyond doubt. Nevertheless, the mobility of trained blind pedestrians falls far short of the mobility of typical sighted pedestrians (Cratty, 1971a).

Many attempts have been made to improve the mobility of blind pedestrians by means of electronic travel aids (ETAs), which provide some of the spatial information lost when vision fails (Clark, 1963, pp. 7–204; Farmer, 1980; Jansson, 1975; Kay, 1974; Mann, 1974; Mims, 1973; Thornton, 1971). Unfortunately ETAs have not worked very well (Airasian, 1972, 1973; Graystone & McLenna, 1968). The reason for this has, in part, been that the signals they display have not been specifically designed to take maximum advantage of the perceptual abilities of auditory observers (Levy & Butler, 1978; Mershon & Bowers, 1979; Sakomoto, Gotoh, & Kumura, 1976). The long cane continues to be the most effective inanimate aid to independent mobility. Though ETAs have not lived up to their expectations, experience with them has been instructive. Analysis of the performance enabled by these aids has made it clear that we do not yet know what spatial information is critical to the successful performance of blind pedestrians, where in the environment that information is to be found, to what perceptual system or systems it should be displayed, and in what manner it should be displayed. In short, effective ETAs have not yet been built because we do not yet understand mobility. It is now evident that the mobility of blind pedestrians will not be further improved until we have made a detailed analysis of the task they perform, and the perceptual and cognitive abilities on which the performance of that task depends. To do this, we must be able to assess the performance of blind pedestrians objectively and quantitatively (Armstrong, 1975; Shingledecker & Foulke, 1978).

THE TASK PERFORMED BY BLIND PEDESTRIANS

Definition

The criteria of mobility include the ability to travel safely, comfortably, gracefully, and independently (Foulke, 1971; Suterko, 1973). The criterion of safety is obvious, but the criteria of comfort and grace may need some explanation. Blind pedestrians lack much of the preview of upcoming situations that enables sighted pedestrians to program their responses to deal effectively with those situations (Barth & Foulke, 1979). Without adequate preview, blind people are often unable to process incoming information in the time that is available for processing (Shingledecker & Foulke, 1978). Furthermore, they must often make decisions involving risks (Foulke, 1971), such as the decision to cross a busy street, without enough information to eliminate uncertainty. For these reasons, they experience stress which, if prolonged, becomes uncomfortable (Heyes, Armstrong, & Willans, 1976; Peake & Leonard, 1971). Blind pedestrians for whom the experience of stress is chronic may avoid traveling. Blind persons who travel independently are, on most occasions, performing in public. If their performances are clumsy or awkward, they will make unfavorable impressions on those around them, thus reinforcing the negative stereotypes concerning blindness with which they must contend (Scott, 1968).

Finally, a satisfactory definition of *mobility* must include the criterion of purposeful mobility (Suterko, 1973). To demonstrate goal-oriented mobility, travelers must know where they are at present, where their goals are, and how to move from their present positions to their goal positions. Acquiring this knowledge depends upon spatial ability; that is, the ability to discover by observation, the shapes and sizes of things, their distances, directions, and relative positions, and their states of motion.

Task Analysis

Mobility is a clear example of what Poulton (1957) has called an open skill. The task performed by blind pedestrians is complex, and it is performed in a continuously changing environment that is not entirely predictable. The sense data needed for performance of the task must be distinguished from a flood of potentially masking sense data and other kinds of interference. Performance of the task is difficult to observe because the sequences of movements made to meet task demands are highly integrated and it is difficult to isolate behavioral events for close observation. One way to begin the analysis of the task performed by blind pedestrians is to try to write the program that would be required by computer-controlled robot equipped with the appropriate sensors to perform the same task. Figure 3.1 is an example of a flowchart that might be drawn in preparation for writing such a program. The process described in this flowchart is both incomplete and schematic but it does indicate several aspects of the problem. To begin with, it describes a purposeful task, a task that is performed to reach a goal. The first four steps in the flowchart indicate that blind pedestrians must have spatial ability to perform the task. They have to be oriented in space, and if they find that they are not, they must acquire spatial information. Steps 5 and 6 indicate that performance is guided, in part, by information obtained from a memorial representation of the space in which the task is performed. Steps 7 through 14 indicate that performance is also guided by contempora-

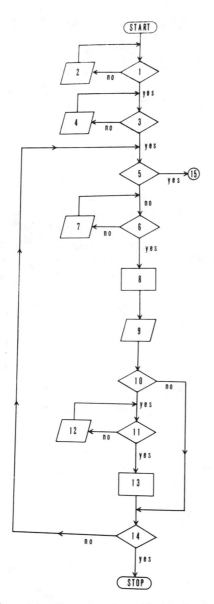
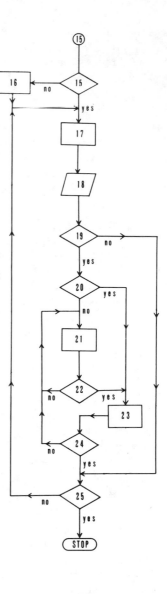

Figure 3.1. The task performed by blind pedestrians:

1. Is goal position known?
2. ACQUIRE INFORMATION
3. Is starting position known?
4. ACQUIRE INFORMATION
5. Is memorial representation of sector of environment containing starting position and goal position sufficient to afford knowledge of one or more routes to goal?
6. Is memorial representation of sector of environment containing starting position and goal position sufficient to permit choice of direction?
7. ACQUIRE INFORMATION

(continued on page 59)

58

neous information; that is, information obtained by observing the space in which the task is performed while the task is in progress. This information supplements spatial knowledge, and functions as feedback. It tests the adequacy of performance, and if the test fails, it indicates the required correction. Although there is no direct indication in the flowchart, it can be inferred that the relative contributions of the information supplied by the memorial representation and the information obtained while the task is in progress depend on the amount of experience with the space in which the task is performed. With little or no experience, blind pedestrians must rely heavily on contemporaneous information. As experience accumulates, they are guided more by information supplied by their memorial representations and less by contemporaneous information.

INFORMATION-PROCESSING COMPONENTS OF LIVING SYSTEMS

The description just given implies some of the components that would have to be included in an information-processing system capable of performing the task. Such a system would, for instance, have to be able to sense the environment, store information, compare items of information, and decide among possible courses of action. The components of the information-processing system described by Miller (1978, Chapter 8) are shown in Figure 3.2.

Transducers receive patterns of energy in one form and transmit analogous patterns of energy in another form. The input and internal transducers of a living system receive patterns of energy—luminous, acoustical, mechanical, and so forth—from the environment of the system and transmit analogous patterns of neural impulses to other components of the system. A *channel* is a route from a transmitter to a *receiver*; a *net* is a set of channels that intersect at points called *nodes*. A net makes it possible for the components of a system to exchange information. *Decoders* convert the patterns transmitted by input and internal transducers to a private code for internal use by the system. Feature detectors, such as

8. ADVANCE IN CHOSEN DIRECTION TO NEXT POSITION
9. ACQUIRE FEEDBACK DATA
10. Do feedback data indicate error?
11. Do error data specify required correction?
12. ACQUIRE DIRECTIONAL INFORMATION
13. MAKE CORRECTION
14. Does present position equal goal position?
15. Is there one and only one route to goal?
16. DISCARD ALL BUT BEST ROUTE
17. ADVANCE TO NEXT ROUTE POSITION
18. ACQUIRE FEEDBACK DATA
19. Do feedback data indicate error?
20. Do error data specify correction?
21. ADVANCE AT RANDOM UNTIL INTERPRETABLE FEEDBACK INFORMATION IS ACQUIRED
22. Is correction specified?
23. MAKE CORRECTION
24. Is route regained?
25. Does present position equal goal position?

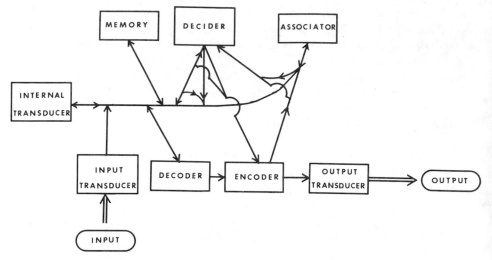

Figure 3.2. An information-processing system. (This is a partial replication of Figure 1.1, in Miller, 1978.)

the edge detectors found by Hubel and Wiesel (1962) are examples of decoders. The *associator* forms associations among the items of information received from input transducers, internal transducers, and *memory*. The functions of memory are the storage and retrieval of the information it receives from input transducers, internal transducers, and from the associator. The *decider* controls the entire system by causing its components to interact in a coordinated manner. It reviews information received from input transducers, internal transducers, and memory, evaluates various alternatives, computes a solution to the problem at hand, and transmits command signals to *encoders* and to other components within the system. Encoders assemble programs of properly sequenced and timed motor impulses that control output transducers. Output transducers change the private code of the system to a public code that is suitable for use in the environment of the system.

Figure 3.2 is not intended to imply that a specific structure is responsible for each of the indicated functions. The figure is simply a convenient way of depicting a set of related functions that must necessarily be included in a system capable of the kind of information processing that characterizes blind pedestrians and other humans.

The Acquisition of Spatial Information

Perception is the acquisition of the information contained in the stimulation that impinges on input and internal transducers. In addition to the functions performed by input transducers, internal transducers, and decoders, perception requires cognitive functions, such as association and memory, and since perceivers are seekers of information it also requires motor functions. Thorough discussions of the structures and functions of perceptual systems are readily available (e.g., Gibson, 1966). I will discuss only properties of perceptual systems that define the capacity for acquiring information about space.

THE VISUAL SYSTEM

The visual system is the spatial system par excellence (see Sedgwick, Chapter 1, this volume; Hermelin & O'Connor, Chapter 2, this volume). It has both a large field of view and selectivity of regard. It is much better at resolving spatial patterns than the other perceptual systems. Physical contact is not required for the excitation of visual receptors. An important consequence of this fact is receptor anticipation (Poulton, 1952) which is the anticipation made possible by the ability to observe the features of a situation in advance of the time at which some action will be required. It is important because it increases the usefulness of visual information. Poulton distinguishes between receptor anticipation and perceptual anticipation, which is made possible by consulting the memorial representation of that situation. It seems to me that Poulton's *receptor anticipation* would be named more accurately *perceptual anticipation,* and that his *perceptual anticipation* would be named more accurately *cognitive anticipation.* Accordingly, I will use the terms *perceptual* and *cognitive anticipation* hereafter to distinguish between the two kinds of anticipation.

As a result of the properties of the visual system just discussed, visual observers are able to answer two kinds of questions about space, what and where (Armstrong, 1975). By visual observation, they know what things are in the space around them and where those things are (Howard, 1973a).

THE AUDITORY SYSTEM

The auditory system is well suited for the analysis of temporal patterns (Green, 1978; Julesz & Hirsch, 1972; Hermelin & O'Connor, Chapter 2, this volume). The information in the acoustic energy that reaches the ears is discovered by interpreting temporal variation. As a consequence, acoustic energy can contain information about the sequence and timing of acoustic events (Julesz & Hirsch, 1972) but no information about the shapes and sizes of objects.

The auditory field is large. Physical contact with auditory receptors is not a requirement for stimulation, and sounds reach the ears from all directions. Within the auditory field, auditory observers can make judgments concerning distance (Batteau, 1968; Coleman, 1963; Fisher & Freedman, 1968), direction (Deatherage, 1966; Simpson, 1972), and state of motion of sound sources (Howard, 1973b; Mills, 1972), but not as accurately as visual observers can (Howard, 1973a). Furthermore, many things in the space encompassed by the auditory field are not sources of sound, and many are sources of sound only part of the time (Foulke & Berlá, 1978). For these reasons, the auditory system is not able to acquire as much information about the positions of things in space as the visual system can. The auditory system is capable of perceptual anticipation, but its value is limited because things in the environment and their positions are not as identifiable by sound as by sight.

Although the auditory stimulus does not contain information about shape and size, auditory observers can often learn to identify objects, or at least categories of objects by the sounds they make. The sound of a car's horn will inform auditory observers who have learned to associate horns with cars that a car is in the vicinity, but they will not know about such spatial qualities as its shape and size.

The auditory system has little or no selectivity of regard. Although auditory observers can, by turning their heads, favor the reception of some sounds over others (Wallach, 1940), auditory selectivity falls far short of visual selectivity (Howard, 1973a). Auditory observers cannot while listening to some sounds exclude other sounds from auditory observation. As a result, the auditory system is more vulnerable than the visual system to interference (Green, 1978).

The auditory system is less able than the visual system to answer "where" questions (Martinez, 1977). The locations of some objects are indicated by sounds, and if they are close enough some objects can be located by echo perception (Ammons, Worchel, & Dallenbach, 1953; Cotzin & Dallenbach, 1950; Juurmaa, 1970a, 1970b; Kohler, 1964; Rice, 1967; Rice & Feinstein, 1965a, 1965b). However, most of the objects with which space abounds escape auditory observation.

THE HAPTIC-PROPRIOCEPTIVE SYSTEM

Observers who learn what things are in the space around them and where those things are by walking to them, reaching for them, and feeling them are informed by both haptic perception and proprioception (Pick, 1980). The transducers on which haptic perception depends include the input transducers that are excited by mechanical stimulation of the skin, and the internal transducers of kinesthesis (Howard, 1973b). Haptic perception provides information about the shapes, sizes, surface textures, and states of motion of things within arm's reach. Proprioception depends on the internal transducers of kinesthesis and the internal transducers of the equilibratory senses (Howard, 1973b). Proprioception provides information about the relative positions of parts of the body and the orientation of the body with respect to the earth. Proprioceptive information is also provided by the visual system, but that will not be considered further here. Because the equilibratory (vestibular) transducers are excited when the body is in motion, they provide information about space that is explored by walking (Beritoff, 1965; O'Keefe & Nadel, 1979). Haptic perception and proprioception employed cooperatively contribute information about the shapes, sizes, surface characteristics, and states of motion of things, and their relative positions in space. This system also provides information about the contours of the surface underfoot and, as will be seen later, this is a particularly important kind of information for blind pedestrians.

Because things must be touched by blind people to appreciate their sizes, shapes, surface textures, and states of motion, the field of the haptic–proprioceptive system is small, and relatively little can be observed at one time. As a consequence, unlike visual observers, haptic–proprioceptive observers are compelled to examine most objects and spaces serially. The serial perception of parts that results from serial examination must then be integrated to achieve the perception of whole objects and spaces (see Hermelin & O'Connor, Chapter 2, this volume).

That physical contact is a requirement for haptic perception, means little perceptual anticipation is possible. Because the position of the body must be changing in order for equilibratory transducers to make their contribution to the perception of space, that contribution is realized as a function of time: Equilibratory transducers are not capable of perceptual anticipation. Of course, their contribution to the perception of space is incorporated in the memorial representation of space that makes cognitive anticipation possible. The haptic–proprioceptive system has more selectivity of regard than either the visual or the auditory system because it can perceive only those things with which it makes contact. There is nothing corresponding to the peripheral vision that provides a minimal awareness of events in the visual field that are outside the focus of attention.

The haptic–proprioceptive system cannot provide as much information about the composition of space as can the visual system. Furthermore, many things that are observable visually are not accessible for observation by touch, because the scale is wrong (they are too large or too small), they cannot be reached, they are too fragile or too flimsy, or too dangerous to touch (Foulke, 1962; Foulke & Berlá, 1978). Nevertheless, in the absence of vision, the haptic–proprioceptive system is the best perceptual system for acquiring information about the composition of space.

The remaining senses can provide some spatial information by way of associative learning (Foulke, 1971; Suterko, 1973). By feeling the heat of the sun on the cheek and by taking into account the time of day, direction can be estimated. The aroma coming from the doorway of a bakery can serve as a landmark. However, the remaining senses contribute relatively little to the perception of space.

PERCEPTUAL ANTICIPATION WITHOUT VISION

Both blind and sighted pedestrians must walk at a reasonable speed, avoid obstacles and other hazards, and follow routes to goals. However, the differential information that is available for mediating behavioral decisions, makes the two tasks quite different. Sighted pedestrians employ a perceptual system that is ideally suited for acquiring the spatial information they need. Blind pedestrians must employ perceptual systems that are not as effective.

Sighted pedestrians have more perceptual anticipation than blind pedestrians. As a consequence, they do not experience the overloads of information that result when perceptual anticipation is inadequate (Shingledecker & Foulke, 1978). They perform mobility tasks with ease and they usually have enough remaining channel capacity to engage simultaneously in other tasks, such as conversing with a friend or observing the contents of a store window. Shingledecker (1976) varied the preview available to blind subjects as they engaged in a mobility task. His results suggest that the inferior performance of blind pedestrians is, in part, a consequence of information processing requirements that intermittently exceed channel capacity. On such occasions, they do not have enough time to process the information available to them and program the appropriate responses. They bump into trees and stumble over curbs even though they may know that the trees and curbs are there.

The information made available to blind pedestrians by the feel of the surface underfoot is a particularly clear example of information that is received without adequate perceptual anticipation. Blind pedestrians are path followers; the surface over which they walk is the source of much of the information they need. The surfaces of paths are composed of materials different from that of the surfaces of the spaces they bound, and those differences can be detected by the way they feel. The concrete or brick sidewalk and the grassy or graveled surface beside it are easily distinguished. The discontinuity in the surface as one passes from sidewalk to street or from street to sidewalk is usually obvious. Outdoor surfaces are rarely uniform and their irregularities, such as changes in texture and slope, serve as landmarks or cues to appropriate action. For example, one blind pedestrian might learn that on a particular route the sidewalk begins to go uphill just before the intersection at which he is to turn left, and that the hump in the sidewalk caused by the tree root underneath marks the place at which he turns to enter his house. Of course, the problem here with using surface information is that it is delivered only when their feet or tips of their canes touch informative regions of the surface. Although they acquire information in this way, the situations that demand changes in behavior are upon them before they have had time to make changes.

One function shared by all travel aids, from the lowly cane to the most sophisticated electronic travel aid, is to increase perceptual anticipation. However, although long canes do give blind pedestrians some of the time they need to prepare their responses, they do not provide enough perceptual anticipation. The range of long canes is approximately 90 cm, the average length of a stride, and at an average walking speed of 4.8 km per hour, long canes afford their users approximately .66 sec in which to prepare responses. However, Barth (1979) found that at an average walking speed, pedestrians need to know about the space 1.5 m ahead of them to avoid errors. At an average walking speed of 4.8 km per hour, 1.5 m of preview provides 1.125 sec in which to prepare responses.

Because ETAs detect obstacles several meters ahead, they should give their users ample time in which to prepare responses. However, the signals displayed by ETAs are often complex and their users may have to spend so much time interpreting them (Levy & Butler, 1978; Mershon & Bowers, 1979) that the time apparently available for preparing responses is lost. Furthermore, the ETAs that have been built so far provide little of the surface information blind pedestrians need.

Cognitive Functions

None of the components in Figure 3.2 are dedicated exclusively to the performance of cognitive functions. Nevertheless, there are components that operate in a coordinated manner to perform a group of related cognitive functions including association, storage, recall, comparison, and decision.

ASSOCIATION AND MEMORY

Because blind pedestrians are less able than sighted pedestrians to identify objects in space and because they cannot observe as much space at one time, they do not experience as much spatial contiguity as sighted pedestrians. They are compelled to observe most objects and spaces serially. The resulting serial perceptions of parts of objects and spaces must then be integrated to achieve perceptions of whole objects and spaces (see Hermelin & O'Connor, Chapter 2, this volume). Is this circumstance reflected in blind pedestrians' memorial representations of space? Do they, as von Senden (1960) suggests, hold conceptions of the external world in which the sizes of objects are defined by the time required to explore their surfaces, and the distances separating them by the time required to move from one object to another; or do they, as the introspective reports of many blind persons suggest, transform temporal experience in a way that restores the spatial coherence of the external world (Révész, 1950; Rowland, 1976)? In either case, their memorial representations of space are likely to be more distorted and to reveal less differentiation of detail than those of sighted pedestrians (Reiser, Lockman, & Pick, 1980). Storage limitations dictate that only some perceptual information can be stored in memory. Such carefully selected information must be entered in an organized way for retrieval to be possible. The organization of information about space should, in some sense, be a reflection of the spatial and temporal coherence of the external world and it is probably the learned consequence of associations resulting from the spatial and temporal contiguity of discriminated experiences. The memorial representations of space that are acquired by pedestrians, blind or sighted, are probably schemata (Schmidt, 1976); that is, they are abstracts that preserve the information needed to travel independently. However, there is more than one way to inform the behavior of a pedestrian performing a spatial task (Pick, Yonas, & Reiser, 1979), and more than one kind of schema is possible. For instance, the ability of some pedestrians to traverse learned routes might depend on a concatenation of stimulus–response associations. Pedestrians who had been programmed in this manner would not be able to indicate, in advance of their trips, the routes they intended to follow. However, as they arrived at choice points, the stimuli they encountered would elicit direction responses. They would, in a sense, be like the participants in a scavenger hunt who find at each choice point the instructions they need to gain the next choice point. Travelers of this type would be inflexible. They would not be able to retrace routes, and if their paths were blocked they would not be able to choose alternatives. If they strayed from correct paths inadvertently they would not know how to find them again. Unlikely as it may seem, there are travelers whose behavior suggests that they solve spatial problems in this way (Kuipers, 1980; Lynch, 1960). Piaget and Inhelder (1967) suggest that such behavior is characteristic of young children.

Travelers who used the "scavenger hunt" strategy could behave more flexibly if the relevant stimuli at choice points and the appropriate responses were represented in memory by linguistic symbols because they could manipulate symbols without actually traversing routes. They could retrace routes, tell others the routes they intended to follow, give directions to others for getting places, etc.

Tolman (1948) referred to the memorial representation of space as a cognitive map. The metaphorical appeal of this term is obvious, and it has been widely accepted and used. However, the term is not well defined and some people (Anderson, 1978; Hart & Berzok, Chapter 7, this volume) object to its use because it implies that a picturelike image of space is stored in memory. For instance, as Kuipers (1980) points out, Lynch's (1960) and Tobler's (1976) use of the term connotes a map in the head that resembles a graphic map. They allow that the map in the head may be a topological transformation of the graphic map that does not preserve a Euclidean representation of space. Lynch (1960), Beck and Wood (1976), and Appleyard (1970) have found that a cognitive map may contain several regions which, though internally well defined, are only loosely related to each other. Nevertheless, they seem to be proposing a picturelike image that can be recalled from memory and consulted for information in the same way one would consult a graphic map or a natural scene. However, a variety of experiments suggest that we do not store images of space in memory, but rather we store spatial knowledge in the form of propositions about space (Pylyshyn, 1973) (see Corballis, Chapter 8, this volume). Though the available evidence is not conclusive about the nature of the memorial representation of space, we may at least conclude that it is selective about the spatial information that is preserved (Appleyard, 1970, 1976; Evans, Marrero, & Butler, 1981; Lynch, 1960; Shagen, 1970).

The space represented in memory is extended space. Extended space includes not only present space but also remembered space that is not presently observable (see Sedgwick, Chapter 1, this volume). The memorial representation of extended space may not be entirely coherent (Lynch, 1960) but neither is it entirely incoherent, and what coherence there is must be achieved by integrating information about remembered spaces that have been observed at different times in the past.

When travelers traverse new routes, they do not have memorial representations of those routes. They must depend on contemporaneous information and aids such as verbal instructions and maps (Bentzen, 1980). Because many of the features of constructed environments are reiterated time and time again, they can also depend on remembered generalizations about space, or what might be called spatial stereotypes. They may know, for instance, that grids of intersecting streets are imposed on cityscapes, that the spaces within these grids are frequently rectangular, that streets are usually at lower levels than the spaces they bound, that sidewalks are usually provided for pedestrians, that these sidewalks are parallel to the streets and often separated from them by strips of grass covered earth, that on both sides of streets and beyond the sidewalks are rows of buildings. Armed with these spatial stereotypes they are able to make many accurate predictions about spaces they are encountering for the first time (Foulke, 1971; Hollyfield, 1981).

Blind pedestrians traversing new routes do so with less safety and certainty than sighted pedestrians because they have to manage with less contemporaneous information and less perceptual anticipation. As experience with routes accumulates, they gradually acquire memorial representations, but because they observe so much less at one time than sighted pedestrians they have more integration to do. They need many more trials than sighted pedestrians to learn new routes and to construct memorial representations of the extended spaces that encompass those routes. Until they have had enough experience to construct these representations, they must often rely on spatial stereotypes for estimates of the structure of present space.

Like sighted pedestrians, blind pedestrians compare contemporaneous information with stored information to maintain orientation in extended space. However, because their representations do not include memories of many things that serve as landmarks for sighted pedestrians it is more difficult for them to maintain orientation.

A HYPOTHESIS CONCERNING THE MEMORIAL REPRESENTATIONS
OF BLIND PEDESTRIANS

Observations made by Appleyard (1970, 1976) suggest that when sighted travelers first learn to find their way in new spaces their memorial representations of space are likely to consist of path structures, but as their experience increases, path structures tend to be replaced by representations consisting of spatially related landmarks. Though the differences in the memorial representations of extended space achieved by blind and sighted pedestrians have not yet been studied in detail, it is a reasonable conjecture that in the early stage of learning about a new extended space blind pedestrians, like sighted pedestrians, will represent that space in terms of a path structure. Unlike sighted pedestrians, however, they will not replace this representation with a representation in terms of spatially related landmarks when they have had more experience. Their memorial representations will come to include landmarks consisting of characteristics of the surfaces of paths. Their representations will include very little of the adjacent landmarks in the spaces bounded by those paths. Like Tolman's (1948) rats, they often grasp the relationships among the paths they know about and can therefore plan novel routes to their goals (James & Swain, 1975; Leonard & Newman, 1967) but they are still maze runners. If they stray from the paths they have learned they are in unfamiliar territory; because their memorial representations do not include much information about adjacent landmarks they easily get lost.

DECISION

Travelers who have goals must decide among routes and possible actions. In order to make these decisions, they must review spatial and nonspatial information from several sources. Travelers depend on contemporaneous information to avoid obstacles, detect discontinuities in the surface underfoot, etc. They must consider the hazards of different routes, their own capabilities (Kaplan, 1973), and their willingness to accept risk (Kogan & Wallach, 1964). They depend on the information in their memorial representations of extended space for orientation. By comparing contemporaneous information with their memorial representations, they find out where they are in extended space. They can then locate spatial goals and plan routes to those goals.

Though the human information-processing system receives inputs of information from many sources, both within and beyond its boundary, there is only one decider. Because all of the information on which decisions are based must be reviewed by the decider, the human information-processing system has a single channel, and because the processing of information requires time the capacity of that channel is limited (Broadbent, 1957). If channel capacity is exceeded, some of the information arriving at the decider cannot be processed, and processing priorities must then be invoked (Keele, 1968). If after priorities have been enforced channel capacity is still exceeded, performance must deteriorate.

INVESTIGATIVE APPROACHES TO THE STUDY OF MOBILITY

The preceding pages describe the tasks of blind pedestrians and suggest some of the perceptual and cognitive functions they must employ to perform the tasks. Although perceptual and cognitive research conducted for other purposes is often relevant, there has been very little research specifically concerned with the mobility of blind pedestrians. Consequently, there is not much experimental evidence in support of this account.

As mentioned earlier, mobility is a complex, open skill performed in a context that is not always predictable and its performance is difficult to observe. Because its performance depends on the coordinated use of a variety of perceptual and cognitive functions it is not surprising that the investigation of mobility has proven to be a difficult undertaking. To date most of the research on mobility has been conducted for the practical purpose of evaluating travel aids and training methods, and because experimental designs have frequently been inadequate, much of this research has been inconclusive.

Researchers have invested countless hours and a great deal of money in the development of ETAs that do not work very well. The mobility of blind pedestrians has been significantly improved by the training methods now in use but it falls far short of the mobility of sighted pedestrians. Further progress will not be realized until we understand the task in which blind pedestrians engage— and for that we must be willing to conduct the kind of basic research that tests theories. An important reason for the scarcity of this kind of research has been the lack of a reliable and valid methodology for measuring the performance of blind pedestrians. This is a prerequisite both for the development of a useful theory of mobility and for the evaluation of the ETAs and training methods currently competing for acceptance. Figure 3.3 presents many of the methodological questions raised by the experimental investigation of the mobility of blind pedestrians (for a more extensive review see Shingledecker and Foulke, 1978) and shows a pair of dichotomized approaches at each of four levels. Any given approach to the measurement of mobility can be described in terms of the position it occupies at each level. Not labels for categories of approaches, the two terms at each level are only indicators of the limits of a continuum. Any given approach may be located at any point on each of the continua.

SUBJECTIVE–OBJECTIVE

Many attempts to measure the performance of blind pedestrians have been subjective. The judgments of expert observers (usually O&M Specialists) con-

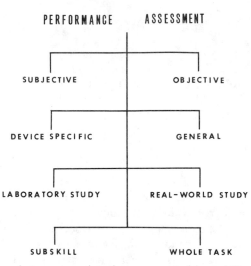

Figure 3.3. A human factors approach to the assessment of the mobility of blind pedestrians. (From Shingledecker & Foulke, 1978.)

cerning the quality of the performance displayed by blind pedestrians as they work through some travel problem have been used as data. In some cases these judgments have been systematized by the use of checklists of subskills (Airasian, 1972, 1973). Checklists increase the likelihood that judges will attend to the same behavior but they imply an objectivity that is specious. Objective assessment of mobility can only be achieved by specifying the behavioral events that are to be observed (Nye, 1973) by specifying the physical operations that are to be carried out to measure those behavioral events and by referring measured behavioral events to criteria in the public domain.

DEVICE SPECIFIC–GENERAL

Most attempts to observe the performance of blind pedestrians have been undertaken to evaluate specific travel aids and measures have usually been chosen specifically to detect differences in aspects of overall performance that were believed to be facilitated by the aid being evaluated (Kay, 1973a, 1973b). This approach fails to provide satisfactory evaluations for at least three reasons. Because experimental tasks are chosen in terms of the physical characteristics of the devices to be evaluated they are unique tasks and it is usually not useful to compare the results of such experiments (Armstrong, 1975); they lack external validity. Device specific experiments do not advance our understanding of the mobility of blind pedestrians very much. Since the measures obtained in such experiments have little bearing on mobility tasks performed in the real world, they are unsuitable for evaluating the adequacy of a general theory of mobility.

In contrast to the device specific approach, the general approach calls for the measurement of performance under stimulus conditions that are analogous in quality, quantity, and distribution to stimulus conditions in the real world. Measures obtained with the general approach have as much potential for internal validity as measures obtained under the device specific approach and can provide information of theoretical interest. They provide a common criterion for the comparative evaluation of different travel aids, under realistic conditions regardless of their modes of operation or the information content of their signals.

LABORATORY-REAL WORLD

The third level of Figure 3.3 expresses a dilemma that must be faced by any investigator who intends to study the mobility of blind pedestrians. If performance is observed in the laboratory there is better control of extraneous variables. However, this control is often achieved at the expense of fidelity to the real world; as fidelity declines there is an increasing probability that laboratory results cannot be generalized to the real world (Chapanis, 1967). However, when performance is observed in the real world, fidelity is not an issue. But there may be loss of the control over extraneous variables that is needed to make an experiment conclusive.

SUBSKILL-WHOLE TASK

At the final level of Figure 3.3, the two terms indicate two approaches to the study of mobility that are familiar to students of human performance. One approach calls for the composite measurement of overall performance on a whole task (Armstrong, 1975; Shingledecker, 1976). The other approach calls for the identification and subsequent separate analysis of the subskills upon which the performance of the whole task depends (Cratty, 1967, 1971a, 1971b; Juurmaa, 1973). Of course, such experiments have the advantages and disadvantages associated with the laboratory approach. In addition, the subskill approach must take into account the possibility that the performance of isolated subskills may be significantly different from the performance of the same subskills when they are components of a whole task. For instance, the performance of a blind pedestrian who is detecting obstacles by means of echo perception in a laboratory experiment where competing stimuli have been deliberately suppressed and the performance of the same skill on a busy city street may be significantly different.

The whole-task approach also requires identification of the subskills that are needed to perform the whole task, but these subskills are exhibited in response to task demands by subjects engaged in the whole task and it is the overall performance of subjects as they engage in the whole task that is assessed. The measure of whole task performance is usually a composite measure that includes measures of the behavioral consequences of the subskills involved in the task. The whole-task approach is functional in its orientation. Successful performance of the whole task is defined in terms of objective criteria related to the components of overall performance that are assessed by the composite measure.

Neither the whole-task approach nor the subskills approach is adequate by itself. The subskills approach permits estimation of the limits of performance. The whole-task approach permits description of manifest performance. Both approaches are needed for the thorough evaluation of travel aids, training methods, and theories of mobility.

SUMMARY

Limited mobility is the most serious handicap experienced by blind persons. They have less ability than sighted persons to travel independently because they cannot acquire enough spatial information. The visual system is uniquely capable of providing information about what things are in space and where they are. Blind pedestrians must travel without much of this information in an environment that is varying continuously and often unpredictably.

Blind pedestrians who master the skills and techniques taught by O&M Specialists can gain some independence in travel, but even with such training their mobility falls far short of the mobility of sighted pedestrians. Blind pedestrians would have better mobility if they could be given access to some of the spatial information they lack. There have been a number of attempts to provide the missing information by means of electronic travel aids that sense the environment and display signals to one or more of the functioning perceptual systems of blind pedestrians. Many of these aids exemplify quite sophisticated engineering but because their designs were guided more by common sense than by knowledge of the perceptual and cognitive abilities of the blind pedestrians who are expected to use them, they do not work very well.

It is now evident that effective ETAs will not be built until their design can be guided by a better understanding of the task that is performed by blind pedestrians when they travel independently and of the perceptual and cognitive functions on which they rely. We need to know what information is critical to mobility, where in the environment to find it, and how to display it. We need realistic assessments of the capabilities and limitations of the perceptual systems on which blind pedestrians depend. How much and in what ways does mobility depend on memory? What demand does the information processing load imposed by the task make on channel capacity?

In short, we need a theory of mobility, and that theory should be firmly rooted in a more general theory concerning the perceptual and cognitive functions of humans. In order to build such a theory we will have to conquer our impatience, set applied research aside for the time being, and conduct the program of basic research that will give us the understanding we need for intelligent applications. The objective of mobility for blind pedestrians that is as effective and efficient as the mobility of sighted pedestrians may not be completely attainable, but we will come much closer to its realization if we are willing to try to understand what we are doing before we do it.

REFERENCES

Airasian, P. W. *Evaluation of the binaural sensory aid.* Washington, D.C.: National Academy of Sciences, 1972.

Airasian, P. W. Evaluation of the binaural sensory aid. *American Foundation for the Blind Research Bulletin,* 1973, *26,* 51–71.

Ammons, C. H., Worchel, P., & Dallenbach, K. Facial vision: The perception of obstacles out of doors by blindfolded and blindfolded deafened subjects. *American Journal of Psychology,* 1953, *40,* 519–553.

Anderson, J. R. Arguments concerning representations for mental imagery. *Psychological Review,* 1978, *85,* 249–277.

Appleyard, D. Styles and methods of structuring a city. *Environment and Behavior,* 1970, *2,* 100–117.

Appleyard, D. *Planning a pluralistic city.* Cambridge: MIT Press, 1976.

Armstrong, J. D. Evaluation of man–machine systems in the mobility of the visually handicapped. In R. M. Pickett & T. J. Triggs (Eds.), *Human factors in health care.* Lexington, Massachusetts: Lexington Books, 1975. Pp. 331–343.

Barth, J. L. *The effects of preview constraint on perceptual motor behavior and stress level in a mobility task.* Unpublished doctoral dissertation, University of Louisville, 1979.

Barth, J. L., & Foulke, E. Preview: A neglected variable in orientation and mobility. *Journal of Visual Impairment and Blindness,* 1979, *73,* 41–48.

Batteau, D. W. Listening with the naked ear. In S. J. Freedman (Ed.), *The neurophysiology of spatially oriented behavior.* Homewood, Illinois: Dorsey Press, 1968. Pp. 109–133.

Beck, R., & Wood, D. Comparative developmental analysis of individual and aggregated cognitive maps of London. In G. T. Moore & R. G. Golledge (Eds.), *Environmental knowing: Theories, research, and methods.* Stroudsburg, Pennsylvania: Dowden, Hutchinson, and Ross, 1976. Pp. 173–184.

Bentzen, B. Orientation aids. In R. L. Welsh & B. R. Blasch (Eds.), *Foundations of orientation and mobility.* New York: American Foundation for the Blind, 1980. Pp. 291–355.

Beritoff, J. *Neural mechanisms of higher vertebrate behavior.* Boston: Little, Brown, & Company, 1965.

Broadbent, D. E. A mechanical model for human attention and immediate memory. *Psychological Review,* 1957, *64,* 205–215.

Chapanis, A. The relevance of laboratory studies to practical situations. *Ergonomics,* 1967, *4,* 35–40.

Clark, L. L. (Ed.). *Proceedings of the international congress on technology and blindness* (Vol. 1). New York: American Foundation for the Blind, 1963.

Coleman, P. D. An analysis of cues to auditory depth perception in free space. *Psychological Bulletin*, 1963, *60*, 302–315.

Cotzin, M., & Dallenbach, K. M. Facial vision: The role of pitch and loudness in the perception of obstacles by the blind. *American Journal of Psychology*, 1950, *63*, 485–515.

Cratty, B. J. The perception of gradient and the veering tendency while walking without vision. *American Foundation for the Blind Research Bulletin*, 1967, *14*, 31–51.

Cratty, B. J. *Enhancing perceptual–motor abilities in blind children and youth.* Paper presented at the Conference of California State Teachers of the Blind, Palo Alto, 1971. (a)

Cratty, B. J. *Movement and spatial awareness in blind children and youth.* Springfield: Charles C Thomas, 1971. (b)

Deatherage, B. H. Examination of binaural interaction. *Journal of the Acoustical Society of America*, 1966, *39*, 232–249.

Evans, G. W., Marrero, D., & Butler, P. Environmental learning and cognitive mapping. *Environment and Behavior*, 1981, *13*(1), 83–104.

Farmer, L. Mobility devices. In R. L. Welsh & B. R. Blasch (Eds.), *Foundations of orientation and mobility*. New York: American Foundation for the Blind, 1980. Pp. 357–412.

Fisher, H. G., & Freedman, S. J. The role of the pinna in auditory localization. *Journal of Auditory Research*, 1968, *8*, 15–26.

Foulke, E. The role of experience in the formation of concepts. *International Journal for the Education of the Blind*, 1962, *12*, 1–6.

Foulke, E. The perceptual basis for mobility. *American Foundation for the Blind Research Bulletin*, 1971, *23*, 1–8.

Foulke, E., & Berlá, E. Visual impairment and the development of perceptual ability. In R. D. Walk & H. L. Pick, Jr. (Eds.), *Perception and experience*. New York: Plenum Press, 1978. Pp. 213–240.

Gibson, J. J. *The senses considered as perceptual systems.* Boston: Houghton Mifflin Company, 1966.

Graystone, P., & McLenna, H. Evaluation of an audible mobility aid for the blind. *American Foundation for the Blind Research Bulletin*, 1968, *17*, 173–179.

Green, D. M. Functional aspects of the auditory sense. In L. D. Harmon (Chair), *Interrelations of the communicative senses*. Conference sponsored by the National Science Foundation, Asilomar, California, 1978.

Heyes, A. D., Armstrong, J. D., & Willans, P. R. A comparison of heart rates during blind mobility and car driving. *Ergonomics*, 1976, *19*, 489–497.

Hill, E., & Ponder, P. *Orientation and mobility techniques: A guide for the practitioner*. New York: American Foundation for the Blind, 1976.

Hollyfield, R. L. *The spatial cognition of blind pedestrians*. Unpublished doctoral dissertation, University of Louisville, 1981.

Hoover, R. E. The Valley Forge story. In I. S. Diamond (Ed.), *Blindness, 1968*. Washington, D.C.: American Association of Workers for the Blind, 1968.

Howard, I. P. Orientation and motion in space. In E. C. Carterette & M. P. Friedman (Eds.), *Handbook of perception* (Vol. 3: *Biology of perceptual systems*). New York: Academic Press, 1973. Pp. 291–315. (a)

Howard, I. P. The spatial senses. In E. C. Carterette & M. P. Friedman (Eds.), *Handbook of perception* (Vol. 3: *Biology of perceptual systems*). New York: Academic Press, 1973. Pp. 273–290. (b)

Hubel, D. H., & Wiesel, T. N. Receptive fields, binocular interaction and functional architecture in the cat's visual cortex. *Journal of Physiology*, 1962, *160*, 106–154.

James, G. A., & Swain, R. Learning bus routes using a tactile map. *New Outlook for the Blind*, 1975, *69*, 212–217.

Jansson, G. *The detection of objects by the blind with the aid of a laser cane* (Report 172). University of Uppsala, Uppsala, Sweden, Department of Psychology, 1975.

Julesz, B., & Hirsh, E. J. Visual and auditory perception: An essay of comparison. In E. E. Davis & P. B. Denes (Eds.), *Human communication: A unified view*. New York: McGraw-Hill, 1972. Pp. 283–340.

Juurmaa, J. On the accuracy of obstacle detection by the blind: Part 1. *New Outlook for the Blind*, 1970, *64*, 65–72. (a)

Juurmaa, J. On the accuracy of obstacle detection by the blind: Part 2. *New Outlook for the Blind*, 1970, *64*, 104–118. (b)

Juurmaa, J. Transposition in mental spatial manipulation: A theoretical analysis. *American Foundation for the Blind Research Bulletin*, 1973, *26*, 87–134.

Kaplan, S. Cognitive maps in perception and thought. In R. M. Downs & D. Stea (Eds.), *Image and environment: Cognitive mapping and spatial behavior*. Chicago: Aldine Publishing Company, 1973. Pp. 63–78.

Kay, L. Sonic glasses for the blind: A progress report. *American Foundation for the Blind Research Bulletin*, 1973, *25*, 25–28. (a)

Kay, L. Sonic glasses for the blind: A progress report. *American Foundation for the Blind Research Bulletin*, 1973, *26*, 35–50. (b)

Kay, L. A sonar aid to enhance spatial perception of the blind: Engineering design and evaluation. *Radio and Electronics Engineer*, 1974, *44*, 605–627.

Keele, S. W. Movement control in skilled motor performance. *Psychological Bulletin*, 1968, *70*, 387–403.

Kogan, H., & Wallach, M. A. *Risk taking*. New York: Holt, Rinehart & Winston, 1964.

Kohler, I. Orientation by aural clues. *American Foundation for the Blind Research Bulletin*, 1964, *4*, 14–53.

Kuipers, B. *Refuting the "map in the head" theory*. (Working Papers in Cognitive Science, TUW-PICS No. 2). Unpublished manuscript. Tufts University, 1980.

Leonard, J. A., & Newman, R. C. Spatial orientation in the blind. *Nature*, 1967, *215*, 1413–1414.

Levy, E. T., & Butler, R. A. Stimulus factors which influence the perceived externalization of sound presented through headphones. *Journal of Auditory Research*, 1978, *18*, 41–50.

Lynch, K. *The image of the city*. Cambridge: MIT Press, 1960.

Mann, R. W. Technology and human rehabilitation: Prostheses for sensory rehabilitation and/or sensory substitution. In J. H. Brown & J. F. Dickson (Eds.), *Advances in biomedical engineering* (Vol. 4). New York: Academic Press, 1974. Pp. 209–353.

Martinez, F. Does auditory information permit the establishment of spatial orientation? Experimental and clinical data with the congenitally blind. *Année Psychologique*, 1977, *77*(1), 179–204.

Mershon, D. H., & Bowers, J. N. Absolute and relative cues for the auditory perception of space. *Perception*, 1979, *8*, 311–322.

Miller, J. G. *Living systems*. New York: McGraw-Hill, 1978.

Mills, A. W. Auditory localization. In J. V. Tobias (Ed.), *Foundations of modern auditory theory*. New York: Academic Press, 1972. Pp. 301–348.

Mims, F. W., III. Sensory aids for blind persons. *New Outlook for the Blind*, 1973, *67*, 404–407.

Nye, P. W. *A preliminary evaluation of the Bionic Instruments—Veterans Administration C-4 laser cane*. Washington, D.C.: National Academy of Sciences, 1973.

Peake, P., & Leonard, J. A. The use of heart rate as an index of stress in blind pedestrians. *Egronomics*, 1971, *14*, 189–204.

Piaget, J., & Inhelder, B. *The child's conception of space*. New York: Norton, 1967. (Originally published in French, 1948.)

Pick, H. L., Jr. Perception, locomotion, and orientation. In R. L. Welsh & B. R. Blasch (Eds.), *Foundations of orientation and mobility*. New York: American Foundation for the Blind, 1980. Pp. 73–88.

Pick, H. L., Jr., Yonas, A., & Rieser, J. J. Spatial reference systems in perceptual development. In

M. Bornstein & W. Kessen (Eds.), *Psychological development from infancy*. Hillsdale, New Jersey: Erlbaum, 1979. Pp. 115–146.

Poulton, E. C. The basis of perceptual anticipation in tracking. *British Journal of Psychology*, 1952, *43*, 295–302.

Poulton, E. C. On prediction in skilled movements. *Psychological Bulletin*, 1957, *54*, 467–478.

Pylyshyn, Z. What the mind's eye tells the mind's brain: A critique of mental imagery. *Psychological Bulletin*, 1973, *80*, 1–24.

Révész, G. *Psychology and the art of the blind*. New York: Longmans-Green, 1950.

Rice, C. E. Human echo perception. *Science*, 1967, *155*, 656–664.

Rice, C. E., & Feinstein, S. H. Echo detection ability of the blind: Size and distance factors. *Journal of Experimental Psychology*, 1965, *70*, 246–251. (a)

Rice, C. E., & Feinstein, S. H. The influence of target parameters on human echo-detection tasks. *Proceedings of the American Psychological Association*, 1965. (b)

Rieser, J. J., Lockman, J. J., & Pick, H. L., Jr. The role of visual experience in knowledge of spatial layout. *Perception and Psychophysics*, 1980, *28*(3), 185–190.

Rowland, W. *Space and blindness*. Unpublished master's thesis, University of South Africa, 1976.

Sakomoto, M., Gotoh, T., & Kumura, Y. On out of head localization and headphone listening. *Journal of Audio Engineering Society*, 1976, *24*, 710–716.

Schmidt, R. A. The schema as a solution to some persistent problems in motor learning theory. In G. E. Stelmach (Ed.), *Motor control: Issues and trends*. New York: Academic Press, 1976. Pp. 41–65.

Scott, R. A. *The making of blind men*. New York: Russell Sage Foundation, 1968.

von Senden, M. *Space and sight: The perception of space and shape in the congenitally blind before and after operation*. Glencoe, Illinois: The Free Press, 1960.

Shagen, J. *Kinaesthetic memory, comparing blind and sighted subjects*. Unpublished doctoral dissertation, George Washington University, 1970.

Shingledecker, C. A. *The development of a methodology for the measurement of blind mobility performance*. Unpublished doctoral dissertation, University of Louisville, 1976.

Shingledecker, C. A., & Foulke, E. A human factors approach to the assessment of the mobility of blind pedestrians. *Human Factors*, 1978, *20*, 273–286.

Simpson, W. E. Locating sources of sound. In R. F. Thompason & G. F. Bosse (Eds.), *Topics in learning and performance*. New York: Academic Press, 1972. Pp. 17–40.

Suterko, S. Life adjustment. In B. Lowenfeld (Ed.), *The visually handicapped child in school*. New York: John Day Company, 1973. Pp. 279–339.

Thornton, W. The binaural sensor as a mobility aid. *New Outlook for the Blind*, 1971, *65*, 324–326.

Tobler, W. The geometry of mental maps. In R. G. Golledge & G. Rushton (Eds.), *Spatial choice and spatial behavior: Geographical essays on the analysis of preferences and perceptions*. Columbus: Ohio State University Press, 1976.

Tolman, E. C. Cognitive maps in rats and men. *Psychological Review*, 1948, *55*, 189–208.

Wallach, H. The role of head movements and vestibular cues in sound by localization. *Journal of Experimental Psychology*, 1940, *27*, 339–368.

Warren, D. H., & Kocon, J. A. Factors in the successful mobility of the blind: A review. *American Foundation for the Blind Research Bulletin*, 1974, *28*, 191–218.

Wiener, W. R. Orientation and mobility come of age. In G. G. Mallinson (Ed.), *Blindness, 1979–1980*. Washington, D.C.: American Association of Workers for the Blind, 1980.

Wiener, W. R., & Welsh, R. L. The profession of orientation and mobility. In R. L. Welsh & B. R. Blasch (Eds.), *Foundations of orientation and mobility*. New York: American Foundation for the Blind, 1980. Pp. 625–651.

THE DEVELOPMENT OF
SPATIAL ABILITIES

J. GAVIN BREMNER

Object Localization in Infancy

A major problem for psychologists studying early abilities can be summarized by asking to what degree the infant's understanding of his surroundings matches that of the adult. Until recently, the prevalent accounts painted a fairly pessimistic picture. It was assumed that young infants had little objective knowledge of the world, being limited to awareness of proximal stimuli rather than distal events and objects. However, over the past two decades or so, increasing evidence has forced the conclusion that even the very young infant's understanding of the world is much more sophisticated than formerly assumed. One consequence of this is a growing tendency to look for the emergence of fairly complex abilities in the first year of life that might not have been expected until the child was older.

One way of tackling the problem of infants' understanding of the environment is to look at the sorts of reference systems that they use to organize the space around them. Indeed, it is difficult to see how they could make much sense of the world without the aid of some organizing principles of this sort. Probably the simplest system that infants could use is one in which objects' positions are defined purely in relation to the subject's body axes. However, this *self-referent* system has one major disadvantage. Whenever the subject moves, all the objects' positions change with respect to the referent.[1] Although

---

[1] See mechanisms of "updating" in Pick and Reiser, Chapter 5, this volume; Potegal, Chapter 15, this volume.

79

SPATIAL ABILITIES
Development and Physiological Foundations

it is clearly important for the infant to appreciate that the subject–object relationship changes during such movements, it is also important that he appreciate the general stability of the environment. Thus, it seems clear that at some point in development, a stable external reference system will be adopted to complement the self-referent system.

At present we know rather little about the reference systems that infants use, and less about the way in which they develop. However, the general developmental literature does not leave us entirely without pointers. In particular, the emphasis Piaget and others have placed on *egocentrism* in infancy must lead us to consider the possibility that young infants may be limited to using a *self-referent* system, that is, a system in which positions of objects are defined relative to the observer (e.g., *straight ahead, to my right,* etc.). If we did find such a limitation in early infancy, we would then be faced with the problem of establishing when and how *external-referent* systems emerge. However, we must also look at recent evidence challenging Piaget's point of view on egocentrism (Bower, 1974), to see if this indicates a different developmental picture.

In this chapter, data from object search tasks are used as the main evidence on which to construct a view of the infant's spatial ability. This is done first at a general level, with the following two sections concentrating on some old and some new evidence about the infant's general concepts of space and objects. Later sections review specific evidence on the spatial reference systems used by infants.

PIAGET'S ACCOUNT OF THE DEVELOPMENT OF THE INFANT'S CONCEPT OF SPACE

Any account of the development of spatial understanding in infancy would be incomplete without a summary of Piaget's (1936/1952, 1936/1954) account of sensorimotor development. In particular, with the central role played by the concept of *egocentrism,* we would expect the theory to be particularly relevant to questions about the form of spatial reference systems used by infants. It seems fair to conclude that one symptom of egocentrism would be the use of a self-referent system to define positions in space. However, Piaget also provides a detailed description of the way in which this egocentrism wanes during the first 2 years of the infant's life. Thus, his theory is also extremely relevant to the development of external-referent systems.

According to Piaget, the newborn is profoundly egocentric, experiencing the world as though it were simply a part of himself, a series of "pictures" brought about by his own activities and without independent material existence. During the sensori-motor period, the infant must progress from this solipsistic state to one of recognition of the independent existence of the surrounding world. In addition, he must come to understand that he is a physical object just like other objects, situated in a common space and governed by the same spatial rules. In other words, the infant progresses from an initial state of viewing the world as an undifferentiated part of himself, through a period of growing *differentiation* between self and environment, to a final stage in which a knowing *integration* is imposed on the differentiated world with the end result that he views himself as a part of the environment just like any other object. In effect, he has to get outside himself; first by realizing that "pictures" experienced within arise from objects outside, and second by relinquishing a privileged egocentric viewpoint to realize that his positions and displacements obey the same rules as those of other objects.

What does this suggest about the nature of spatial reference systems used in infancy? If Piaget's description is correct, it is difficult to see how the very young infant's coding of positions could be anything but self-referent. Indeed, it may not even make sense to talk about understanding of *position* at this stage, in view of the suggestion that this only develops gradually from an elementary association between a "picture" and a specific action. However, Piaget's concept of egocentrism is considerably more complex and far reaching than a simple statement about the type of spatial code used by infants.[2] Only a fuller analysis of his account will clarify whether he makes any specific predictions about these spatial codes.

In *The Construction of Reality in the Child* (1936/1954) Piaget claims that the development of the concepts of objects and space are inextricably linked. "Space . . . is not at all perceived as a container but rather as that which it contains, that is, objects themselves . . . [Piaget, 1936/1954, p. 98]." In other words, space is an ordering of objects, so when we talk about space we unavoidably talk about objects. As in the case of the object concept, Piaget describes six stages in the development of the concept of space, spanning the period from birth to around 2 years.

In the first stage (0–1 month) and second stage (1–4 months), space is not differentiated from the actions that the infant makes in it. Rather than taking place in space, action creates space, and displacements of objects are experienced purely as extensions of the subject's acts. Consequently, the infant does not distinguish changes in position from changes in state, and it really does not make sense to talk of the infant understanding positions in any way. On a similar note, writing about the infant's understanding of depth, Piaget says: "the child certainly perceives at various depths, but there is no indication that he is conscious of these depths or that he groups the perceived displacements on different planes [1936/1954, p. 111]."

[2]To prevent confusion I have avoided using the term *egocentric* to describe coding positions relative to the self.

During stage III (4–8 months), understanding of space takes a large step forward. The infant begins to observe herself in action, and this coordination of vision and prehension allows the first step to be made toward objectification (and hence externalization) of positions and displacements in space. However, these positions and displacements are not yet interpreted independently of her own activity. The dissociation between self and environment is only beginning. At this stage the infant still does not search for an object hidden beneath a cover, and Piaget believes that this is further indication that the object's position is only related to the subject and not to other objects, in this example, the cover. In other words, objects are now assigned positions in space, but these positions are self-referent ones with no evidence that the infant can relate an object's position to an external referent. Thus, if Piaget is correct, the earliest point at which it makes sense to say that the infant codes an object's position is stage III, and at this stage coding is self-referent.

During stage III, the infant's understanding of depth relations advances considerably. Through observing his own actions, the zone of space in which activity takes place is externalized and elaborated as a system of relations in depth. However, regions of space out of reach are not elaborated, taking a form similar to the naive concept of astronomical space; as if the objects within it are arranged on the inside surface of a hemisphere bounding the tactual zone, all at the same distance relative to the subject. In stage III, for the first time, near space is now definitely external. Also, distant space, although poorly elaborated, is now external to the extent of being at the boundary of near space.

Stage IV (8–12 months) is probably the crucial stage in the sequence for the development of both space and object concepts. The main feature of this stage is the emergence of search for the hidden object. After watching a toy being hidden under a cover, the stage IV infant will lift the cover and retrieve it successfully—a reaction absent during stage III. Apart from its presumed implications for the development of the concept of the permanent object, this new ability is seen as a significant factor in development of understanding spatial relations. Piaget (1936/1952) emphasizes that in searching for an object hidden under a cover, the infant is coordinating two separate schemes (cover lifting and grasping the object) to achieve the goal. This new higher order scheme is specifically suited to obtaining hidden objects and would not be required if the object were invisible but directly accessible. In fact, Piaget notes that stage III infants will reach for an object that is outside the visual field. This does not qualify as stage IV ability, because the same action scheme suffices as when the object is visible.

In stage III, means–ends differentiation emerged, and consequently space became external. Now in stage IV, two separate schemes are combined in an ordered sequence to reach a goal, and the spatial implication is that this coordination of schemes is accompanied by a dawning understanding of the spatial relationship *between* the objects acted upon. In the case of search for a hidden object, the ordering of cover lifting and grasping is accompanied by understanding of the spatial relationship between object and cover, and hence, by realization of the material existence of the object under the cover.

This new ability to relate objects to each other opens the door to the possibility of the infant using external spatial reference systems. However, Piaget points to a curious search error made by stage IV infants as evidence that their thinking retains a considerable subjective element. Whereas infants have no difficulty in obtaining an object hidden consistently in one place, they run into difficulties if the object is hidden in a new place. Although they watch attentively while the object is hidden there, they show a strong tendency to search at the original hiding place. According to Piaget, although the infant can relate object and occluder, this capability is limited to a single context relating to a particular activity that has been successful in the past. Although this means that the link between environment and activity is not completely severed, the connection between action and object is now one step removed. The object is no longer "at the disposal of the act" (stage III), but "at disposal in the place where the action has made use of it [Piaget, 1936/1954]." Thus, *place* is now the middle term. The crucial question for our present purposes is how the infant defines that place, but I think that Piaget's account of this leaves us in some doubt. Is search directed to a constant place relative to the subject, or to a constant place relative to some features of the environment? Arguments in terms of egocentrism might seem to favor the first alternative. However, as we have seen, Piaget believes that the stage IV infant can relate objects to each other, so definition of this position relative to an environmental feature is also a possibility.

In addition to searching for hidden objects, the stage IV infant is better at using his whole body to reach out for more distant objects, and hence the zone of tactual space is expanded as well as elaborated. However, a poorly elaborated distant space still surrounds the semi-objective near space. Strangely enough, Piaget does not mention crawling as an activity likely to affect the infant's understanding of space, an ability that most infants gain during this stage.

The onset of stage V (12–18 months) is marked by the disappearance of the stage IV error. Apparently, the infant now views space in an objective manner, as an interrelationship of objects independent of his activities. Objects can now be mentally represented as well as perceived in their spatial relationships to each other. However, Piaget claims that the infant does not yet understand his own position in space. This is surprising since most infants can crawl by the time they enter stage V, and hence have plenty of firsthand experience of the consequences of their own displacements. Do they believe that in crawling they bring desired goals to them rather than the reverse? We shall come to some evidence later that suggests that even stage IV infants may be more aware of their own positions in space than Piaget argues.

Although the stage V infant can represent the positions of objects, he still experiences problems in representing their displacements. If an object is moved to a new hiding place while hidden in a transport container, the infant will only search in the transport container, failing to deduce that the object's absence there implies its presence at the second hiding place. Thus, the infant only searches where the object has been *seen* to go, and cannot represent its movements mentally.

During stage VI (18–24 months) both of the limitations characteristic of stage V are overcome. The infant solves the invisible displacement problem. In movements about the room she gains the ability to make detours around obstacles in order to reach goals, which indicates foresight of the consequence of her own displacements. Hence, she can understand and represent her own positions and displacements in the same way as she does those of other objects.

RECENT EVIDENCE AND REINTERPRETATIONS OF PIAGET'S THEORY

Most of the work relating to Piaget's theory has been directed more to the object concept than to the concept of space. In addition to formation of standardized tests of sensori-motor intelligence (Casati & Lezine, 1968; Escalona & Corman, 1967; Uzgiris & Hunt, 1966) and attempts at replication of the developmental sequence (Kramer, Hill, & Cohen, 1975; Miller, Cohen, & Hill, 1970), alternative accounts of the development of the object concept have been formulated (Bower, 1974; Harris, in press; Moore, 1975; Moore & Meltzoff, 1978). The work of Bower and his colleagues is particularly relevant. In contrast to Piaget's picture of extreme egocentrism in the young infant's concepts of objects and space, Bower claims that the very young infant understands the substantial permanence of objects but lacks understanding of the rules governing positions and displacements of objects in space.

Surprise and Visual Tracking as Measures of the Infant's Knowledge

Bower's main findings can be summarized as follows: First, even at 20 days, infants who had just seen a stationary object being occluded by a vertical screen showed surprise if the object was absent when the screen was removed; a reaction that did not occur in the case of normal reappearance (Bower, 1966a). However, this reaction only appeared in these very young infants if the occlusion time was less than 6 sec suggesting a fairly fragile effect. Nevertheless, Bower concluded that as early as Piaget's stage I or II, infants understand the permanence of the hidden object.

When we add to this Bower's (1966b) claim that infants of much the same age understand simple depth relationships and size constancy, it looks as if we must accept that the young infant's understanding of the world is quite advanced. However, subsequent studies of tracking behavior (Bower, Broughton, & Moore, 1971; Bower & Paterson, 1973) reveal some limits to the young infant's knowledge. Bower *et al.* (1971) found that if young infants had to track an object that passed behind a screen on the middle part of its trajectory, they would continue to track undisturbed if the object stopped behind the screen. More surprisingly, they did the same thing even if the object stopped in full view. Of course, both findings could have arisen because the infant simply could not arrest the tracking process. However, more detailed investigation of the tracking records indicated that tracking did not actually continue undisturbed. When the object stopped in view, subjects arrested tracking momentarily, only to resume the same activity a moment later (Bower & Paterson, 1973). Bower and Paterson concluded that infants noted the presence of the object when it stopped, but did not recognize the identity between it and the one that they had been tracking. Hence they resumed tracking in order to recapture the original object.

In addition, Bower and Paterson found that it was only around the age of 6 months that infants began to show surprise if the object being tracked was exchanged behind the screen so that a different one emerged at the usual time. However, younger infants were disturbed if an identical object emerged earlier than it should have if traveling at uniform velocity. This fits with the conclusion by Bower *et al.* (1971) that before the age of 6 months, a moving object is treated more or less as a movement sensation. Its features are not identified and it is dissociated from its stationary counterpart. Hence, when an object moves it becomes a different object. As a consequence, even when it stops again it is treated as a different object since there is no dynamic link between it in its first and second places. Bower's description seems to fit quite well with the spatial half of Piaget's account. For instance, Bower's claim about the infant's inability to understand the displacements of objects fits well with Piaget's claim that below stage III the infant fails to differentiate between changes in position and changes in state. However, unlike Piaget, Bower credits the young infant with considerable objective knowledge of space and objects.

Some workers (Goldberg, 1976; Nelson, 1971, 1974) have interpreted anticipation of an object's reemergence from a screen in terms of ability to predict event sequences rather than in the stronger terms adopted by Bower. According to Goldberg, the 4-month-old's surprise at early emergence of the object need only mean that the infant is sensitive to violation of a hitherto constant trajectory. On the other hand, the 6-month-old's surprise at reemergence of a different object is explained in terms of event prediction in conjunction with object discrimination. The infant discriminates between different objects in motion and expects (from past experience) to see ones that match at both sides of the screen. Alternatively, Moore, Borton, and Darby (1978) suggest that a concept of *object identity* is a sufficient basis for the sort of responding exhibited by the older infants in the work by Bower *et al.* (1971). In this case, object identity is specified by two rules (Moore, 1975): a rule that confers identity on an object moving along a given trajectory, and a rule specifying an object seen at two successive times as the same object if it possesses the same features.

Evidently, Bower's data can be interpreted in different ways. However, none of the alternatives seem to cope with the evidence of continued tracking after the object stops in view. Although it has been shown that 4-month-olds can generalize a response from a stationary to a moving object (Hartlep & Forsyth, 1977), Bower's evidence suggests quite strongly that they do not treat them as a single entity.

Spatial Factors and Search Failure by Six-Month-Olds

Given Bower's evidence, why do 6-month-old infants show no attempt to search for hidden objects in the traditional Piagetian stage III task? As we have seen, Bower *et al.* (1971) found that difficulties with object displacements were overcome by infants aged 5 or 6 months, and they claim that 6-month-olds have a good understanding of objective permanence and the rules governing movements of objects. Bower offers a spatial explanation of the search deficit. Infants do not understand the relationship between object and cover, so that when an object is hidden they conclude that it has been mysteriously replaced by or transformed into the cover.

Again, Bower's argument seems to fit in well with Piaget's account of the infant's spatial knowledge. Piaget also claims that the stage III infant does not understand the relationship between object and cover, but he argues that knowledge of this relationship is inseparable from a general understanding of the object's continuing existence. In contrast, Bower claims that infants have this general understanding and only run into problems with certain types of occlusion. He suggests that the relation *inside,* used in the Piagetian task, is more complex than the relation *behind.* Indeed, this qualification is necessary if the account is to be reconciled with his earlier argument that infants as young as 1 or 2 months understand the continued existence of an object behind a screen.

It seems quite likely that infants should have particular trouble with the relation *inside,* since the cover or container overlaps the object's position, and hence could be identified as occupying the same position as the object. Support for the validity of the distinction between *behind* and *inside* was provided by Brown (1973) who found that infants who failed to search in the traditional task (involving the *inside* relation) were successful if the object was hidden *behind* a screen. However, this explanation must be modified in the light of more recent findings. Neilson (1977) found that the degree of separation between object and screen was a crucial factor determining whether or not infants would search. If the object–screen separation is small infants will not search, whereas if it is large they will search. This result raises the possibility that in the *inside* case it is the small separation between object and occluder that gives the infant problems, rather than the complex enclosure relationship.

One fascinating aspect of Neilson's results was the appearance of this seperation effect even when the object was placed in full view in front of the screen instead of behind it. Infants did not reach for the object if it was placed directly in front of the screen, but they did reach if it was placed some distance in front. Thus, it seems that infants around 6 months old have general difficulties in perceiving or understanding relations between objects that are placed close together. The simplest interpretation seems to be that if the infant does not perceive a clear separation between two objects, he does not treat them as separate entities.

Taken together, the results in this section suggest that we should look harder at spatial factors in search tasks. Apparently, infants have difficulty in treating objects as distinct if the separation between them is small. This may explain why they do not search in the Piagetian task. Failure to perceive a clear separation between object and cover at the point of disappearance may lead to failure to represent them as separate objects once hiding is complete.

Spatial Factors and Search Errors by Nine-Month-Olds

If the spatial problems studied by Bower and his associates are overcome by the end of stage III (8 months), a new problem emerges. Although the stage IV infant knows how to search for a hidden object, whether behind or inside an occluder, he makes the error of searching in only one place, even when he has just seen the object hidden elsewhere. I would argue that this is a phenomenon that requires a separate explanation from those presented so far. Although a fragile grasp of the relationship between object and occluder may contribute to the stage IV infant's problems, it is not clear why he should search at one place as a consequence. Hence, an additional factor must be identified if we are to account for this perseverative error.

We might hope to find some clues to the nature of this second factor by considering some of the recent work on the phenomenon. Most studies have investigated the error for what it can tell us about the stage IV infant's object concept. However, we cannot be at all sure what the infant's search behavior means in terms of his notions about the object that he seeks. What does the infant think he does with the object when he searches? Does he know that it is under the cover and that his action simply makes it accessible, or does he think that his action re-creates it anew (an argument that Piaget used to explain behavior during stage III but not during stage IV)? An empirical solution to a question pitched at this level is most unlikely, and I believe that experimentalists should adopt an approach that avoids questions of this sort.

Alternatively, a spatial analysis may be more useful. Whatever the stage IV error may indicate about the infant's object concept, we should remember that it is a change in *position* that presents the difficulty for the infant. In the first spatial interpretation Evans and Gratch (1972) found that infants still made the stage IV error if the object hidden in the second position (B) was different from the one hidden at the first position (A). This conflicts with Piaget's claim that the object is treated as existing only in the place where it has been found before. Since the second object had never been hidden at position A, the infant had no reason to believe in its presence there when out of sight. Consequently, Evans and Gratch decided that the stage IV error must be a "place going error" that tells us nothing about the infant's object concept. Unfortunately, however, a recent replication (Schuberth, Werner, & Lipsitt, 1978) revealed quite different results.[3] Infants made very few B trials errors when the object was changed prior to hiding at position B; a result much more in line with Piaget's argument. Even without these conflicting results, Evans and Gratch's conclusion was questionable. As Gratch (1975) has since noted, infants may really have been looking for the old toy when they searched at position A after the toy switch.

Harris (1973) adopted a different approach to the problem, pointing out that the object movement in the standard stage IV task involved a change in the object's *absolute* position *and* a change in the left–right relationship between hiding and empty container (*relative* position). Harris (1973) used an ingenious technique to separate absolute and relative changes. The infant was seated at a table on which there were two hiding positions, one in front of him and the other to his right. Once he had found the object at the central position several times, the positions were moved laterally so that the central position was now at the infant's left and the right-hand position was at his midline. After this

---

[3]A methodological difference may account for this disparity. Schuberth *et al.* (1978) modified the "standard" condition by removing the toy after the last A trial and reintroducing it before the first B trial. This alteration made "standard" and "toy change" conditions identical in all respects apart from the factor under investigation, and it seems to have had the effect of increasing B trial errors in the "standard" condition.

change, the object was hidden either in the central position or in the left-hand position.[4] In the first case, the absolute position of the object was the same as before, whereas the relative position was reversed. In the second case, the relative position was the same as before, whereas the absolute position of the object changed.

Harris obtained low error rates after both sorts of change. However, it seems likely that this happened because he did not impose a delay between hiding and search, a condition under which few errors occur on the standard task (Gratch, Appel, Evans, LeCompte, & Wright, 1974; Harris, 1973). In a follow-up study, Butterworth (1975) rectified this problem by including a standard condition for comparison, and imposing a delay between hiding and search. The results of this study indicated that a change in either the relative position or the absolute position of the object led to errors at much the same level as when both changes took place together. Butterworth concluded that the infant uses both of these systems to define the position of the object. Furthermore, he concluded that the stage IV error cannot be explained as a tendency to repeat a successful response, since such a strategy would have led to success whenever the absolute position of the object remained the same.

This last point could be taken as a challenge to Piaget's account. However, I believe that a hard look at what he says about the stage IV error indicates that this new evidence presents not a challenge as much as a new problem that his account does not deal with. In Piaget's (1936/1954) terms, the object is no longer "at the disposal of the act [p. 24]" as in stage III, but instead is "at disposal in the place where the action has made use of it [p. 50]." While this may mean that action would spatially define the place of search there is another possibility—namely, that successful action specifies the appropriate place without defining it in spatial terms. Hence, the infant might take account of external information such as the container relationship, but might rely egocentrically on past successful action to specify which of the two positions is appropriate for search.

In summary, our problem is to determine whether the infant errs by repeating a successful response, or whether successful past action simply provides a basis for choice between the two positions, or indeed, whether it has any effect at all. Butterworth's evidence that infants use external relational codes casts doubt on the first interpretation. In any case, whichever interpretation is adopted, the crucial factor determining the error seems to be the information that the infant uses to guide search, rather than the spatial system within which she locates the object. After all, if she searched where she saw the object

---

[4]This is very similar to the technique used by Harrison and Nissen (1941) when studying absolute versus relative position responses by chimpanzees in delayed response problems.

go on B trials, she would be correct no matter what system she used to locate it. It is her reliance on out of date information (some sort of past experience at A) that leads to the error.

There have been a number of attempts to assess the importance of action in the stage IV task (Butterworth, 1974; Evans, 1973; Landers, 1971). In all cases, the technique involved a condition in which the subject was a passive observer during all, or most (Landers, 1971), of the trials at the first position. The object was hidden and revealed several times without the infant being allowed to search. Despite this limited experience, roughly the usual number of infants made an error when they were allowed to search on the first B trial.

Although these results allow us to conclude that overt manual action is not a necessary condition for error, this does not necessarily weaken Piaget's position. He notes that infants also make the same sort of error in their visual search habits, looking for an object where they have seen it appear before. Hence, in passive versions of the stage IV task, the specific visual fixation associated with appearance of the object may have been the action used to specify the appropriate place for subsequent search. This sort of interpretation is interesting because it suggests that if action has any effect on the direction of search, it is really at quite an abstract level. In addition, the presence of the error in looking habits reinforces the idea that if we adopt Piaget's point of view, a spatial analysis of the error must be divided into two components—one relating to understanding the object–occluder relationship through coordination of two manual action schemes, and another relating to the perseverative error itself, which still occurs when coordination of schemes is not a task requirement.

The evidence presented in this section has helped us to eliminate some explanations of the error. It is not simply perseveration of a manual response (Butterworth, 1974; Evans, 1973; Landers, 1971). At a higher level we have Butterworth's evidence that the infant relies to some extent on the relationship between containers when deciding where to search. However, we still need to find out why the infant chooses to rely on past instead of present information. There are plenty of explanations of how position A is specified, but very little information to explain why more recent information from B trials should be neglected.

I believe that we can learn most from the stage IV search error by drawing a clear distinction between two questions.

1. Why does the stage IV infant treat only one position as the appropriate place for search?
2. Given that he is likely to search at only one place, how does he define that place spatially?

At present we do not seem close to an answer to the first question, either in terms of the infant's object concept or in terms of his concept of space. However, later on I shall present an alternative explanation, based on the infant's ideas about the functions of places rather than on the physical laws governing objects and space. First, however, I shall deal with evidence bearing on the second question.

SPATIAL REFERENCE SYSTEMS USED BY INFANTS

Evidence from the Stage IV Error

Having separated the questions of cause of error and the form of the infant's spatial codes, we can approach the stage IV error in a new way. Instead of performing experimental manipulations designed to eliminate the error, we can hope that the error will remain but that our manipulations will reveal something about the way or ways in which infants code the position of search.

Butterworth (1975) has shown that 9-month-old infants use both absolute and relative position codes when locating a hidden object. What does this mean in terms of spatial reference systems? Absolute responding could have meant two things. A fixed position relative to the subject was also a fixed position relative to stable environmental features, since the infant was in the same position throughout the task. Thus, absolute responding could have meant that the infant was using a self-referent system as Butterworth suggests, or it could have meant that she was using an external system based on stable environmental features. This second system can be totally independent of the subject's position; for instance, a position that is close to a landmark remains so, whatever the infant's position. Second, a relative position code involves use of external features, but the relational terms *left* and *right* are self-referent. Although this system can be used to specify a given position despite a degree of subject movement, it breaks down in this role if the subject's perspective changes too much. For instance, if the baby approaches the problem from the opposite side of the array, the left-hand container becomes the one that was on the right, and vice versa.

Hence, there are three systems that the infant might use; self-referent, left–right relative, or external-referent. Self-referent (e.g., to my left) and relative systems (e.g., the left-hand container) both depend on the subject's position. Hence, they will both be classed as self-referent definitions to contrast them with external-referent definitions based exclusively on external landmarks and independent of the subject's position (e.g., the container near to landmark x).

We attempted to find out whether 9-month-olds could use an external reference system (Bremner & Bryant, 1977): Each infant was seated in front of a small table containing hiding positions to his left and right. To provide external spatial cues, the table was painted so that one position lay on a black background, the other lay on a white background. After five A trials the infant was moved to the opposite side of the table, and the object was hidden again with him watching from his new position. In the case of one group (A), the object was hidden in the opposite position from before, whereas in the case of the other group (B), it was hidden in the same position as before (see Figure 4.1).[5]

If infants responded to the same self-referent position as before, those in Group A would be correct whereas those in Group B would make errors. However, if they responded to the same position relative to external cues (for instance, the table side differentiation cues), those in Group A would make errors and those in Group B would be correct.

The results were quite clear. Very few infants in Group A made an error on the first trial after the rotation, whereas most of the infants in Group B did err. Thus, after rotation there was a clear tendency for infants to change the side of the table at which they searched, but no evidence that they changed their response. These results were backed up by those from two other conditions in which we rotated the table instead of the infant. Few subjects erred when the object was hidden at the same self-referent position as before, although this was on a different background, and a large number of errors occurred when the object was hidden on the same background but at the opposite self-referent position.

As a whole, these results suggest that most infants were using some form of self-referent definition of the search location. It is not clear whether they were using a simple self-referent system (maybe related to response repetition) or one based on the left–right relation between the containers. However, there was certainly no evidence that infants were using an external system based on cues contained in the table or in the rest of their surroundings.

Nevertheless, I was not convinced that this was the whole story. It seemed possible that the cues differentiating the sides of the table were not adequate, and that in the presence of more salient cues, infants might respond differently. Consequently, I performed a similar experiment (Bremner, 1978a) in which the cues were enhanced by exchanging the black and white table and gray covers for a gray table with a black and a white cover.[6] This simple alteration produced a dramatic change in infants' performance. In Group B, only a small

[5] This technique was derived from the method used by Tolman, Ritchie, and Kalish (1946) to study place versus response learning by rats.

[6] This assumption is backed up by the finding that animals acquire a discrimination more quickly if there is spatial contiguity between stimulus, response, and reinforcement (Cowey, 1968).

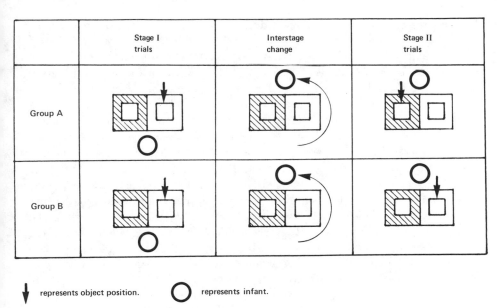

	Stage I trials	Interstage change	Stage II trials
Group A			
Group B			

↓ represents object position.   ◯ represents infant.

Figure 4.1. The important conditions from Bremner and Bryant's (1977) experiment.

number of infants made an error; a significant improvement over the background cue case. In Group A, infants performed at much the same level as before, although this involved lifting a differently colored cover after the transformation. Thus, it seems that success in Group B was not simply the result of responding to a particular cover rather than searching for the object. Apparently, the infant is capable of defining the position of search relative to external features, but only if these cues are sufficiently strong.

However, it turned out that these cover cues did not have such a strong effect in the equivalent condition in which the transformation involved reversal of the covers instead of rotation of the infant. Thus, provision of stronger cues was not sufficient on its own to produce a shift from a self-referent to an external-referent system. A movement of the infant was also necessary, maybe because it served some sort of prompting function.

Unfortunately, there is a problem in interpreting these data. The assumption has been that most infants were searching at the same position (however defined) both before and after the transformation, and that when infants were successful after the transformation this was because the addition of strong cues had made them change their coding of the "same place" from self-referent to external-referent. That is, the experimental manipulation did not remove the error but redefined the place at which they searched. However, there is another possibility. The infant-movement condition may simply have led to nonperseveration. After movement, infants may have searched where they saw the object go instead of where they found it before. It is not clear why this should only happen when very strong spatial cues were provided; however, the ambiguity of these findings gave me reservations about the further usefulness of the stage IV task as a tool for finding out about spatial understanding. I required a task that would remove any chance of the infant using post-transformation experience to guide his search, but at the same time I wanted to retain the useful features of the old task. The solution was to interpose the rotation between hiding and search. The infant would see the object hidden as usual, but would be prevented from searching until after he had been moved (or until after the table had been rotated). Then I had to decide whether to include preliminary search trials before these rotation trials commenced. Was search experience necessary if the infant was to code the object's position at all, or did this sort of experience strengthen the tendency to use the self referent code, maybe at the expense of a more satisfactory strategy? The data supported the second alternative. Self-referent errors increased on rotation trials when preliminary search experience was included (Bremner, 1978b).

With preliminary search experience omitted from the new task, 9-month-olds performed in much the same way as they had on the stage IV task. When cover cues were used, infants were very good at finding an object that had been hidden before they moved, but were less successful when background differentiation was the cue. The same cue effect appeared in conditions in which the table was rotated instead of the infant, but performance was significantly worse overall. Thus, the same two effects emerged. The first effect is simple. Stronger cues enhanced performance. The second effect is not so easily explained and is certainly more interesting. Infants performed better at the new spatial transformation task when the transformation involved movement of themselves rather than movement of the array.

Of course, there is an alternative explanation of the movement effect. Instead of superior performance being the result of moving the infant between stages, it could have arisen because the object remains stable relative to the surroundings in the *infant movement* case but not in the *array movement* case. Thus, it might really be an *object stability effect*. However, if this were the case, we would have difficulty explaining the absence of a difference between infant and array movement conditions in the first study (Bremner & Bryant, 1977). If the less immediate surroundings were used as referents, one would expect object stability to have a stronger effect when the immediate cues were less strong, rather than the reverse. Thus, the idea that movement of the subject serves some sort of prompting function seems by far the simplest explanation of the data.

The movement effect suggests that infants are better at taking account of their own movements than the movements of objects. This runs counter to Piaget's claim that the infant does not understand his own displacements until late in the sensori-motor stage, so his theory may need modification here.

Evidence from Other Sources

Our results were gathered from only one age group (9-month-olds). However, another study using a formally similar task has been performed with subjects tested at three different ages (Acredolo, 1978). In this case, only an *infant rotation* task was presented. Infants were placed in a small rectangular room with windows to their left and right. They were trained to anticipate the appearance of an experimenter at one of the two windows, the cue being the sound of a centrally placed buzzer. After a consistent response to one window was established, subjects were rotated through 180 degrees about the midpoint of the room, so that the window that had been on their left was now on their right and vice versa. The crucial observation was the direction of the infant's response when the buzzer sounded after this movement.

Acredolo presented this task both with and without spatial cues to 6-, 11-, and 16-month-old infants. In the "landmark" condition the correct window was surrounded by a large star; in the other condition no such cue was provided. At 6 months the majority of infants responded to the same self-referent position after they had been moved, and thus anticipated the event at the wrong side of the room. The landmark did not seem to have any effect at this age. At 11 months performance was still bad in the "no landmark" group; however, only half of the "landmark" group made an error after rotation. Thus, at this age infants seem to derive some benefit from the landmark. Nevertheless, even performance of the "landmark" group was rather poor, and the only clear change in performance took place between 11 and 16 months, with the majority of 16-month-olds performing correctly from their new position whether or not a landmark was provided. It is interesting that spatial cues seemed to be unnecessary by 16 months, despite the hint of an effect at 11 months. Maybe strong spatial cues are really important only while the infant is acquiring the ability to use external reference systems.

It is worth noting that 11-month-olds did not perform as well as 9-month-olds in the strong cueing condition of my task (Bremner, 1978b). This difference could have arisen for a number of reasons. First, Acredolo gave infants a number of training trials before rotation, a procedure that produced strong self-referent responding in my task. Second, the disparity may be due to a difference in cue salience between studies. Acredolo and Evans (1980) found that presentation of stronger landmarks (stripes and flashing lights around the correct window) enhanced performance on Acredolo's task considerably. Under these conditions, most 9- and 11-month-olds performed correctly after rotation, and even 6-month-olds performed in a very mixed way compared to their clear self-referent performance in the original study.

Another potential source of difference has been identified. Acredolo (1979) found that babies tested at home were more likely to use an external referent in searching for an object after movement than were infants tested in the laboratory. She suggests that this factor may contribute to the difference between her results in the laboratory and my work in the infant's home. Infants may be better at orienting in their own home surroundings than elsewhere, or they may perform better at home simply because they are more relaxed there (cf. Hart & Berzok, Chapter 5, this volume).

In summary, although it is clear that infants are capable of using external cues to specify positions, we can identify quite a number of factors that may affect whether or not the infant will display this competence. Perhaps 6-month-olds would perform as well as older infants if all these factors were in their favor.

EARLY EXPERIENCE AND INFANTS' SEARCH STRATEGIES

Possibly the simplest way of accounting for some of the more puzzling results is to look for an explanation in terms of the infant's general experience up to and around the age of 9 months. Are there constraints on her everyday experience that might lead to her perseverative errors? Similarly, could her choice of spatial reference system be constrained or influenced by her limited experience?

A Way of Explaining the Stage IV Error

Piaget (1936/1954) reports that it is not until around 9 months that infants spontaneously engage in hiding and finding games. However, infants have probably been finding things in customary places (e.g., the toy box) for some time before this. It is possible that before they have a full understanding of physical relationships between objects, they make simplifying assumptions about the places where hidden objects are usually found. These assumptions may be based more on finding objects than seeing them hidden, for although the young infant often takes an object out of a container, he very rarely puts it in one.

A tentative explanation of the stage IV error can be built up from this. Once the infant learns to search for an object that he sees being hidden, he may still be influenced by his old strategy of searching where he has come to find objects rather than where he has just seen them hidden. Thus, on the first A trial of the stage IV task, he searches where he sees the object disappear (his new strategy), because he had no prior hypothesis about where it should be found. However, when he finds it, this experience establishes that as the place where the object is to be found, and henceforth the old strategy dominates so that the infant searches at the first place, no matter where the object is hidden.

All this may leave some doubt about the reason for the infant relying on his idea about the "appropriate place" for search, even when he can use the newer strategy of searching where he sees the object go. The answer may be that the new strategy is not well established, or is at least more of a strain on the infant's mental resources. Under such circumstances, the natural course would be to fall back on an old strategy that worked well in the past.

It should be evident that this explanation is very much like an extended version of the "place going error" explanation offered by Evans and Gratch (1972). However, the difference is that in this case the question is left open of whether an appropriate place for search is treated as a general source of objects, or whether its appropriateness is limited to the particular object hidden there. At present, there seems to be no conclusive evidence suggesting an answer.

Early Experience and Spatial Reference Systems

The increasing mobility of the infant during her first year seems likely to have a fairly direct effect on her developing spatial abilities. The basic argument[7] is that initially the infant is quite likely to find a self-referent definition of positions fairly adequate in her dealings with the world, but that later, and particularly when she begins to crawl, such a strategy will become very much less successful. A self-referent code breaks down as a system for specifying stable features of the environment every time the infant moves. However, the basis of my argument is that the young infant does not link up the different views of the world that she obtains from different positions. From each single viewpoint, stable objects maintain constant positions relative to the self, hence, within this fragmented view of the world, the self-referent system is adequate.

The essential factor is that the immobile infant is limited to positions in her environment that her parents select for her, and that in many of these habitual positions she will be placed in a consistent orientation. The infant is presented with a relatively small number of different views of the world that are inseparable from the context in which they arise (e.g., lying in her cot, having her diaper changed, etc.). Within a particular context, stable objects and even her parents attending her are likely to appear in consistent positions relative to her.

[7]Originally suggested to me by Peter Bryant.

There will be some cases in which the same object can be seen from more than one of her habitual positions. However, she is unable to link up the separate perspectives that she obtains from these different positions. The context in which she finds herself determines where she looks, and the fact that she finds the same object from different positions does not necessarily indicate a proper understanding of space, or indeed the realization that it is the same object that she sees in both cases (cf. Bower's argument [1974] that the young infant fails to confer identity on one subject seen in different places; see the section entitled "Surprise and Visual Tracking as Measures of the Infant's Knowledge," p. 84).

Not only will stable objects maintain their self-referent positions within a particular context, but unstable objects will disappear and reappear at similarly stable places: The infant's parents will disappear and reappear at particular places, and toys are likely to be found in the same box (maybe in a slightly mysterious way, because it is rarely the infant who puts them there).

This argument predicts that up to a certain age the infant will find self-referent definition of positions adequate, but that when he becomes mobile he is likely to find this strategy less satisfactory. He will find himself in many more positions and orientations than before. The context will no longer determine his position, and in order to cope with the large number of new perspectives with which he is faced, he will have to develop a way of linking them. The obvious way of doing this would be to relate positions to a stable external system instead of to the unstable self. It should be noted that this increase in perspectives will also occur before the infant begins to crawl, as a result of changes in the way the infant's parents treat him as he gets older. Particularly when he is able to sit upright, he is likely to be given more freedom of position, being placed in a number of positions on the floor and in chairs. Thus, although the onset of crawling is likely to produce the major increase in perspectives, other factors will also contribute to this. This suggests that around the time that the infant is beginning to find the self-referent strategy generally unsatisfactory, he may be aware that it is particularly after he has just been moved that some other strategy is required. This would explain the good performance of infants in the "infant movement" conditions of my experiments (Bremner, 1978a, 1978b).

This explanation also suggests a direction for future research. If crawling is important for the development of spatial abilities, maybe we should be devising tasks in which the infant has to crawl to reach her goal. Corter, Zucker, and Galligan (1980) have found that the stage IV error occurs when the infant has to crawl in order to find her mother, and this technique could easily be adapted to tackle the spatial problems that we are concerned with here. If the present explanation is correct, the infant is most likely to use an external-referent system in a task of this sort.

SUMMARY OF THE EVIDENCE COVERED SO FAR

In the introduction, I said that a spatial system for object localization based on environmental cues might emerge some time in infancy, particularly around the time that the infant is becoming more mobile. What does the current evidence tell us about this?

I have argued that two separate relationships are contained in object search tasks used with 9-month-olds. First of all, there is the simple external relationship between object and cover. If search for a hidden object could be taken as good evidence for understanding of the object–occluder relationship, we would have strong evidence for this sort of spatial ability fairly early in infancy. Infants will search for hidden objects at around 7 or 8 months, and if Bower is right, they understand simpler object–occluder relationships much earlier. But the snag is that infants may search for a hidden object without understanding its relationship to the occluder, believing that their action re-creates the object.

However, Brown's (1973) evidence shows that simplification of the object–occluder relationship leads to search at an earlier age. This supports the idea that search indicates understanding of this simpler relationship, and that infants understand simple relationships before more complex ones, or ones that are harder to perceive (Neilson, 1977).

Second, search tasks present the infant with a choice between alternative locations. This also applies in visual anticipation tasks (Acredolo, 1978; Acredolo & Evans, 1980), and the problem has been to determine how he defines the position at which he searches. So far, we know that there is a gradual progression from self-referent to external-referent definition of this position between 6 and 16 months (Acredolo, 1978), and that strong landmarks enhance external coding around the middle of this period (Acredolo & Evans, 1980; Bremner, 1978a, 1978b). In addition, infants are better at taking account of their own movements than movements of objects. I argued that this may be because everyday experience leads them to believe that an external-referent system should be used whenever they have moved to a new position.

EXTENSIONS OF THE ACCOUNT

Topological Principles in External Reference

Although the work reviewed so far tells us a good deal about the development of external-reference systems and the circumstances under which infants will use them, little has been said about the organizing principles involved in these systems. Piaget and Inhelder (1948/1956) claim that the infant is initially restricted to understanding *topological* relationships, particularly the relationship of *neighborhood,* and that understanding of projective relationships like *in front* and *behind* only begins to emerge in limited form during stage IV. It may be necessary to question this conclusion, since it is based in part on Piaget's (1936/1954) now doubtful claim that infants do not understand the principle of size constancy until the later sensori-motor stages (cf. Bower, 1966b).

Nevertheless, Lucas and Uzgiris (1977) have obtained evidence that stage IV infants are still influenced by topological principles. Testing at weekly intervals from the age of 6 months, they found that if infants were given a task in which an object placed in front of a landmark screen was then hidden and displaced behind a second screen, they often erred by searching at the landmark screen, and took an average of five sessions longer to solve this task than one in which the landmark screen was omitted. Lucas and Uzgiris concluded that this was because these infants coded the object as *in the neighborhood* of the landmark before occlusion, and that they did not code it as *in front,* so saw nothing wrong in searching behind the screen for an object last seen in front of it.

In a second experiment, they found that errors of this sort were significantly reduced if the object was placed to one side of the landmark screen instead of in front of it, and they concluded that a strong determinant of perceived neighborhood is overlap between contours of object and landmark. This is an interesting result, and investigation of this sort of detail seems to be a good way of finding out more about the infant's method of coding positions.

One other finding supports this line of argument. As I mentioned, Acredolo and Evans (1980) found that 11-month-old infants could use a very noticeable landmark to guide visual search after movement. However, they also noted that if a separation was produced between event and landmark by presenting the latter at the opposite window, infants showed much less evidence of relying exclusively on the external referent. This suggests that the infant is only competent at specifying positions in terms of their neighborhood or proximity to landmarks.

On the surface, these results seem in conflict with those obtained by Neilson (1977) with younger infants. Lucas and Uzgiris found that if the contours of object and landmark did not overlap, the neighborhood effect disappeared. However, Neilson found that younger infants failed to reach for an object when separation between it and the screen was small, but succeeded when it was larger. In one case, proximity prevented search, whereas in the other it promoted search at a landmark. Maybe this can be explained in terms of a complementary process of differentiation and integration in spatial organization. Initially, the infant may only identify an object if he can differentiate it from its surroundings, so he thinks something rather strange happens to it if it moves so that he can no longer perceive separation between it and some other object. Hence, he does not reach for an object placed close in front of a screen, and does not search for it if a similar relationship exists just before occlusion. The infant's problem at this stage is to understand that objects remain distinct even in the absence of clear separation between them. However, once this differentiation is established, cases will arise in which it is important for the infant to relate objects that she knows are separate, and it is in these cases that neighborhood may be the organizing principle. This is just the sort of description that Piaget gave to the changing relationship between infant and environment during sensori-motor development, absence of differentiation leading to differentiation, with a consequent requirement for knowing integration.

Links with Spatial Abilities of Older Children

In addition to extending our knowledge of infants' spatial abilities, it would be useful if we could relate this to what we know about spatial abilities in older subjects. The bulk of the literature in this area has stemmed from Piaget and Inhelder's (1948/1956) claim that children are unable to adopt the spatial perspective of another until around the age of 7 or 8 years. However, much of this work has been concerned more with the question of whether or not young children can adopt the perspective of another than with a detailed breakdown of the spatial components of the task (e.g., Cox, 1975; Flavell, Omanson, & Latham, 1978; Houssiadas, 1965; Laurendeau & Pinard, 1970; Salatas & Flavell, 1976). Nevertheless, considerable similarity exists between the sort of task used to measure spatial perspective taking and the rotation tasks used with infants.

Piaget and Inhelder's (1948/1956) main task was the well-known "three mountains problem," in which the child was presented with a three-dimensional model of three mountains and was asked to select or construct the perspective appropriate to a doll observer viewing the scene from a number of different positions. Children under 7 or 8 years of age had considerable difficulty with this problem, and showed a strong tendency to select or construct their own perspective in representing the doll's view, even when it was placed at the opposite side of the display from the child. Two components have been identified in this task (Coie, Costanzo, & Farnill, 1973; Piaget & Inhelder, 1948/1956). The child must appreciate that a person in another position has a different spatial perspective from his own, *and* he must be able to construct this perspective from the available evidence.

The second component of the task relates directly to the child's spatial ability, and may just be a more complex version of the problem faced by infants in rotation tasks. In this case the child does not move around the array, but it has been argued that he has to do the mental equivalent if he is to put himself in the observer's position (Huttenlocher & Presson, 1973). Some studies have investigated the child's ability to predict how a display will look from a new position that he actually adopts (Huttenlocher & Presson, 1973; Schantz & Watson, 1970, 1971). Huttenlocher and Presson found that rotation tasks were solved earlier than the sort of perspective task used by Piaget and Inhelder, and Schantz and Watson found that children as young as 3 or 4 years old could work out how an array should look after adopting a position at the opposite side of it.

Despite the striking similarity between these tasks and those used with infants, two major factors make them more complex. First, children are given far less information on which to base their reconstruction. In perspective tasks they have to imagine their own movement (if this is how they solve the task) and in rotation tasks the array is screened before rotation. In contrast, infants can use any cues that are provided, both during and after the rotation. Using a task in which the child had to search for an object after rotation of the array in which it was hidden, Lasky, Romano, and Wenters (1980) found that performance was better than chance at 4 years if the array rotation was witnessed by the child. But when the child did not witness the rotation and had to infer it from the changed position of a landmark, performance did not exceed chance until 7 years. Although they did not carry out a direct comparison of these conditions in the case of subject movement, it seems likely that the same sort of results would emerge.

Second, in perspective tasks children are asked to locate several objects rather than to identify one position, and in many cases the arrays have been organized in two dimensions; near–far as well as left–right. Hence, reconstruction often requires the use of a coordinate reference system with objects located relative to near–far and left–right dimensions. According to Piaget, this demands the ability to represent spatial relationships, a step beyond the infant's ability to impose some sort of spatial order on what she perceives. He argues that initially the preoperational child's representations are based on topological principles, but that later, egocentric thought is used to differentiate these relationships. More concretely, the infant will intially locate objects in proximity to landmarks but later will begin to differentiate between *near and to the left of* and *near and to the right of* the landmark.

This system has two limitations that are overcome only gradually. First, the self-referent component of the system does not cope with changes in perspective; a position is *near and to the left of* a landmark for all viewing positions. Second, the child is unable to perform logical multiplication of positions on the near–far and left–right dimensions, with the result that she locates the object in terms of one dimension alone. Pufall and Shaw (1973) investigated these points by asking children to place an object on a rotated display in the same position as that of a similar object on a fixed comparison display. They found that 4-year-olds used the sort of self-referent system mentioned above and failed to coordinate dimensions, so that the position was specified as *near* or *left,* but not *near left*. This led to placing errors even when the display was not rotated. However, the 6- and 10-year-olds were capable of coordinating dimensions and compensating for different perspectives quite well. But, when landmarks were omitted and the child had to represent a position with respect to an abstract frame of reference, both age groups fell back on the self-referent system. [8]

Despite these interesting results it is also important to ask how the child would perform under circumstances with which he is more familiar. A change in perspective due to rotation of a display is not something that a child is often faced with, and even infants were more successful in subject rotation conditions than array rotation conditions (Bremner, 1978a, 1978b). In addition, this sort of small scale array is more like a model than a real space.

One study overcomes both of these problems. Acredolo (1977) tested 3-, 4-, and 5-year-olds on a task similar to the one she used with infants (Acredolo, 1978). The child was placed at the midpoint of one side of a rectangular room and was trained to find an object in one of two cups placed at the midpoints of the walls to his left and right. The same cup was always used, but hiding was done in such a way as to prevent the child from detecting which cup had been "baited." Once he had been successful on five consecutive trials, he was tested again but this time starting from the opposite side of the room. Roughly half of the 3- and 4-year-olds made errors by searching at the same self-referent position as before, whereas none of the 5-year-olds made an error after movement. In addition, provision of landmarks enhanced performance, leading to a majority of correct performers in all age groups. Surprisingly enough, landmarks were effective in this way even when they were quite indirect. Differentiating the front and back walls of the room had the same effect as differentiating the immediate surroundings of the hiding places. This is interesting because it looks as if children were doing something quite complex in using these indirect landmarks. Differentiation in the near–far plane only provides information about a left–right choice if it is applied within a coherent coordinate reference system based on orthogonal dimensions.

In summary, there is a good deal of overlap between the questions being asked in these two areas; however, we still have little evidence about how spatial abilities in infancy develop into the abilities of the young child. Although the work covered here has begun to answer some of the questions, at least as many new questions have emerged. Future research in spatial development should have two objectives. First, we need much more information about the precise forms of spatial organization used by infants and children, and the

---

[8]The sort of self-reference observed by Pufall and Shaw is really a combination of self-reference and external-reference. The object is related first and foremost to the landmark, but the details of this relationship are based on the self as referent. In fact, this is rather similar to the "relative" system that Harris (1973) and Butterworth (1975) investigated with infants (see pp. 88–90).

conditions under which one system is used rather than another. Second, we must attempt to elucidate the developmental relationship between the spatial systems used by infants and those used by children.

REFERENCES

Acredolo, L. P. Developmental changes in the ability to coordinate perspectives of a large-scale space. *Developmental Psychology*, 1977, *13*, 1–8.

Acredolo, L. P. Development of spatial orientation in infancy. *Developmental Psychology*, 1978, *14*, 224–234.

Acredolo, L. P. Laboratory versus home: The effect of environment on the 9-month-old infant's choice of spatial reference system. *Developmental Psychology*, 1979, *15*, 666–667.

Acredolo, L. P., & Evans, D. Developmental changes in the effects of landmarks on infant spatial behavior. *Developmental Psychology*, 1980, *16*, 312–318.

Bower, T. G. R. Object permanence and short-term memory in the human infant. Unpublished manuscript, University of Edinburgh, 1966. (a)

Bower, T. G. R. The visual world of infants. *Scientific American,* December, 1966, 80–92. (b)

Bower, T. G. R. *Development in infancy.* New York: Freeman, 1974.

Bower, T. G. R., Broughton, J. M., & Moore, M. K. Development of the object concept as manifested in changes in the tracking behavior of infants between 7 and 20 weeks. *Journal of Experimental Child Psychology*, 1971, *11*, 182–193.

Bower, T. G. R., & Paterson, J. G. The separation of place, movement and object in the world of the infant. *Journal of Experimental Child Psychology*, 1973, *15*, 161–168.

Bremner, J. G. Spatial errors made by infants: Inadequate spatial cues or evidence of egocentrism? *British Journal of Psychology*, 1978, *69*, 77–84. (a)

Bremner, J. G. Egocentric versus allocentric spatial coding in nine-month-old infants: Factors influencing the choice of code. *Developmental Psychology*, 1978, *14*, 346–355.

Bremner, J. G., & Bryant, P. E. Place versus response as the basis of spatial errors made by young infants. *Journal of Experimental Child Psychology*, 1977, *23*, 162–171.

Brown, I. *A study of object permanence.* Unpublished honours thesis, University of Edinburgh, 1973.

Butterworth, G. *The development of the object concept in human infants.* Unpublished D.Phil. thesis, University of Oxford, 1974.

Butterworth, G. Object identity in infancy: The interaction of spatial location codes in determining search errors. *Child Development*, 1975, *46*, 866–870.

Casati, I., & Lezine, I. Administration manual: The stages of sensory-motor intelligence in the child from birth to two years. Revision by C. B. Copp & M. Sigman, unpublished manuscript, U.C.L.A., 1968.

Coie, J. D., Costanzo, P. R., & Farnill, D. Specific transitions in the development of spatial perspective taking ability. *Developmental Psychology*, 1973, *9*, 167–177.

Corter, C. M., Zucker, K. J., & Galligan, R. F. Patterns in the infant's search for mother during brief separation. *Developmental Psychology*, 1980, *16*, 62–69.

Cowey, A. Discrimination. In L. Wiezkrantz (Ed.), *Analysis of behavioral change.* New York: Harper & Row, 1968.

Cox, M. V. The other observer in a perspective task. *British Journal of Educational Psychology,* 1975, *45*, 83–85.

Escalona, S., & Corman, H. The validation of Piaget's hypothesis concerning the development of sensori-motor intelligence: Methodological issues. Paper presented at the meeting of the Society for Research in Child Development, New York, 1967.

Evans, W. F. *The stage IV error in Piaget's theory of object concept development: An investigation of the role of activity.* Unpublished dissertation, University of Houston, 1973.

Evans, W. F., & Gratch, G. The stage IV error in Piaget's theory of object concept development: Difficulties in object conceptualization or spatial location? *Child Development,* 1972, *43,* 682–688.

Flavell, J. H., Omanson, R. C., & Latham, C. Solving spatial perspective-taking problems by rule versus computation: A developmental study. *Developmental Psychology,* 1978, *14,* 462–473.

Goldberg, S. Visual tracking and existence constancy in 5-month-old infants. *Journal of Experimental Child Psychology,* 1976, *22,* 478–491.

Gratch, G. Recent studies based on Piaget's view of object concept development. In L. B. Cohen & P. Salapatek (Eds.), *Infant perception: From sensation to cognition* (Vol. 2). New York, Academic Press, 1975.

Gratch, G., Appel, K. J., Evans, W. F., LeCompte, G. K., & Wright, N. A. Piaget's stage IV object concept error: Evidence of forgetting or object conception. *Child Development,* 1974, *45,* 71–77.

Harris, P. L. Perseverative errors in search by young infants. *Child Development,* 1973, *44,* 28–33.

Harris, P. L. Object perception and permanence. In P. Salapatek & L. B. Cohen (Eds.), *Handbook of infant perception.* New York: Academic Press, in press.

Harrison, R., & Nissen, H. W. The response of chimpanzees to relative and absolute positions in delayed response problems. *Journal of Comparative Psychology,* 1941, *31,* 447–455.

Hartlep, K. L., & Forsyth, G. A. Infants' discrimination of moving and stationary objects. *Perceptual and Motor Skills,* 1977, *45,* 27–33.

Houssiadas, L. Coordination of perspectives in children. *Archiv fur die gesamte Psychologie,* 1965, *117,* 319–326.

Huttenlocher, J., & Presson, C. C. Mental rotation and the perspective problem. *Cognitive Psychology,* 1973, *4,* 277–299.

Kramer, J. A., Hill, K. T., & Cohen, L. B. Infants' development of object permanence: A refined methodology and new evidence for Piaget's hypothesized ordinality. *Child Development,* 1975, *46,* 149–155.

Landers, W. F. The effect of differential experience on infants' performance in a Piagetian stage IV object–concept task. *Developmental Psychology,* 1971, *5,* 48–54.

Lasky, R. E., Romano, N., & Wenters, J. Spatial localization in children after changes in position. *Journal of Experimental Child Psychology,* 1980, *29,* 225–248.

Laurendeau, M., & Pinard, A. *The development of the concept of space in the child.* New York: International Universities Press, 1970.

Lucas, T. C., & Uzgiris, I. C. Spatial factors in the development of the object concept. *Developmental Psychology,* 1977, *13,* 492–500.

Miller, D., Cohen, L., & Hill, K. A methodological investigation of Piaget's theory of object concept development in the sensory-motor period. *Journal of Experimental Child Psychology,* 1970, *9,* 59–85.

Moore, M. K. Object permanence and object identity: A stage-developmental model. Paper presented to the Society for Research in Child Development, Denver, April, 1975.

Moore, M. K., Borton, R., & Darby, B. L. Visual tracking in young infants: Evidence for object identity or object permanence? *Journal of Experimental Child Psychology,* 1978, *25,* 183–198.

Moore, M. K., & Meltzoff, A. N. Object permanence, imitation, and language development in infancy: Toward a neo-Piagetian perspective on communicative and cognitive development. In F. D. Minifie & L. C. Lloyd, (Eds.). *Communicative and cognitive abilities. Early behavioral assessment.* Baltimore, Maryland: University Park Press, 1978.

Neilson, I. E. *A reinterpretation of the development of the object concept in infancy.* Unpublished Ph.D thesis, University of Edinburgh, 1977.

Nelson, K. E. Accommodation of visual tracking patterns in human infants to object movement patterns. *Journal of Experimental Child Psychology,* 1971, *12,* 182–196.

Nelson, K. E. Infants' short-term progress toward one component of object permanence. *Merrill-Palmer Quarterly of Behavior and Development,* 1974, *20,* 3–8.

Piaget, J. [*The origins of intelligence in children*] (M. Cook, trans.). New York: Basic Books, 1952. (Originally published in French, 1936.)

Piaget, J. [*The construction of reality in the child*] (M. Cook, trans.). New York: Basic Books, 1954. (Originally published in French, 1936.)

Piaget, J., & Inhelder, B. [*The child's conception of space*] (F. J. Langdon & J. L. Lunzer, trans.). London: Routledge & Kegan Paul, 1956. (Originally published in French, 1948.)

Pufall, P. B., & Shaw, R. E. Analysis of the development of children's spatial reference systems. *Cognitive Psychology,* 1973, *5,* 151–175.

Salatas, H., & Flavell, J. H. Perspective taking: The development of two components of knowledge. *Child Development,* 1976, *47,* 103–109.

Schantz, C., & Watson, J. S. Assessment of spatial egocentrism through expectancy violation. *Psychonomic Science,* 1970, *18,* 93–94.

Schantz, C., & Watson, J. S. Spatial ability and spatial egocentrism in the young child. *Child Development,* 1971, *42,* 171–181.

Schuberth, R. E., Werner, J. S., & Lipsitt, L. P. The stage IV error in Piaget's theory of object concept development: A reconsideration of the spatial localization hypothesis. *Child Development,* 1978, *49,* 744–748.

Tolman, E. C., Ritchie, B. F., & Kalish, D. Studies in spatial learning IV. Place learning versus response learning. *Journal of Experimental Psychology,* 1946, *36,* 221–229.

Uzgiris, I., & Hunt, J. McV. An instrument for assessing infant psychological development. Mimeographed paper, Psychological Development Laboratory, University of Illinois, 1966.

HERBERT L. PICK, JR.
JOHN J. RIESER

5

Children's Cognitive Mapping[1]

Surely one of the most remarkable achievements in human spatial orientation unaided by artificial technical navigational aids are the voyages of the Pulawat Islanders. They travel in outrigger canoes across vast stretches of the Pacific. Their trips sometimes extend for a thousand miles and most of them are concluded successfully. It is instructive for us to examine the procedures by which such navigation is accomplished for, on the one hand, many of the processes seem to be the same as those used for more ordinary (but perhaps no less remarkable) orientation tasks. On the other hand, some of the elements in the Pulawat accomplishments differ markedly from what we are used to, and this contrast may help us identify the common underlying processes. With such reasoning in mind we would like to describe briefly, on the basis of Thomas Gladwin's (1970) analysis, how the Pulawat Islanders find their way from island to island, identifying what we think are some of the crucial features.

[1]The preparation of this chapter was supported by Program Project Grant HD-050207 from the National Institute of Health to the Institute of Child Development of the University of Minnesota and by the Center for Research in Human Learning of the University of Minnesota. The work of the Center is supported by Research Grants from the National Science Foundation and from the National Institute of Health. Reiser was supported in part during preparation of the chapter by Grant HD-04510 awarded to the John F. Kennedy Center Institute on Mental Retardation and Intellectual Development, George Peabody College.

SPATIAL ABILITIES
Development and Physiological Foundations

Then we would like to review some of our own research on the spatial orientation of children and adults that implicates some of the same features.

The Pulawat Islanders may set out on a journey as something of a lark after an evening of celebration. First they must decide the direction in which to sail. While their home island is in sight this is no problem. They start by sailing directly away from one side or another of their island and landmarks apparently are lined up to maintain azimuth. However, they are quickly out of sight of their own island and then must depend on other information for heading. Various cues and combinations of cues are useful. Directions of stars are useful unless occluded by clouds, particularly the directions of the points of rising and setting of the various stars. For moment to moment maintenance of heading, the angle of the boat with waves and wave patterns is valuable, and the Pulawat navigators seem to sense this on the basis of motion of the boat and of the sound of the water against the boat. The angle of heading to the wind also provides valuable information. Fortunately, the wind direction and wave patterns are highly predictable in this region of the sea. For example, there are three different wave trains that typically occur. Their directions differ, but they are distinguishable by period and by shape. As the journey proceeds, there are numerous subtle landmark (or should we say, seamarks) that indicate more or less precisely where one is. These include changes in the wave patterns as one passes in the shadow of an island and changes in the color of the water due to reefs. The flying ranges of birds of different types are known, and some species fly toward their home islands at dusk. This information indicates direction and maximum distance to some land.

Besides using these physical cues for maintaining heading and pinpointing locations when a specific seamark can be seen, the Pulawat Islanders conceptualize, that is, mentally represent, their journey in an intriguing way. The term *Etak* is used to refer to this conceptualization. A *hypothetical* island is "placed" over the horizon off to the side of the route between the home and destination islands as illustrated in Figure 5.1. At the beginning of the journey the direction of the hypothetical island is specified by its stellar direction. It may, for example, lie in the direction of the rising of Gamma Aquilae. When they are at the destination, the Etak island lies in the direction of the setting of the Little Dipper. In between, various points along the journey are marked by the reference island lying in the direction of the rising and setting directions of other stars. Indeed the journey is marked off in apparently equal units as the Etak Island is thought of as moving past the boat first lying in one direction and gradually shifting its bearing. In short, *place is kept* during the journey by the imagined relative movement of this hypothetical island in the *opposite* direction to the actual movement of the boat. By comparison with what we do in order to keep track of where we are as we move about our world, this may seem very strange, but it is a completely reasonable system and works quite well.

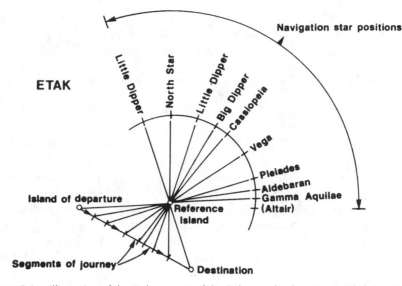

Figure 5.1.   Illustration of the Etak concept of the Pulawat Islanders. (From Gladwin, 1970.)

The focus of our discussion is on two aspects of spatial behavior, particularly as reflected in developing children. First we discuss the ways in which children guide their travel by reference to landmarks, just as the Pulawat use landmarks flexibly, guiding their course according to stellar positions, wave characteristics, and visible terrain as each is available and relevant to their course. Second, we consider the general problem of how people relate their locomotion to their knowledge of a general locale and particular destination, and we discuss the early development of this ability and the experiences facilitating its development. The Pulawat Islanders accomplish this by noting their movements over the water and relating these movements to knowledge of their destinations, conceptualizing these movements in terms of the Etak.

USE OF LANDMARKS

Encoding and remembering the position of an object or event in terms of its relation to a single landmark is a powerful technique. The Pulawat Islanders can tell that they are on the correct route by noting the presence of a reef marked by a particular change in water color, or that they are passing an island that is marked by a change in wave pattern. The effective guidance of travel by reference to landmarks presupposes a number of skills: (*a*) the ability simply to associate the marker and destination; (*b*) discriminative capacity to select as landmarks features that are distinctive and perceptually available within a particular locale versus those which are not; and (*c*) knowledge of spatial concepts, so that alternative spatial relations between landmark and destination can be remembered and used.

The simple association of a distinctive marker and the expected location of an event emerges quite early in life. For example, one of us (Rieser, 1979) taught 6-month-old infants to look, prompted by the sound of a bell, at one of several distinctively patterned windows placed symmetrically around their line of sight. This is illustrated in Figure 5.2, which depicts an infant learning to find the event at the window above line of sight. During training the sound of the bell was followed by the experimenter appearing in a particular window, smiling, and saying "peekaboo." After training, the babies were turned and the buzzer was again sounded. The babies looked reliably at the same distinctive window as they had in training. In another condition the windows were not distinctive, but the babies were trained in the same way to look at one particular window. In this case, after training when they were rotated around their line of sight, they looked at the window that was in the same location relative to their body as during training; in short, they made spatially egocentric responses. Thus, under some conditions, young infants can encode the position of an event in terms of a particular relation to their own body. It should be noted that Bremner (with infants, Chapter 4, this volume) and Hart and Berzok (with older children, Chapter 7, this volume) also contrast egocentric and allocentric strategies in spatial orientation in their chapters in this book.

Little is known about infants' capacities to discriminate features that are more versus less appropriate landmarks in different settings. However, Acredolo (1976) has studied similar discriminative capacities in older children, ranging from 3 to 7 years in age. She adduced that given a choice of distinctive

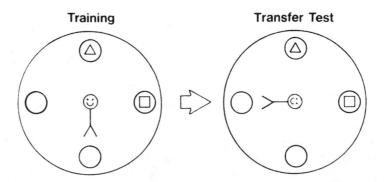

Figure 5.2.   Schematic of Reiser's (1979) experiment in which babies were trained in a supine position to look upward at a distinctive window. After training they were rotated 90 degrees for a transfer test.

features the older children tended to select as landmarks features that are more remote from the destination and more permanently placed in position, whereas the younger children tended to select as landmarks features that were closer to, the destination and moveable. Acredolo used an elaboration of the classical place versus response learning problem in the following way. Children were brought into a small, nondistinctive room that had a door at one end, a window at the other end, and a table along one wall, as illustrated in Figure 5.3. They were walked up to a corner of the table and blindfolded. While blindfolded, they were walked in a circuitous route back to the door or to the window end of the room. The blindfold was removed and the children were asked to return to the place at which they had been blindfolded. Unknown to the children, sometimes the table was silently moved across the room during their blindfolded walks. Children who remembered the starting place in terms of its relation to the room's walls would be expected to return to the right place, regardless of the table's location and whether they needed to turn right from the door or left from the window. This is what the older children tended to do. The youngest children, the 3-year-olds, tended to be spatially egocentric, turning in the direction they originally turned during the procedure. They tended, for example, to turn right no matter where they started from or where the table was. Children in the middle of the age range tended to return to the location of the table even if it had been moved. Thus, in this small room, one can see a progression in preference of features used as landmarks, from relatively moveable features that were adjacent to the destination, to permanently fixed features that were more distant from the destination. This progression may result from children's learning to select as landmarks permanently fixed features of the environment.

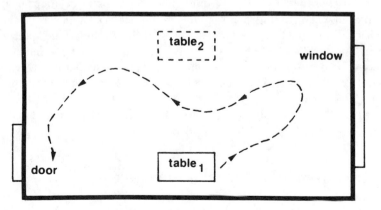

Figure 5.3. Schematic of procedure in Acredolo's (1976) experiment in which blindfolded children are walked to one end of a room and then were asked to return to the place where blindfold was put on.

Whether a feature can serve as an effective landmark depends on its context. For example, people sometimes encode their automobile's location in large parking lots in terms of its proximity to another car (e.g., beside the bright red car). Later searching for their car, such unfortunates realize that there are many similar "bright red cars" in the particular array. This ability, to discriminate what features are distinctive within particular contexts, is a key to the effective use of landmarks. Although it has not been widely investigated in children's use of landmarks in spatial tasks, it has been studied with children in referential communication tasks. For example, Glucksberg and Kim (cited in Glucksberg, Krauss, & Higgens, 1975) asked nursery school children to instruct a listener how to stack colored blocks on a peg. The array in front of the children contained blocks that had two colors, for example, one block was red and blue, another was red and green. If the children described the blocks with only one color name in situations like this, they were equally likely to identify the particular block with the redundant color (in this case, red) as with the informative color (in this case blue or green). The efficiency with which children select informative referential labels within communication tasks varies with the particulars of the task and context as well as with age from 3 to 14 years (Flavell, 1977; Glucksberg, Krauss, & Higgins, 1975). Similar trends probably occur in children's selection of landmarks for future reference in spatial orientation tasks.

The use of single features as landmarks is often mediated by one's knowledge of spatial relations. For example, remembering that a particular book is somewhere "above" another book instead of simply "nearby on the same bookcase" can make quite a difference in search time. Likewise, remembering that a particular store is "to the left" of a particular view of a landmark instead of "in the same part of the city" can save time. The use of spatial relations by infants and children when guiding their search by reference to a landmark has not been systematically studied. For example, the work by Bremner (Chapter 4, this volume) and by Rieser (1979) with infants involved the use as landmarks of distinctive patterns which largely coincided spatially with the desired target object. However, there has been a great deal of work concerning children's learning of spatial relations. Much of it has involved tasks where children are asked to identify similar shapes that vary only in their orientation. This work is reviewed in detail by Rudel (Chapter 6, this volume).

REFERENCE SYSTEMS AND SPATIAL INFERENCE

Knowing the location of an object or oneself relative to a single landmark is particularly valuable in contained spaces where the landmark can be detected easily. It is of limited value in larger, less contained spaces because of two general problems. The first problem involves knowledge of one's own location.

For example, knowing that one is a certain distance from a specific landmark is directionally ambiguous. The situation is similar to that of people without a compass at sea mired in deep fog, or blind people at a particular known intersection who have lost track of their facing direction. In cases like these, people "know" their location, but not knowing their headings or facing directions are unable to get anywhere else. The second problem involves the locations of different objects or events, each encoded in terms of its spatial relation to *different* landmarks. In situations in which this occurs, people have no basis for inferring the spatial relations among the various objects or events, and thus, cannot know how to travel from one to the others. A reference system, or frame of reference, is needed through which the location of all objects or events of interest is specified. We focus here on one type of reference system, a set of common landmarks with respect to which all locations are known.

The use of a reference system to relate the positions of two or more objects generally requires that the position of each object be known with respect to the given reference system and that an inference be made relating these two positions. Thus, for example, if we know in a Cartesian coordinate system that something is located at $x_1 y_1$ and something else is located at $x_2 y_2$, we still have to calculate the distance between these two positions and the direction to one from the other if we want to know them. We know of no experimental studies that explicitly teach children to use a reference *system* as opposed to individual landmarks. However, there are a number of studies in which it seems plausible to assume that children have learned locations of objects with respect to some reference system and that indicate that children can make this kind of inference about relations between locations.

In an illustrative experiment of this kind, 4- to 5-year-old children and adults were taught to go from home base to each of 10 locations, as illustrated in Figure 5.4 (Kooslyn, Pick, & Fariello, 1974). The subjects were trained so that they could go rapidly and directly from a home position to each location but were given no direct experience moving from one location to another. After training, the subjects were asked to consider in turn each location and order in closeness to that location all the other locations. (This is in some ways an analogue of the classical Maier 3-Table maze used with rats.) These data were subjected to multidimensional scaling analyses, and the best fitting "map" of the data was constructed. As seen in Figure 5.5, the maps of both the children and the adults conform fairly closely to the true layout. It is apparent that even young children can make inferences about spatial relations with which they have had no direct experience. It is interesting to note that, although the adults made more accurate inferences than the children, the children's references were remarkably good. Inferences made by children and adults conformed more closely to the Euclidian straight line distances than to the distances they had traveled.

Perhaps an even more interesting example of children's developing ability to

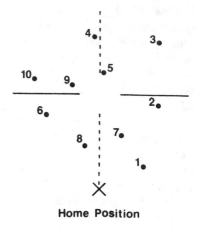

Figure 5.4. Layout of the space used in study by Kosslyn, Pick, and Fariello (1974) in which children were trained to go from a "home base" to each of 10 locations. Solid lines represent opaque barriers, dashed lines represent transparent barriers.

make spatial inferences is provided by a study of three-dimensional spaces recently completed in our laboratory (Lockman & Pick, in preparation). The basic question of the study concerned how well children and adults knew the layout of their own apartments. They were asked to point, or more specifically aim, a sighting tube directly at objects in their duplex apartments which were out of sight from their immediate location by virtue of being behind walls,

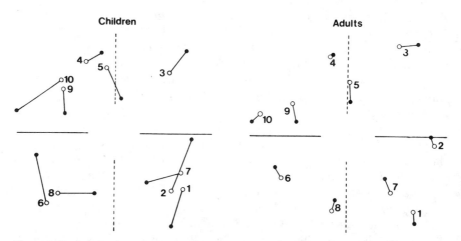

Figure 5.5. Relation between layouts generated from judgments of distances between all pairs of locations and the actual layout of space in study of Kosslyn, Pick, and Fariello (1974). Filled dots represent actual location; unfilled dots represent locations based on judgments.

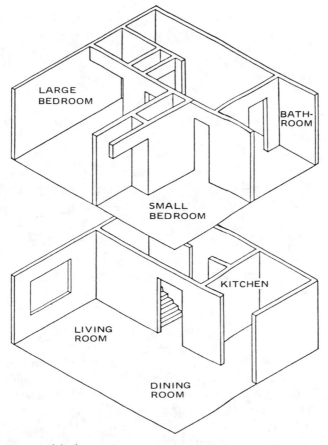

Figure 5.6. Layout of duplex apartments.

beneath floors, or above ceilings. The sighting tube apparatus could be read to measure both azimuth and elevation information. The layouts of the apartments in a housing complex were all identical and the common layout is portrayed in Figure 5.6. An example of the azimuth and elevation measure for a typical target location is illustrated in Figure 5.7. Children who were 4 to 6 and 8 to 9 years of age and their parents participated in the study. They were taken from station point to station point around their apartment and were asked to aim the sighting tube at various target locations. Sometimes the target locations were on the same floor; sometimes they were on the floor above or below. Since all the target locations were out of sight from the station points, the aiming had to be done on the basis of some kind of inference about the direction of the object. That is to say, to point accurately at a target, people had to go beyond

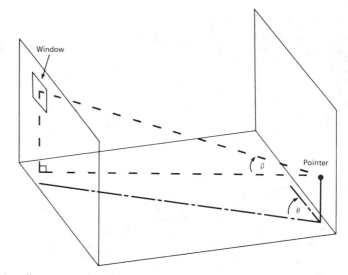

Figure 5.7. Illustrations of angles of azimuth (θ) and altitude (β) measured when subjects pointed at various out of sight locations.

their direct spatial experience. They could not have directly experienced the direction of the target from the sighting station.[2] We want to focus on results involving a comparison of the accuracy in aiming at targets on the same floor with that of aiming at targets on different floors. We will examine just the azimuth measure (θ), as the elevations (β) for targets on the same floor were all essentially equal and zero. The azimuth errors in aiming at targets on the same floor and on different floors are presented in Figure 5.8. It is apparent that even the younger children do quite well in aiming at targets on the same floor, but they appear to be at a considerable disadvantage in making judgments about the directions of objects on different floors. Adults, by contrast, do just as well in judging direction of objects on different floors as on the same floor. The older children's performance is between that of the younger children and that of the adults. The reason for the greater difficulty in making inferences between floors than within floors for the children may be due to the fact that their attention when going up and down stairs is on the vertical movement and they

---

[2]Nothing more is implied by the assertion that subjects have made an inference than that they have gone beyond the spatial information that they have directly experienced. From the data of this study it is not clear when such inferences are made, that is, at the time of testing or earlier as spatial knowledge is built up. It is also possible that what are called spatial inferences here are in fact provided by some higher order variable of the spatial stimulation such as Gibson (1966) has suggested as an alternative to other constructivistic perceptual hypotheses. If such is true, then the term *inference* would, of course, be a misnomer.

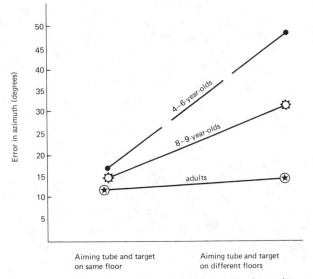

Figure 5.8. Azimuth (θ) errors in pointing at out of sight locations when subjects were on same floor as target and when they were on different floors from the target location.

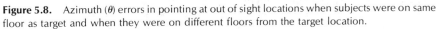

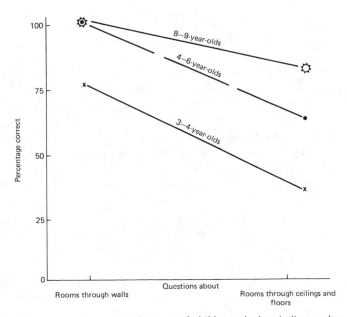

Figure 5.9. Percentage of correct judgments of children asked to indicate what rooms lay behind walls (i.e., on same floors), above ceiling, or below floors (i.e., in different floors).

117

do not notice that there is a horizontal displacement as well. This is quite consistent with Piaget's observation that young children have difficulty in coordinating changes in two dimensions at once, which underlies some of the difficulties they have with the standard conservation tasks. If our hypothesis about the failure to notice the horizontal displacement in moving vertically from one floor to another is correct, we might expect not just an average increase of errors in aiming but a high proportion of "reversals" in which targets actually in the front of the house are judged to be in the back, etc. To get an idea of the extent of such large errors, the proportion of errors over 90 degrees was calculated for each age group. The adults made none, the 8–9-year-olds, 5%, and the 4–6-year-olds, 21%. Finally, to assess this sort of ability in even younger children without the complexity of the aiming task, we asked some younger 3 4-year-old children, as well as the 4–6-year and the 8–9-year-old children to indicate simply what rooms lay above or below the rooms they were in and what rooms lay behind walls. The performance on these questions is depicted in Figure 5.9. Again, it is evident that all children, even the very youngest, do quite well with judgments on the same floor, but performance drops off progressively on between floor judgments with decreasing age.

COGNITIVE MAPPING AND MENTAL OPERATIONS

So far we have discussed orientation with respect to single landmarks and making inferences about spatial relations among locations defined in terms of a set landmarks or a reference system. Knowing the set of spatial relations among all the objects in a space, whether by direct perception or by inference, is an important constituent of the *process* of "cognitive mapping." However, there are additional processes. We would like to suggest that cognitive mapping implies being able to operate on spatial knowledge in a way that is analogous to viewing the space from different station points. The kinds of operations we have in mind are all equivalent to "updating" the spatial relations between all the objects in a space and an observer whose position relative to the objects changes while locomoting. Such updating can occur when a person is moving through a space in which she can actually see all the objects in question. In that case she can *perceptually* update the spatial relations between herself and the objects in the space. Cognitive mapping is implicated when the updating is done without perception, when, for example, all the spatial relations cannot be seen because the space is too big or complex, or it is dark, or one is blind, or the movement is imaginary; that is, when the updating is done mentally (one implication of such mental updating is being able to plan alternative routes or make detours when necessary).

What can we say about the development of this ability to update spatial relations between oneself and one's environment in children? We know that a simple form of such updating occurs in young children. In a recent study by Rieser (Heiman & Rieser, 1980; Rieser & Heiman, in preparation), 18-month-old toddlers were presented with a spatial problem in a small room. The room was round and had eight identical windows at about child's eye height. In training, the children started from the center of this room, always facing the same direction. With help from a parent, they quickly learned on a bell signal to walk from the centered starting position over to a particular target window (for example, the window directly to their left). This window was the only window that would automatically open when touched by the toddlers, whereupon they were entertained by watching a toy display. This procedure is schematized in Figure 5.10. In the test situation the toddlers were always centered in the round room, but the parent would then turn the child into a required test position. Then the same bell signal would occur, prompting the child to find the target window. When the toddlers had been turned just a bit past the target, they reliably backtracked, reversing that direction of turn to follow the shortest route to the target. When the children had been turned well past the target, they turned in the same direction to find the target, still selecting the shortest route but they were imprecise in their search for the target, usually choosing one of seven, nontarget windows. However, they selected the shortest-route direction of turn significantly more often (more than 85% of their trials) than a longer-route direction. Heiman and Rieser (1980) obtained similar results with a highly selected group of 14-month-old toddlers. This experiment illustrates an ability in very young toddlers to keep track of their locomotion with respect to their knowledge of a single target location. Of course, for mature spatial orientation and for evidence of sophisticated cognitive mapping ability, a person must be able to update his location with respect to all the object locations in a space. Does this ability develop with age?

In a study by Hardwick, McIntyre, and Pick (1976) the ability of subjects of different ages to update the spatial relations between themselves and several objects in a large familiar room was investigated. As in the study of people's ability to aim at different locations in their apartment, earlier described, the subjects in this study were asked to aim a sighting tube at three different target objects in a large room. In this case they could not see the locations of the targets because they were aiming from behind an occluding screen. After aiming from one station point, the subjects were taken to a second station point, where they again aimed the sighting tube at the same targets and, finally, they repeated the process from a third station point. Thus, for each target, each subject provided three direction lines—one from each station point. These direction lines could be plotted on an outline of the room to yield a "cognitive map" derived from the subject's knowledge of the layout of the room. An example of such a representation of a cognitive map is portrayed in Figure 5.11.

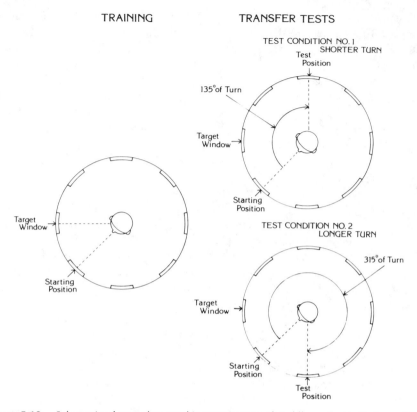

TRAINING   TRANSFER TESTS

Figure 5.10. Schematic of procedure used in experiment with toddlers asked to find a particular unmarked window after being turned a given distance. Training is depicted on the left side of the figure. All subjects were trained to find the display at the target window, located 45 degrees of turn from starting position. The two transfer test conditions appear on the right. Subjects were always turned toward and past the target window. Sometimes, subjects were turned 135 degrees into test position; on these trials, the shorter route to the target was in the reverse direction of movement into test position. At other times, subjects were turned 315 degrees into position; on these trials, the shorter route to the target was in the same direction as the movement into test position.

The subjects could not see the target locations in the room when they were pointing and to perform the task they had to update their spatial relations as they moved from station point to station point. It is important to note that subjects more or less demonstrated the ability to perform this updating operation on spatial information. As experimenters, we constructed the cognitive map. The subjects in this study were first graders (about 6 years old), fifth graders (about 11 years old), and college students. There were slight quantitative differences as a function of age in accuracy of aiming, but again the

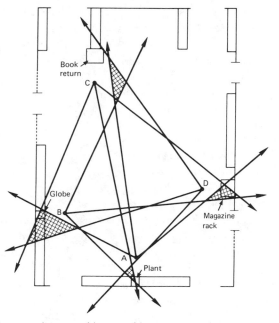

Figure 5.11. Schematic diagram of layout of large room and representation of a "cognitive map." Subjects were asked to point at various targets from three station points. Black dots beside letters A, B, C, and D indicate positions of the station points. Lines with arrows indicate pointing directions. Small labeled rectangles indicate target objects: plant, globe, etc. The shaded triangles are the intersections of the direction lines for a given target object and can be taken to define the subjective position of the target.

remarkable result was that even the young children performed well. First graders made an average error of 9.6 degrees, fifth graders, 5.8 degrees, and college students, 6.5 degrees. It is tempting to say that their cognitive mapping ability in such a situation is high. However, in moving from station point to station point, the subjects were able to see their changing relations, so it is possible that subjects were perceptually updating their spatial relations rather than operating on their spatial knowledge.

As a better test of subjects' ability to operate on this spatial knowledge, a subsequent condition was added. Subjects remained behind an occluding screen and were asked to imagine that they were moving from station point to station point and to aim at the various targets as if at the different imagined station points. Under this condition, there were quite marked quantitative and apparently qualitative age differences. Some of the first graders responded egocentrically; that is, they aimed at all the targets as if from their actual station point rather than from the imagined station points. The others responded nonegocentrically but chaotically as if they knew that responding egocentrically was wrong but did not know how to decide what was correct. some of the fifth graders were using this latter nonegocentric strategy while others were able to figure out in a *general* way what was correct but were not precise. They could get in the right ball park, so to speak, but were unable to fine tune their judgments. They were egocentrically biased by prominent target objects. Similarly the college students made these generally correct judgments or else their aimings were fine tuned and very accurate. Thus, we see that young children, particularly, are at a disadvantage when the task involved at least this type of mental operation on spatial knowledge—a kind of mental updating. According to our criterion, they do not yet exhibit processing ability characteristics of cognitive mapping. The work Olton describes (Chapter 14, this volume) suggests an ability of rats to operate on spatial knowledge in a way analogous to the taking of different perspectives.

What factors facilitate the kinds of updating operations that characterize cognitive mapping? First, the *absence* of conflicting spatial information eases the process of mentally taking different viewpoints. In the imagined-station-point study just described, another condition was added in which the occluding screen was removed when the subjects were required to imagine that they were at the different station points. Under this condition, they had to imagine themselves in one place when the immediate perceptual information was that they were in another place. This was more difficult for subjects of all ages. Second, the physical movement from a familiar station point to a novel station point facilitates judgments of target directions from that novel station point— even if no direct perceptual information about the spatial relations is available during the movement. This has been demonstrated in recent work by Rieser (in preparation). Rieser tested adults in familiar and unfamiliar rooms. At the start, subjects stood in one place and pointed repeatedly at objects they could see around them in the room. Sometimes subjects were then asked to pretend they were standing across the room facing a different direction and to point quickly and accurately at the target objects as if they occupied this novel station point. To be sure that the contradictory sights and sounds would not interfere with their performance, subjects were always tested while equipped with blindfold and with earphones, so that they could not see or hear their actual positions. The subjects reported that this task was difficult, and their pointing responses were quite slow and relatively inaccurate. This result is not surprising, since the task required subjects to mentally transform their spatial relations to the target objects that they had seen in order to take into account their imaginary position change. More surprising are the results obtained in different conditions, conditions that *logically* require the same mental operations. In these conditions, subjects stood at one station position and repeatedly pointed at the target objects they could see. Then they were equipped with blindfolds and earphones, so that they could not see or hear where they stood, and were walked across the room to stand in a novel station point. Under these latter conditions, the subjects reported that the task was quite easy; their speed and error scores were significantly better than the scores obtained in the previous condition, and equivalent to their baseline scores collected while they stood in the original station point. The task in both conditions required people to operate mentally on their knowledge of the layout of target objects, transforming the spatial relations they had seen. Logically identical products of these mental operations were needed. However, psychologically, the mental operations were sharply different. Subjects reported the operations were effortless and automatic when they actually moved, which is substantiated by their good, baseline-level speed and error scores; conversely, subjects reported the operations were difficult and required strategic thinking when they merely imagined moving, which is substantiated by their relatively slow and inaccurate responses. Some properties of the physical movement seem to potentiate people's abilities to accomplish these mental operations to update their positions relative to their knowledge of the spatial layout in which they operate. The exact, sensory properties of the physical movement which accomplish this are still under investigation (see Potegal, Chapter 15, this volume).

Spatial updating, the ability to relate one's own movements to knowledge of the spatial layout in which one operates, has an interesting developmental history. As discussed at the beginning of this chapter, infants do not show evidence of this ability during the first year of life. In the absence of landmarks, infants younger than 12 months search for hidden objects in a spatially egocentric manner, failing to differentiate their own movement from the stability of the general environment. However, the spatial behavior of toddlers during the second year of life is qualitatively different. Under similar conditions, these toddlers show their use of a viable self-reference system, relating their own movements to their knowledge of single target locations. However, this ability appears to be relatively imprecise in toddlers, and much more precise in school-age children. Thus, the later development of this ability may involve growth in its precision, as well as growth in the ability to relate one's movement to knowledge of increasingly larger, more complex spatial layouts.

What might influence the early development of these mental operations which seem to emerge during the second year of life? Neural maturation is one possibility, whether its site is the relatively later-maturing parietal lobe as suggested by some work of Beritoff (1965) and Semmes, Weinstein, Ghent, and Teuber (1963) or the earlier-maturing hippocampus as suggested indirectly by other work (O'Keefe & Nadel, 1978; Olton, 1977). Also relevant is the work Potegal (Chapter 15, this volume) discusses on vestibular–striatal involvement in updating. Another possibility involves the *timing* of the qualitative change from infant spatial egocentrism to toddler self-reference. This supports the suggestion that particular classes of experience play a role in the early development of these mental operations.

The specific classes of experience would be those provided by independent locomotion. It is certainly true that infants during the first year, although they may crawl, and toddlers during the second year, who ordinarily walk, differ in type and amount of self-directed locomotion. It is plausible that this difference provides the older group with more frequent opportunities to discover that their own movements alter their spatial relations to things that are in fact spatially fixed (see Bremner, Chapter 14, this volume).

We propose that the change to bipedal locomotion provides an additional class of experience that may play a role in the development of updating. We refer to a class of visual experience, seeing the simultaneous transformation of the network of spatial relations among self and visible objects during locomotion. Although quite indirect with regard to questions concerning the normal early development of this ability, there is evidence to suggest that early vision plays a role in the developmental history of updating. Previous research suggests that, on some spatial, locomotory tasks the performance of early-blinded adults is deficient compared to performance by late-blinded and sighted adults (Warren, 1979). Recent evidence collected by one of us (Rieser & Guth, in preparation) suggests that early-blinded adults do not automatically relate their physical movements to knowledge of a spatial layout, although late-blinded and sighted adults do so. The relevant evidence has been collected in a different version of the study described earlier in this chapter—adults pointed at familiar objects from novel station points after actually moving to the new place versus after only imagining that movement. These procedures were modified, so that all learning and testing was conducted without vision and without auditory locational cues. Under these completely nonvisual conditions all six sighted and all six late-blinded subjects (who were blinded after the age of 8 years) responded as in the previous experiment. Namely, these subjects reported difficulty and the use of explicit strategies when asked to imagine moving to the new station point, and described their responses as easy and automatic after they physically walked to the novel station point. These introspective reports were substantiated by the response speeds and errors. Qualitatively different results were obtained from the six early-blinded subjects (all subjects were blinded before the age of 4 years). Similar to the sighted and late-blinded, these subjects all reported using explicit strategies when asked to imagine moving to a new station point; their speed and error scores were slightly better than those produced by the late-blinded, but insignificantly better. Different from the sighted and late-blineded, these subjects all reported using similar, explicit strategies after they physically walked to a new station point. The introspective reports were substantiated by statistical analysis of their speed and error scores across these conditions. This pattern of results suggests that early visual experience does play a role in the developmental history of updating, the ability to relate one's physical movement per se, without auditory reference information, to knowledge of the relevant spatial layout. The exact nature of the role played by experience and the precise impact of its timing awaits further investigation. See Hermelin and O'Connor (Chapter 2, this volume) for further discussion of the role of vision in spatial orientation.

CONCLUSIONS

We have made a long journey—at least we have covered a lot of ground, from actual navigation through the Pacific to imagined movement around a campus room. We have tried to indicate that many of the grander problems of navigation are, in principle, not much different from the more mundane kinds of orienting that even young children engage in while getting around their world. In particular, we have tried to show that very young children can use landmarks for orientation at least under certain conditions as in the study with 6-month-old infants, and by 14 or 18 months they can update their own position relative to a specific target location when they move, as in the study with toddlers. In that case, the children also seemed to understand the identity transformation of turning 360 degrees in real space. We also saw that 4- and 5-year-olds can make spatial inferences about distances with which they have had no direct experience. Where there appears to be most room for development in children's cognitive mapping is, on the one hand, in the integration of spatial information in complex, three-dimensional spaces, as demonstrated in the study of the layout of the children's apartments. On the other hand, there seem to be radical changes in children's abilities to operate mentally on their spatial knowledge.

There are a number of potentially practical implications of these abilities and their development in children. First, notwithstanding the developmental changes which will occur, the surprising sophistication of young, normal children suggests that they are solving practical spatial problems that are more complex than any problems they are yet able to handle in other domains. Might there be a way of translating nonspatial problems into a spatial form that could capitalize on this cognitive power? Second, the identification in these spatial tasks of some component processes such as orientation to landmarks, orientation in terms of frames of reference, spatial inferences, operations on spatial knowledge suggests the possibility of analytically diagnosing people with spatial deficits and developing training techniques that focus on their specific deficiencies. Such techniques could be used especially for certain retarded children said to have particular difficulties with practical spatial problems such as finding their way. (Others apparently show spatial abilities that are far beyond their achievements in other domains.) These techniques might be helpful with girls with Turner's syndrome, a genetic anomaly, which is reported to be associated with selective deficits in spatial abilities, their other intellectual abilities falling within the normal range (Alexander, Walker, & Money, 1964; McGee, Chapter 9, this volume). Blind persons are confronted with extremely difficult problems of spatial orientation (Foulke, Chapter 3, this volume; Shingledecker & Foulke, 1978) and, in fact, are given formal mobility training that accentuates attention to landmark cues mediated by nonvisual sense modalities and attempts to facilitate the learner's spatial inference and updating behavior. A less severe form of sensory deficit (from the point of view of orientation) is loss of the vestibular system. Under conditions of reduced visibility, persons with this deficit may need special help in orienting. The research described above is basic—not directly concerned with these important problems—but it points the way to an analysis of the practical problems of spatial orientation that should be quite salutary.

We would like to conclude on three theoretical notes. First, earlier we regarded the use of egocentric reference information as more primitive in some sense than the use of landmarks. However, when surveying the various kinds of studies that we have described it seems obvious that under some conditions the use of a self-reference system can be accomplished on a very sophisticated level as in the studies of updating information when one moves about. So rather than consider egocentric responding, in general, as being primitive we would like to suggest there are immature and mature levels of self-reference information. Second, earlier in our thinking we treated a single landmark as a sort of minimal reference system. Now, on reflection, the concept of frame reference or reference system seems, as the term suggests, to imply a systematicity or integration that is not captured by a single landmark. As noted above, locating objects with respect to a common frame of reference permits the possibility of relating the locations to each other although this may require an additional cognitive operation. Again, use of a single landmark may reflect immature functions whereas use of a reference system may be a more mature level of functioning. Some of these distinctions are summarized in Table 5.1. Finally,

TABLE 5.1
Levels of Reference Information

Level	Self	Geographic
Immature	Egocentric	Landmarks
Mature	Self-reference (updating)	Frames of reference

we spoke earlier of criteria for establishing the existence of cognitive maps. We tended to use the term *cognitive map* to refer to the *set* of spatial relations that people knew, and we were particularly interested in the spatial relations people knew by virtue of making spatial inferences. As our work progressed, we have become more and more uncomfortable with this reification of such mental representations. (Hart and Berzok in Chapter 7 also share some of our reservations about the "map" metaphor.) Rather, we feel that the focus should be on the *process* of maintaining spatial orientation. It does seem that some level of sophistication in that process may warrant the name of cognitive *mapping*. This level of sophistication includes not only *making* spatial inferences about relations between locations but *updating* the relation between one's own position and all the other locations in a space. Although this shift may appear small, we feel it is important since it changes the emphasis from the thing to the process.

REFERENCES

Acredolo, L. P. Frames of reference used by children for orientation in unfamiliar spaces. In G. Moore & R. Gooledge (Eds.), *Environmental knowing*. Stroudsberg, Pennsylvania: Dowden, Hutchinson, and Ross, 1976.

Alexander, D., Walker, H. T., Jr., & Money, J. Studies in the direction sense—1: Turner's syndrome. *Archives of General Psychiatry, 1964, 10, 337–339.*

Beritoff, I. S. Neural mechanisms of higher vertibrate behavior. Boston: Little, Brown & Co., 1965.

Flavell, J. H. *Cognitive development.* Englewood Cliffs, New Jersey: Prentice–Hall, 1977.

Gibson, J. J. *The senses considered as perceptual systems.* New York: Houghton–Mifflin, 1966.

Gladwin, T. *East is a big bird.* Cambridge: Harvard University Press, 1970.

Glucksberg, S., Krauss, R., & Higgins, E. The development of referential communication skills. In F. D. Horowitz (Ed.), *Review of child development research* (Vol. 4). Chicago: University of Chicago Press, 1975.

Hardwick, D. A., McIntyre, C. W., & Pick, H. L., Jr. Content and manipulation of cognitive maps in children and adults, *SRCD Monographs, 1976, 41*(3), Serial No. 166.

Heiman, L., & Rieser, J. Spatial orientation at eighteen months of age: Search mediated by self-movement. Paper presented at the International Conference on Infant Studies, New Haven, Connecticut, April 1980.

Kosslyn, S. M., Pick, H. L., Jr., & Fariello, G. R. Cognitive maps in children and men. *Child Development, 1974, 45, 707–716.*

Lockman, J., & Pick, H. L., Jr. Developmental comparison of knowledge of a three-dimensional layout, in preparation.

O'Keefe, J. and Nadel, L. *The hippocampus as a cognitive map.* Oxford: Clarendon Press, 1978.

Olton, D. S. Spatial memory. *Scientific American*, June, 1977, 236, pp. 82–98.

Pick, H. L., Yonas, A., & Rieser, J. J. Spatial reference systems in perceptual development. In M. Bornstein & W. Kessen (Eds.), *Psychological development from infancy.* Hillsdale, New Jersey: Erlbaum, 1979.

Rieser, J. Reference systems and the spatial orientation of six month old infants. *Child Development, 1979, 50, 1078–1087.*

Rieser, J. J. Locomotion facilitates the mental transformation of spatial knowledge. Article in preparation, 1982.

Rieser, J. J., & Guth, D. The role of visual experience in mental transformations of spatial knowledge. Paper presented at the Society for Research in Child Development, Boston, April, 1981.

Rieser, J. J., & Heiman, M. L. Spatial self-reference systems and the shortest-route behavior in toddlers, in press.

Semmes, J., Weinstein, S., Ghent, L., & Teuber, H. Correlates of impaired orientation in personal and extrapersonal space. *Brain*, 1963, *86*, 747-772.

Shingledecker, C. A., & Foulke, E. A human factors approach to the assessment of the mobility of blind pedestrians. *Human Factors*, 1978, *20*, 273-286.

Warren, D. H. Perception by the blind. In E. C. Carterette & M. P. Friedman (Eds.) *Handbook of perception* (Vol. 10). New York: Academic Press, 1979.

RITA G. RUDEL

The Oblique Mystique:
A Slant on the Development of
Spatial Coordinates

There is by now a substantial literature dedicated to the paradox that vertical and horizontal lines separated by 90 degrees are readily discriminated, whereas oblique lines separated by the same angle are not (for review see Appelle, 1972). If we consider that paradox in terms of the complexity of language required to describe the four different orientations, we may be able to clarify, if not explain it.

The differentiation of opposite obliques demands retention of a compound spatial label. We use *vertical* and *horizontal* to distinguish up–down from left–right axes, which are 90 degrees apart, but there is no distinguishing term for opposite obliques, also 90 degrees apart. The vertical–horizontal axis describes four directions, up, down, left, and right (Hebb, 1949), and although four directions are also described by opposite obliques, there is no distinct lexical designation for any of them (Olson & Hildyard, in preparation). They can be described only by combining the *vertical–horizontal* vocabulary. Thus, a vertical goes up–down and a horizontal left–right, but an oblique can only be described as "up to the right" (or "down to the left") in distinguishing it from one that is "up to the left" (or "down to the right"). Adding to the confusion, verbal descriptions can begin with the left–right rather than with the up–down axis (left–up to right–down), the serial ordering of language in this instance being completely arbitrary inasmuch as obliques occupy both coordinates of space simultaneously.

129

SPATIAL ABILITIES
Development and Physiological Foundations

That it is not a shared lexicon but the shared space that creates the difficulty with obliques can be seen from the comparability of data on animals and children, from acuity experiments with normal adults and infants, as well as from clinical findings with brain damaged patients. Electrophysiological and acuity data as well as studies of learning and behavior after restricted rearing suggest that the oblique may not be built into the nervous system but is somehow built up in the course of development. It seems to appear later phylogenetically as well as ontogenetically. It never achieves the stability of vertical or horizontal orientations and is more readily lost with brain damage. Teuber (1975) once said, "Clinical and experimental aproaches are not mutually exclusive . . . it remains true that the scotoma, in addition to being a sign of impairment of visual pathways, can give us access to the role of specific neural structures in visual function—blindness illuminating sight . . . [p. 457]." Clinical data presented in conjunction with research in this chapter may similarly shed light on the issues.

PERCEPTION OF OBLIQUES

The difficulty of oblique orientations transcends memory or language. Relative to the vertical or horizontal it is an unstable form, difficult to detect or to sustain perceptually. The first account of this was published in 1893: Jastrow reported that subjects required to reproduce visually presented lines or to set lines to particular orientations, did so with reduced accuracy when the lines were oblique than when they were vertical or horizontal. There have been several subsequent studies demonstrating a decrease in accuracy as a line target approached 45 degrees of tilt and an increase in accuracy as it moved toward a true vertical or horizontal (Burns, Mandel, Pritchard, & Webb, 1969; Liebowitz, Myers, & Grant, 1955). Recently, Olson has demonstrated: (a) that normal adults took longer to make same–different judgments about oblique lines than about vertical or horizontal lines (Olson & Hildyard, 1977); and (b) that 14-year-old children found it significantly easier to search for arrows bearing a horizontal or vertical orientation than for those bearing an oblique orientation. Olson concluded, "The hypothesis that obliques must be matched against a more complex representation than horizontal and vertical lines is very compatible with the data [Olson & Hildyard, in preparation]." In transformations of letter-like forms (Gibson, Gibson, Pick, & Osser, 1962), 45 degree rotations were nearly twice as difficult for children to recognize as up–down reversals (Schaller & Harris, 1974).

Craig and Lichtenstein (1953), testing the effect of extended periods of fixation on lines in varying orientation, reported maximal disappearance the closer they came to the 45 degree oblique positions. Consistent with this,

Figure 6.1. Target detection test: Subjects are instructed to cross out the geometric form shown at the top of the page wherever it appears. (From Rudel, Denckla, & Broman, 1978.) The particular data sheet reproduced here is that of a 53-year-old patient 1 year after removal of a life occipital-parietal meningioma. (From Holtzman et al., 1977.)

131

stabilized images of obliques disappear much more readily than lines at right angles (McFarland, 1968).

For a fine wire to be visible in an oblique orientation, it must be twice as wide as in a vertical or horizontal orientation (Ogilvie & Taylor, 1958). The distance between lines on an oblique grating have to be wider than on either horizontal or vertical grating to be perceived as a grid and not as a solid gray (Emsley, 1925). Diminished acuity for obliques is probably not the consequence of early experience in a "carpentered" environment consisting primarily of verticals and horizontals (as suggested by Annis & Frost, 1973·), since Leehey, Moskowitz–Cook, Brill, and Held (1975) found the identical acuity deficit in infants, age 6–50 weeks. There was an acuity *shift* with age, so that as the children grew older the distance between the lines (spatial frequencies per degree) could be decreased before the grids were responded to as if they were a uniform gray; but at every age, the lesser acuity for the oblique grid was significant.

The instability of the oblique appears to extend as well to the diamond shape, a figure made up of four obliques, ◇, (which, if rotated 45 degrees, becomes a square). Asked simply to cross out the figure wherever it appears on the page, many young children select the square instead, despite the presence of the diamond target at the top of the page. This occurs more frequently among children with congenital brain damage, particularly those with oculomotor disturbance, and in some brain injured adults (Rudel & Teuber, 1971). A 53-year-old male engineer, following removal of a left occipital-parietal meningioma did the task as shown in the reproduction of the data sheet (Figure 6.1). He crossed out 12 squares instead of 14 diamonds on this timed target detection test. His postoperative Wechsler Adult Intelligence Scale IQ at the time he took this test was 130 (Holtzman, Rudel & Goldensohn, 1977).

DISCRIMINATION OF OPPOSITE OBLIQUES

Sutherland (1957, 1958, 1960) reported that the octopus, an advanced invertebrate, could readily learn to choose a line oriented vertically rather than one oriented horizontally but not opposite obliques. This was also true for children under the age of 6 (Rudel & Teuber, 1963) and for almost half (40%) of those between 6 and 9. Further, although children could easily learn to discriminate an oblique from a horizontal or vertical, they chose the opposite oblique almost as often as the original in a test of transfer, again attesting to the equivalence in memory of the two orientations. Bryant (1969) confirmed these findings by comparing performance on discrimination of obliques with other "mirror image" stimuli and concluded that although obliques are recognized as clearly different from vertical or horizontal, the difference in direction of their slope is not encoded in memory and the difference is more difficult to recognize than it is for other mirrored images (e.g., opposite ⊔ figures) or the vertical and horizontal. These findings have by now been confirmed for species as divergent as goldfish (Mackintosh & Sutherland, 1963), octopus (Sutherland, 1957), and young human beings, in "carpentered" (Rudel & Teuber, 1963) and in "uncarpentered" environments (Olson, 1970; Serpell, 1977).

Immediate context, if not the broader environment, may, however, alter the difficulty of discriminating opposite obliques. It has been shown that how mirror image stimuli (obliques or left–right inverted U-figures) are displayed affects the difficulty of discrimination. Placing one of the pair above the other, instead of side by side, makes them easier for children to discriminate (Huttenlocher, 1967; Stein & Mandler, 1974). Such altered settings reduce the difficulty of opposite oblique relative to vertical–horizontal discriminations but do not reverse it. However, there has been a study demonstrating that opposite obliques, presented successively in isolation from any external context (through a viewing aperture) and each provided with a verbal label (*mine* versus *yours*) were discriminated even by preschool children (Fisher, 1979). Lacking any spatial context, the children may, like Lashley's rats (1938), do this by looking at only a small part of the visual field. A strategy of restricted attention may make learning a discrimination with successive presentations, that is, one stimulus at a time, simpler than simultaneous presentations of both obliques. Thus, the lower left segment of the area containing the stimulus will always have something on it when the oblique runs from lower left to upper right; in contrast, the lower right will have something on it when the stimulus runs the other way (Figure 6.2). A child responding in this way to a detail may always make the correct choice simply by consistently choosing whatever appears on

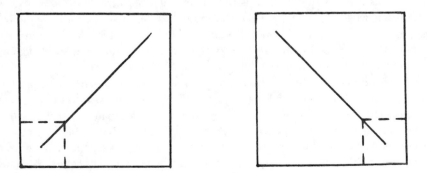

Figure 6.2. Oblique lines on a square presentation card. Possible effect of visually isolating either end of each line is indicated.

the lower left (or lower right) corner. He or she need not, and at the preschool level could not, use the compound directional representation (lower-left) if he or she restricts his or her attention to the lower edge of the area. In this case the child need only remember "this side" versus "that side," or "this side is mine," "that side is yours." Indeed, another study by the same author (Fisher, 1980) appears to confirm this. Whereas children as young as 5:6 to 6:5 years could discriminate successively presented opposite obliques, they could do this only when the obliques were shown in a "constant frame location," that is, when a segment of the positive oblique always appeared in the same corner of the stimulus card. Changing the stimulus–frame relationship significantly depressed learning scores. Whatever the context, most research has agreed with the conclusion of Bryant (1969) that the encoding in memory of the direction of visually presented oblique slope is particularly difficult for children.

Such encoding seems to be even more difficult by touch, for without vision the possibility of utilizing restricted attention or a strategy of location within the frame is reduced. Attention to one part of the stimulus is much more difficult, since the entire extent (or most) of it must be palpated in order to get the direction of slope. There is also less opportunity and it is much more complicated to utilize a "constant frame relationship" which Fisher (1980) found essential to making the oblique discrimination. The importance of using orthogonals as context, or frame, is apparent from the research of Millar (1976) with the blind. Rotating to oblique directions was more difficult for blind children than for blindfolded sighted children, a finding which Millar attributes to their failure to use orthogonals as referents or as a constant frame. Blind children with visual experience prior to the age at which opposite obliques are discriminated were not better at this task than the congenitally blind (Hermelin and O'Connor, Chapter 2, this volume).

In the Braille alphabet, the E is represented by two obliquely oriented dots, ⋱, whereas the I is the identical stimulus in the other direction, ⋰. These as well as other left–right mirror-image letters in that haptic alphabet were distinguished (using a "same–different" paradigm) less accurately by children (age 7–14) than were pairs of Braille letters that differed in the number of dots or the distance between them (Rudel, Denckla, & Hirsch, 1977).

The difficulty with discrimination of obliques is clearly, then, not restricted to visual display and, in fact, is more difficult without it. Differentiating one from another demands some representation of two coordinates of space simultaneously. In drawing, constructing, or walking an oblique line, one must simultaneously pass through the vertical and horizontal coordinates, which are distinct and clearly differentiated in the mammalian (and inframammalian) visual system. To differentiate opposite obliques demands holding one dual coordinate in mind while comparing it with another.

THE MOTOR COMPONENT:
DRAWING, CONSTRUCTING, AND WALKING OBLIQUE LINES

According to norms developed on the Stanford–Binet Intelligence Scale (1960), a 4-year-old is expected to copy a square, but it is not until age 7 that he or she is expected to copy a diamond. Similarly, on the Beery–Buktenica Developmental Test of Visual–Motor Integration (1967) the diagonal line is expected to be drawn later on the developmental scale than the vertical or horizontal, the triangle (a three-sided figure) later than the square. The vertical diamond, ◇, which has the same number of sides as the square appears on the test at age 8 and a diamond lying on its side, ◇ , at age 10.

Olson (1970) has documented cross-culturally the difficulty inherent in constructing the diagonal across a checkerboard [Figure 6.3(a)]. Even in imitation, children tend to avoid the hypotenuse, and take right angles instead

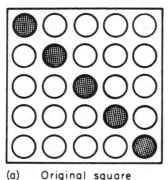

(a)   Original square

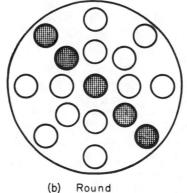

(b)   Round

Figure 6.3.   Checkerboards used by Olson (1970) for constructing obliques. (a) The rectangular, vertical–horizontal. (b) The circular boards. (From Olson, 1970.)

(Figure 6.4). The use of a round board without vertical–horizontal edges and with the diagonal clearly running through the center [Figure 6.3(b)] improves performance but not to the level of vertical or horizontal line construction. One cannot invoke different effects of carpentered versus uncarpentered environments to explain this, since the skyscraper architecture of children in New York could not have been anything like that of Olson's subjects or Serpell's (1971) in Africa where the Rudel and Teuber study (1963) was replicated.

The shortest distance between the distant endpoints of two lines at right angles is the hypotenuse. Yet, taking the shortest route constitutes a problem for some children. To traverse the maps shown in Figure 6.5(a) (Semmes *et al.*,

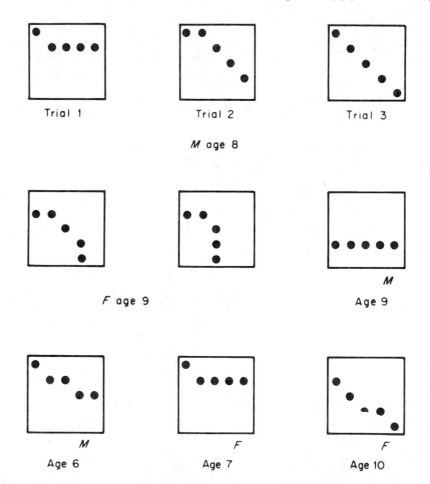

Figure 6.4. The patterns produced by children ages 6–10, including three attempts by one child to place five checkers in an oblique orientation. (From Olson, 1970.)

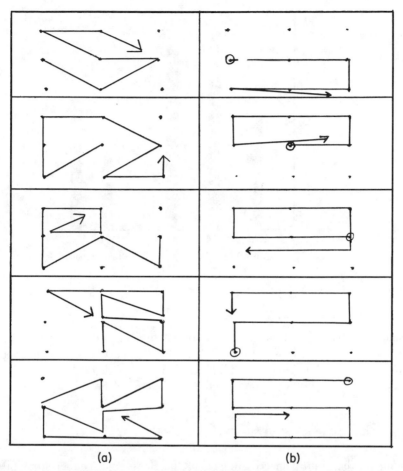

$$(a) \qquad\qquad (b)$$

Figure 6.5. (a) Five of the 10 maps used in studies of children's map-walking performance. (From Rudel & Teuber, 1971; based on Semmes, Weinstein, Ghent, & Teuber, 1955.) (b) Performance of a 15-year-old girl with right hemisphere damage and severe oculomotor apraxia.

1963), it is essential to make some oblique moves (i.e., to walk diagonally from one dot across to another). The route to be taken is marked on a map held by the subject; the nine unconnected dots are laid out in a 3 × 3 horizontal–vertical matrix on the floor. Of the 88 moves (on 10 maps), 34 are oblique. In a study of normal and brain damaged children (Rudel & Teuber, 1971), the latter were, as expected, relatively more impaired in terms of total number of correct moves, but much more deviant in terms of the total number of diagonal moves made, correctly or incorrectly. Brain damaged children with a mean CA of 11:3 years

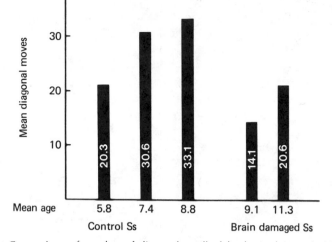

Figure 6.6. Comparison of number of diagonals walked by brain-damaged children and younger normal controls. (From Rudel & Teuber, 1971.)

walked only as many diagonals as did normal subjects with a mean CA of 5:8 (Figure 6.6). Similarly Keogh and Keogh (1967) reported that "educationally subnormal" boys had greater difficulty walking a triangle or diamond than any other shape (i.e., circle or square) on an unmarked floor.

THE ROLE OF EYE MOVEMENTS

Among the brain damaged subjects of the Rudel and Teuber study (1971) were some with oculomotor disturbances, defined in terms of impairment of one or more of the following: tracking, coordination of head and eyes, upward gaze, convergence, or nystagmus. Some had strabismus. The spatial orientation test was performed better by tbose brain damaged subjects who had apparently undisturbed oculomotor function, the 34 oblique moves contributing heavily to the difference within the brain damaged group (i.e., between those with and those without oculomotor disturbances).

Hebb (1949) proposed that the perception of pattern, of which the oblique seems to be one of the most complex, is built up out of voluntary eye movements around contours, permitting the organism to transpose (and therefore to recognize) the stimulus under different conditions. There has been considerable controversy about this proposal (for review see Bond, 1972). From recent studies, Hein, Vital-Durand, Salinger, and Diamond (1979) conclude that although surgical immobilization of a kitten's eye prevents the acquisition of visually coordinated behavior, the immobilized eye is capable of mediating whatever visually guided behaviors had been acquired prior to surgery. This is consistent with the observation that eye movements need not occur during the execution of visually guided behavior. Haptic discrimination of spatial contours (including Braille letters) by the blind or blindfolded, also involves movement but of the hands or fingers rather than of the oculomotor system which, from Millar's studies (1976), appears to do so much more effectively.

Two studies of children (Denckla, Rudel, & Broman, 1980; Rudel & Teuber, 1971) have made apparent that route-finding ability is more adversely affected by right than by left hemisphere damage, the side of damage inferred from neurological signs on the contralateral side. Those with imputed right hemisphere damage also traversed fewer diagonals than either a left hemisphere or bilaterally impaired group. The two factors, oculomotor and right hemisphere impairment, may not be independent. It has long been known that patients with right hemisphere damage suffer unilateral neglect (Brain, 1941) which may, in fact, be related to scanning difficulties (Weinberg, Diller, Gordon, Gerstmann, Lieberman, Lakin, Hodges, & Ezrachi, 1977). McFie (1969) has noted oculomotor impairment after right hemisphere damage, and Twitchell (personal communication, 1975) has confirmed that although the oculomotor system is represented bilaterally, most children with oculomotor abnormalities show neurological signs of right hemisphere damage.

The oculomotor system may help build perception of contour and space, as suggested by Hebb (1949), contributing to the spatial capacity of the right hemisphere. For 8 years we have followed a patient of M. B. Denckla. She is now 15-years-old and has a verbal (WISC-R) IQ of 85 and Performance IQ of 51 (and, incidentally, reads very well). She has neurological signs of right hemisphere damage and severe oculomotor apraxia (the inability to dissociate head from eye movements) with *pupen augen* (dolls' eyes) on passive head movement. Recently retested on walking 5 maps (of the 10), she failed to walk a single diagonal [Figure 6.5(b)], nor could she reproduce a block design with even this simple oblique pattern ▟ . Limited as she is by severe impairment of the oculomotor system, there would appear to be no obliques in her world. With cogenital or early damage to the oculomotor system, response to the unstable diagonal form may fail to develop. The capacity to process oblique lines may also deteriorate with some types of late damage affecting left or right occipital–parietal regions. Damage to the oculomotor system appears to be associated with inadequate development (or the deterioration) of some right hemisphere skills, one of which may be the capacity to process oblique orientations.

Educators and others involved in remediation have been alert for some time to the difficulty oblique movements pose for children with slow or otherwise anomalous development. They have ascribed this to a difficulty with "crossing the midline," presumably a form of perceptual neglect (Ayres, 1977). Since these children readily "cross the midline" when they draw, walk, or construct horizontals, this appears more a restatement of the problem than an explanation. The vulnerability of oblique orientation to early damage may at least in part be due to the complex mental representation it demands.

LINE ORIENTATION AND HEMISPHERIC ASYMMETRY

The clinical data of others have supported our findings of a relationship between judgments of line orientation and right hemisphere function (Warrington & Rabson, 1970; DeRenzi, Faglioni, & Scotti, 1971). In the normal, right-handed adult, however, the evidence appears to be somewhat more equivocal. Umiltà (1974) found that four lines, a vertical, horizontal, and two obliques, were discriminated more rapidly in the right visual field (by the left hemisphere). With an increased number of stimuli in intermediate oblique positions superiority shifted to the left visual field.

For the adult in a discrimination situation involving four orientations, verbal mediation was clearly a feasible strategy, but it ceased to be so with an increasing number of stimuli. Umiltà concluded that discrimination of line orientation is performed by the left hemisphere when the task is easily verbalized but by the right when the task is more difficult, not readily verbalizable. For children, the oblique discrimination is difficult because they cannot verbalize compound labels, and therefore, it is possibly more vulnerable to right hemisphere impairment.

The ongoing studies of Hubel and Wiesel (1962, 1968) and Levick (1961) on neuronal line detectors in the visual cortex of the kitten, monkey, and rabbit suggest that discrimination of line orientation in human beings may also depend upon these detectors during at least the early stages of the discrimination process. Ultimately, however, higher level processes take over and what Umiltà calls the *discriminative decision* (1974, p. 172) is lateralized (cf. Ratcliff, Chapter 13, this volume).

An interesting model of how children come to discriminate mirror images is provided in a study by Aaron and Malatesha (1974). Five asymmetrically reversible letters and five triangular forms in different left–right orientations were presented tachistoscopically at two different exposure times, 80 msec and 3–5 sec. Paradoxically, children younger than 8 were more accurate in their choices after the briefer exposure. These data, the authors believe, fit Noble's hypothesis (1968) that mirror-image confusions are due to "interhemispheric reversal" which provides antagonistic information to the two hemispheres, in contrast to the situation when the stimulus is presented to only one hemisphere at a time. One would have to hypothesize that after age 8 there is some suppression of one side on bilateral presentations and/or that verbal mediation makes the original stimulus more accessible in recall. Appealing as this model may be, there is still controversy about "interhemispheric reversal" and the study awaits replication.

THE DUAL REPRESENTATION OF THE OBLIQUE

A child required to consistently discriminate an oblique line from a vertical or horizontal line can do this as readily as he or she can distinguish vertical and horizontal lines from each other (Rudel & Teuber, 1963), and in this ability differs from the octopus (Sutherland, 1960), possibly because the child possesses language. He or she will tell you "I can do it because I always pick the slanty one" (or "the straight one"). Sometimes the oblique is characterized as the "crooked one." The difficulty arises only when the child must choose one of two opposite diagonals.

The oblique problem does not represent a unique case of dual mental representation. There are many instances in which a single stimulus partakes simultaneously of two relationships, and Piaget and Inhelder (1958) have described the difficulty children have with these as egocentric or preoperational levels of functioning. Thus, the child is not only himself but simultaneously a brother. Asked how many brothers he has, he includes himself, unable to separate his being a brother from his position as a possessor of brothers. Conservation problems may also be seen in this context, for when an object changes shape it is seen by the child as also changing size. Preoperational thought is not a fixed developmental stage but depends upon the child's grasp of the relationships involved and the amount of information he or she is given. Pufall (1975) found that egocentric responding on spatial tasks depended upon the symmetry of relations between space and self. Children as old as 10 responded "egocentrically" when topographic information was reduced (Pufall & Shaw, 1973), but after they were sensitized to distinctive features, asymmetry predominated and responses became less egocentric or preoperational. Ives (in preparation) in a study of spatial perspective (the three-mountain problem), found that when the correct orientation demanded a two-part linguistic label (i.e., "front" and "side"), kindergarten children tended to leave out one component of the description, this type of error accounting for 96% of the nonegocentric errors at that level.

The difficulty with dual representation is also apparent in experiments on the intermediate-size problem (Rudel, 1960). The child of less than 3 years can readily learn to choose the larger or smaller of two stimuli, but it is not until he or she is 6 that he or she can consistently choose the same stimulus presented (in randomly varying spatial order) with a smaller and a larger stimulus. The middle-size stimulus, which is simultaneously "larger than" and "smaller than" the other two, partakes of two relationships, thus demanding dual representation. Before they grasp the idea of "middle-size" children sometimes choose the stimulus in the positional middle, the compound designation *middle-size* being more difficult. That concept, however, is aided in the course of development by innumerable stories involving triads—sisters, brothers, pigs, or bears—each set of which must have one that is middle-size, not to mention the middle-size bowls of porridge, chairs, and beds. The oblique path is rarely chosen by the protagonists in mammalian (or, apparently, inframammalian) threesomes, all of whom, if I have these stories right, prefer the "straight and narrow." The diagonal, as we have seen is a difficult path to take, walk, draw, or construct, possibly because in the pre-Euclidean, nonabstract natural world it has little survival value except when it may have been the shortest path. In this respect, however, there may be differences within the human species. Held (1977) notes, citing the research of others, that Chinese adults and American Cree Indians show less acuity reduction for oblique edges than do Caucasians. The American Cree live on flat terrain but build homes with oblique contours. Held concludes that the genetic component may be fairly significant, since both Chinese and American Indians have a common Mongoloid origin.

The artist in our own culture uses horizontals and verticals to divide up space and to depict objects, but uses the diagonal (along with other techniques) to suggest distance. Appreciation of this sophisticated abstraction comes late in development and is not essential to survival in the natural, three-dimensional world where distance is "given" by convergence and bilateral retinal disparity.

CONCLUSIONS

In a world dominated by letters of the alphabet, plane geometry, maps, and other two-dimensional representations, the oblique orientation and discrimination of opposite obliques take on survival value very different from what is involved in the Natural environment. Psychologists and educators have armed themselves with tests that exploit the form. Out of 36 items on the Raven's Coloured Progressive Matrices, 9 (or 25%) require a choice between opposite oblique orientations. The Bender–Gestalt has the diamond form, diagonals, or angles juxtaposed at 45 degrees on 7 out of 9 items. The Beery–Buktenica Test of Visual–Motor Integration includes every possible variation of the diagonal form, and, of course there are the diagonal blocks and designs on the WISC-R and WAIS. The use of obliques in testing is probably out of proportion to the practical daily need to discriminate, construct, draw, or walk them. Their mystique lies not in any distinctive property they possess but in the insight they provide into the development and functioning of the nervous system. The capacity to process obliques is a measure of the extent to which that form has been developed, or, in the case of an adult with brain damage, to which it has been retained.

Additionally, the ability to distinguish opposite obliques to some extent characterizes the related capacity to conceptualize and maintain dual representations simultaneously, not only of dual *spatial* relationships. Much recent research on cerebral hemispheric asymmetry suggests that parallel processing is dependent upon the integrity of the right hemisphere. From the results of research, as well as from clinical observations, that integrity, in turn, is at least, in part, dependent upon the spatial information provided by an intact voluntary oculomotor system. Tbe reciprocal relationship of right hemisphere functions and eye movements has been implicated also in the scanning difficulties of adult patients with right hemisphere impairment.

The fragile oblique, with its poor discriminability, so difficult to sustain in preception or memory, nonetheless sheds light on these mechanisms. It remains a symbol, a tool for gauging levels of neuronal functioning and the complexity of mental operations of which the organism under scrutiny is capable.

REFERENCES

Aaron, P. G., & Malatesha, R. N. Discrimination of mirror image stimuli in children. *Neuropsychologia*, 1974, *12*, 549–551.

Annis, R. C., & Frost, B. Human visual ecology and orientation anisotropies in acuity. *Science*, 1973, *182*, 729.

Appelle, S. Perception and discrimination as a function of stimulus orientation: The "Oblique Effect" in man and animals. *Psychological Bulletin*, 1972, 78 (4), 266–278.

Ayres, A. J. The development of perceptual–motor abilities: A theoretical basis for treatment of dysfunction. In H. A. Solan (Ed.), *The psychology of learning and reading difficulties*. New York: Simon and Schuster, 1977. Pp. 385–392.

Beery, K. E., & Buktenica, N. A. *Developmental test of visual–motor integration*. Chicago: Follett, 1967.

Bond, E. K. Perception of form by the human infant. *Psychological Bulletin*, 1972, 77, 225–245.

Brain, R. Visual disorientation with special reference to the lesions of the right cerebral hemisphere. *Brain*, 1941, *64*, 244–272.

Bryant, P. E. Perception and memory of the orientation of visually presented lines by children. *Nature*, 1969, *224*, 1331–1332.

Burns, B. D., Mandel, G., Pritchard, R., & Webb. C. The perception of briefly exposed point-sources of light. *Quarterly Journal of Experimental Psychology*, 1969, *21*, 299–311.

Craig, E. A., & Lichtenstein, M. Visibility–invisibility. Cycles as a function of stimulus orientation. *American Journal of Psychology*, 1953, *66*, 554–563.

Denckla, M. B., Rudel, R. G., & Broman, M. The development of spatial orientation skill in normal, learning-disabled, and neurologically-impaired children. In D. Caplan, (Ed.), *Biological studies of mental processes*. Cambridge: MIT Press, 1980. Pp. 44–59.

DeRenzi, E., Faglioni, P. and Scotti, G. Judgements of spatial orientation in patients with focal brain damage. *Journal of Neurology, Neurosurgery, and Psychiatry*, 1971, *34*, 489–495.

Emsley, H. H. Irregular astigmatism of the eye: Effect of correcting lenses. *Transactions of the Optical Society*, 1925, *27*, 28–41.

Fisher, C. B. Children's memory for orientation in the absence of external cues. *Child Development*, 1979, *50*, 1088–1092.

Fisher, C. B. Children's memory for line orientation: A re-examination of the "oblique effect." *Journal of Experimental Child Psychology*, 1980, *29*, 446–459.

Gibson, E. J., Gibson, J. J., Pick, A. D., & Osser, H. A developmental study of the discrimination of letter-like forms. *Journal of Comparative and Physiological Psychology*, 1962, *55*, 897–906.

Hebb, D. O. *The organization of behavior*. New York: Wiley, 1949.

Hein, A., Vital–Durand, F., Salinger, W., & Diamond, R. Eye movements initiate visual–motor development in the cat. *Science*, 1979, *204*, 1321–1322.

Held, R. Early deprivation and meridional variation in visual acuity. *Neurosciences Research Progress Bulletin*, 1977, *15*, 467–469.

Holtzman, R. N. N., Rudel, R. G., & Goldensohn, E. S. Paroxysmal alexia. *Cortex*, 1977, *14*, 592–603.

Hubel, D. H., & Wiesel, T. N. Receptive fields, binocular interaction, and functional architecture in the cat's visual cortex. *Journal of Physiology*, 1962, *160*, 106–154.

Hubel, D. H., & Wiesel, T. N. Receptive fields and functional architecture of monkey striate cortex. *Journal of Physiology*, 1968, *195*, 215–243.

Huttenlocher, J. Discrimination of figure orientation: Effects of relative position. *Journal of Comparative and Physiological Psychology*, 1967, *63*, 359–361.

Ives, S. W. Children's ability to coordinate spatial perspectives through structured linguistic descriptions. In D. R. Olson & E. Bialystock (Eds.), *Exploration in inner space: Aspects of the nature and development of spatial cognition*, in preparation.

Jastrow, J. On the judgment of angles and positions of lines. *American Journal of Psychology*, 1893, *5*, 214–248.

Keogh, B. K., & Keogh, J. F. Pattern copying and pattern walking performance of normal and educationally subnormal boys. *American Journal of Mental Deficiencies*, 1967, *71*, 1009–1013.

Lashley, K. S. The mechanism of vision: XV. Preliminary studies of the rat's capacity for detail vision. *Journal of General Psychology*, 1938, *18*, 123–193.

Leehey, S. C., Moskowitz–Cook, A., Brill, S., & Held, R. Orientation anisotropy in infant vision. *Science*, 1975, *190*, 900–902.

Levick, W. R. Receptive fields and trigger features of ganglion cells in the visual streak of the rabbit's retina. *Journal of Physiology*, 1961, *188*, 285–307.

Liebowitz, H. W., Myers, N. A., & Grant, D. Radial localization of a single stimulus as a function of luminance and duration of exposure. *Journal of the Optical Society of America*, 1955, *45*, 76–78.

Mackintosh, J. K., & Sutherland, N. S. Visual discrimination by the goldfish. The orientation of rectangles. *Animal Behavior*, 1963, *11*, 135–141.

McFarland, J. H. "Parts" of perceived visual forms: New evidence. *Perception and Psychophysics*, 1968, *3*, 118–120.

McFie, J. The diagnostic significance of disorders of higher nervous activity. *Handbook of Clinical Neurology*, 1969, *4*, 1–11.

Millar, S. Spatial representation by blind and sighted children. *Journal of Experimental Child Psychology*, 1976, *21*, 460–479.

Noble, J. Paradoxical interocular transfer of mirror image discrimination in the optic chiasm sectioned monkey. *Brain Research*, 1968, *10*, 127–151.

Ogilvie, J. C., & Taylor, M. M. Effect of orientation of the visibility of fine wires. *Journal of the Optical Society of America,* 1958, *48,* 628–629.

Olson, D. R. *Cognitive development: The child's acquisition of diagonality.* New York: Academic Press, 1970.

Olson, D., & Hildyard, A. On the mental representation of oblique orientation. *Canadian Journal of Psychology,* 1977, *31,* 3–13.

Olson, D., & Hildyard, A. Proving the Whorfian hypothesis with octopi, white rats, and children: The perception and lexical representation of oblique lines. In D. R. Olson & E. Bialystock (Eds.), *Exploration in inner space: Aspects of the nature and development of spatial cognition,* in preparation.

Piaget, J., & Inhelder, B. *The growth of logical thinking from childhood to adolescence.* London: Routledge, 1958.

Pufall, P. B. Egocentrism in spatial thinking: It depends on your point of view. *Developmental Psychology,* 1975, *11,* 297–303.

Pufall, P. B., & Shaw, R. E. Analysis of the development of children's reference systems. *Cognitive Psychology,* 1973, *5,* 151–175.

Rudel, R. G. The transposition of intermediate size by brain-damaged and mongoloid children. *Comparative and Physiological Psychology,* 1960, *53,* 89–94.

Rudel, R. G., Denckla, M. B., & Broman, M. Rapid silent response to repeated target symbols by dyslexic and nondyslexic children. *Brain and Language,* 1978, *6,* 52–62.

Rudel, R. G., Denckla, M. B., & Hirsch, S. The development of left-hand superiority for discriminating Braille configurations. *Nuerology,* 1977, *27,* 160–164.

Rudel, R. G., & Teuber, H. L. Discrimination of direction of line in children. *Journal of Comparative and Physiological Psychology,* 1963, *56,* 892–898.

Rudel, R. G., & Teuber, H. L. Spatial orientation in normal children and in children with early brain injury. *Neuropsychologia,* 1971, *9,* 401–407.

Schaller, M. J., & Harris, L. J. Children judge "perspective" transformations of letterlike forms as different from prototypes. *Journal of Experimental Child Psychology,* 1974, *18,* 226–241.

Semmes, J., Weinstein, S., Ghent, L. and Teuber, H.-L. Spatial orientation in man after cerebral injury. I. Analyses by locus of lesion. *Journal of Psychology,* 1955, *39,* 226–244.

Semmes, J., Weinstein, S., Ghent, L., & Teuber, H. L. Correlates of impaired orientation in personal and extrapersonal space. *Brain,* 1963, *86,* 747–772.

Serpell, R. Discrimination of orientation by Zambian children. *Journal of Comparative and Physiological Psychology,* 1977, *75,* 312–316.

Stein, N. L., & Mandler, J. M. children's recognition of reversals of geometric figures. *Child Development,* 1974, *45,* 604–615.

Sutherland, N. S. Visual discrimination of orientation by octopus. *British Journal of Psychology,* 1957, *48,* 55–71.

Sutherland, N. S. Visual discrimination of the orientation of rectangles by *Octopus vulgaris* Lamarck. *Journal of Comparative and Physiological Psychology,* 1958, *51,* 452–458.

Sutherland, N. S. Visual discrimination of orientation by octopus: Mirror images. *British Journal of Psychology,* 1960, *51,* 9–18.

Teuber, H.-L. Effects of focal brain injury on human behavior. In H. Tarver (Ed.), *The clinical neurosciences* (Vol. 2: The Nervous System). New York: Raven Press, 1975. Pp. 457–480.

Twitchell, T. E. Personal communication, 1975.

Umiltà, C., Rizzolatti, G., Marzi, C. A., Zamboni, G., Franzini, C., Camarda, R. and Berlucchi, G. Hemispheric differences in the discrimination of line orientation. *Neuropsychologia,* 1974, *12,* 165–174.

Warrington, E. and Rabson, P. A preliminary investigation of the relation between visual perception and visual memory. *Cortex,* 1970, *6* (1), 87.

Weinberg, J., Diller, L., Gordon, W. A., Gerstmann, L. J., Lieberman, A., Lakin, P., Hodges, G., & Ezrachi, O. Visual scanning training effects on reading-related tasks in acquired right brain damage. *Archives of Physical Medicine and Rehabilitation*, 1977, 58, 479–486.

ROGER HART
MAXINE BERZOK

7

Children's Strategies for Mapping the Geographic-Scale Environment

An extremely important aspect of children's development of competency in the world is the ability to establish a knowledge of its spatial qualities. A child travels between home, school, friends' houses, shopping streets, and sometimes goes on trips out of the neighborhood. A child walks to some places, rides in a car to some places, and usually takes school trips by bus. Out of these experiences develop sets of expectancies about what will be encountered and the temporal and spatial relations among certain places and events. In effect, the mental equivalent of a "map," or sets of "maps," develops.

A body of literature has grown in developmental psychology, particularly over the past 10 years, to describe and explain the development of this kind of cognitive activity. Unfortunately, the focus of theory and research has been on the ontogenesis of the competence underlying the production of geometrically correct "maps." Such sophisticated cognitive mapping ability is undoubtedly valuable. It is important in some practical way-finding problems such as planning movement between a number of widely separated points that have not been encountered in sequence before. We do not believe, however, that it accounts for all of children's cognition of geographic space, or that it is a higher order strategy that subsumes other simpler systems.

There are many different mental strategies one may adopt for maintaining orientation in geographic space. For example, one may observe that even small children will anticipate a series of markers on long car trips and be aware of

147

SPATIAL ABILITIES
Development and Physiological Foundations

ISBN 0-12-563080-8

changes in order. They can give directions on how to drive home that may not employ the correct verbal labels but nevertheless are usually accurate. They can give in-context "go this way, go that way" directions with appropriate motions. Whether they can produce this information out of context and in correct geometric relation involving representation of direction and metricized distance can be questioned, but they clearly have some useful mapping strategies.

The use of the Euclidean, geometrically-correct map as the model for "mapping" research has had a long history, going back at least as far as Tolman's classic paper (1948). Roger Downs has very recently written a critical account of the history and current usage of the cartographic "map" as a metaphor in spatial cognition (Downs, 1981). Piaget's investigations of a child's knowledge of space have provided a model for the investigation of children's knowledge of geographic space (see reviews by Hart & Moore, 1971; Seigel & White, 1975). He focused on the development of a child's ability to represent the spatial world in reduced scale on a Euclidean geometric base (Piaget, 1947/1967, 1948/1970). It is not at all surprising, then, that the idea of such a map has heavily influenced our thinking about the structure and functions of the mental processes behind spatial behavior. Such Euclidean maps are commonly mistaken for the real world and used as standards against which the quality of people's cognitive maps (as revealed in drawings, models, etc.) are measured. As Downs has pointed out, there is no single type of cartographic map (Downs, 1981). Maps are designed for specific purposes.

All maps may be thought of as distorted representations of the "real world" to some degree. It is a question of designing the best map for the job. The same is true for "mental maps." People use mapping strategies or transformations that are suited to their particular purpose or problem. Our methods of research and analysis must recognize this. In particular, we should not use the traditional Euclidean map as "the" standard of analysis, but as something that scores performance relative to the particular goals of the person at the time.

This chapter will present some relevant issues in current research on the range of strategies children employ in solving practical way-finding or location-establishing problems in macro (or geographic) space. Touching on traditional cognitive mapping problems, our major purpose is to begin the development of a taxonomy of mapping strategies. Drawing from field studies and personal observations, we will show that the mapping strategy employed by a person is influenced by purpose, environmental structure, varied experience, and the like. Our hope is that this might help to encourage a different kind of research which will attempt to elicit mapping strategies under a variety of conditions. Only this kind of research will help us discover the elements and rules of organization children use to order their everyday spatial experiences. More descriptive studies of the varieties of mapping strategies people use will enable us to discover if (and then how) mapping strategies develop and if (and then when) they do become hierarchically ordered. (For a similar approach to the problem of cognitive mapping see Wapner, Kaplan, & Cohen, 1973.) This would lay the foundation for a discussion of the finer points of children's performances in rather difficult, abstract, laboratory spaces. Such a research program should lead us to form hypotheses and develop theory more relevant to the problem area.

A CONSIDERATION OF CURRENT THEORY AND RESEARCH

The Purpose of Cognitive Mapping

People have many purposes for mentally constructing the environment. Developmental research on spatial cognition rarely discusses these purposes, proceeding as though this is either an obvious or irrelevant issue. We think neither is the case. Research commonly seems to be implicitly concerned with problems of way-finding. But this is only one of many reasons we have for mentally representing environments in some spatially organized ways. In the Appendix of his ground-breaking early study, *The Image of the City* (1960), Kevin Lynch discusses other reasons and suggests that the overriding one is our fear of being lost. More basically, we could say that we mentally map the environment to surround ourselves with a known, and hence, more secure or safe world. Spatial imagery is also probably valuable for storing information. By mentally locating phenomena, in some way or other, we know how they are distributed and how they correlate with the spatial distribution of other phenomena, so that we can make decisions about such things as where is the nearest place to buy milk, or which dog-walking place is the least exposed to wind on a cold day (cf. Foulke, Chapter 3, this volume).

In this chapter we focus upon way-finding not because we think this is the most important reason for cognitive mapping but because this is the issue upon which most research seems to have focused and because we think it is ripe for some theoretical reorientation. We shall begin by briefly summarizing the current state of theory in the development of spatial cognition for way-finding in children.

Current Research

Almost all research on way-finding has been conducted in laboratory settings. This has resulted, we believe, in an underestimation of the spatial problem-solving competences of children. Most of this research focuses on the acquisition of concepts of correct measurement of distances and their integration into projective and Euclidean coordinations, proportions, and symbolic conventions. This research has involved nonpurposeful tasks, commonly suggesting lower levels of spatial competence than we believe young children possess.[1] Some recent work by Hazen, Lockman, and Pick (1978) and Pick and Lockman (1979) supports this contention. In the earlier study, conducted in a bleached laboratory setting of small, empty, white-walled rooms, with center-placed toy animal landmarks, they found that 4-year-olds performed poorly with a spatial inference task. In the later study, conducted in the home environment, children of the same age could make spatial inferences.

There are a number of reasons why we might expect children to have a better organization of geographic space than laboratory research indicates. A primary one has to do with the modes of representing and communicating one's knowledge. For example, if a child were asked, "Can you show me another way or a shorter way to school," (rather than to *tell* or *draw*) we hypothesize that real paths constructed from alternate starting points would give evidence of "mapping" integration of three or more nonsequential points which could not be well represented in traditional measures of spatial cognition such as drawn maps or constructed models. Equally important factors in laboratory situations have been the use of unsuitable landmarks and loss of control over decision making, both related to the common meaninglessness of the tasks to the children. We should expect children to organize more complex information in the "real world" because they can select and use personally relevant landmarks; they can freely explore extensive large-scale areas (instead of being lead through experimental environments); and they have very good reasons for needing to mentally map their surroundings. It is these factors that probably account for the high degree of cognitive mapping competence found by Hart (1979, 1981) in children's home town mapping.

The General Paradigm of Current Research

Any mapping strategy requires some kind of reference system that allows us to orient ourselves to the environment. In an earlier paper Hart and Moore (1971/1976) discussed three categories of reference systems—egocentric, fixed, and coordinated (see Figure 7.1). Much of the research has adopted a similar scheme to this drawn from the theory of Piaget and his colleagues.

THE EGOCENTRIC SYSTEM OF REFERENCE

This is a very simple and limited system. A child uses his own body as the sole system for ordering the position of things in space (Howard &

---

[1]In a related vein, Nelson (1978, 1979) and Nelson and Greundel (1979) have found that young children give well-ordered accounts of their everyday experiences, contrary to Piagetian hypotheses of the memorial fragmentation of young children's experience.

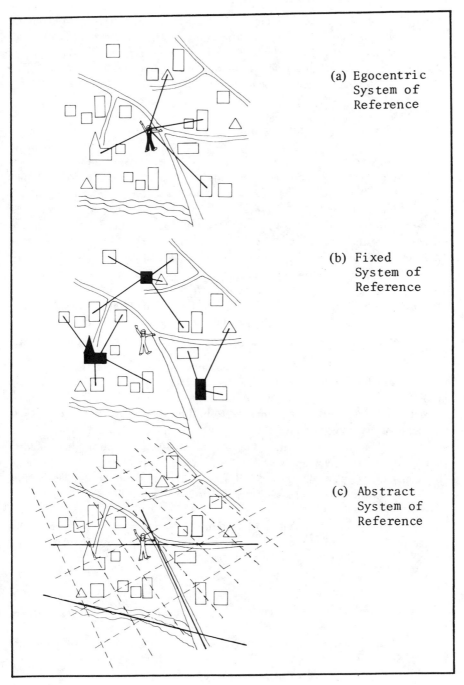

(a) Egocentric
    System of
    Reference

(b) Fixed
    System of
    Reference

(c) Abstract
    System of
    Reference

Figure 7.1. Three reference systems for children's orientation in the landscape.

151

Templeton, 1966). It would seem to offer serious computational problems for navigational orientation because as the child moves, the system of reference moves (cf. Pick and Reiser, Chapter 5, Potegal, Chapter 15, this volume). For more efficient geographic orientation a child must establish some kind of system of reference independent from the self.

THE FIXED SYSTEM OF REFERENCE

In this system a child orients in terms of fixed elements in the environment rather than relying solely upon orientation to the body. It is a more efficient system for localization when the child can view her environment from many different locations. It is believed to develop subsequent to the egocentric system of reference. Piaget found that children using this system could relate places to each other in clusters around certain landmarks or along particular routes. However, they could not coordinate these clusters or routes with each other using this system because they are each dependent upon a single viewpoint or journey. Much literature has pointed to "route maps" as evidence of this second stage. Piaget provides a theoretical account of why these route maps precede co-ordinated, or survey maps, hypothesizing that a child cannot represent space beyond the immediately visually present space until she is able to logically coor-dinate the relationship of the independent journeys or viewpoints.

THE COORDINATED SYSTEM OF REFERENCE

This system depends upon the transition from a series of fixed points or fixed routes for reading spatial relations to a freely transferable point of read-ing. The coordinated system of references need not be the cardinal (compass) directions used in cartographic maps. A child may select roads as a coordinate system, for example, although they may not be such clean, ready-made coordi-nate systems as the one we have in New York City. In Geneva, Piaget found children used the river and roads. The degree of suitability of environments for constructing some kind of coordinate system undoubtedly varies and is probably one major reason why some cities are more difficult to orient in than others.

This three-fold categorization of systems is quite elementary, of course, for there are many different reference systems lying between purely egocentric and the most sophisticated of coordinated systems.

AN ALTERNATIVE APPROACH

Mapping strategies may be grouped under two broad headings: sequential strategies, or route mappings, and simultaneous display strategies, or survey mappings (see Hart & Moore, 1971; Moore, 1973; Shemyakin, 1962; Siegel &

White, 1975). Previous discussion of these terms in the literature has largely focussed on when these strategies appear ontogenetically—route maps have been thought to precede survey maps. We wish to assess the respective utility of route maps and survey maps for particular problems. We believe that such a perspective opens the way to investigate the degree to which differences in mapping strategy reflect not only ontogenesis of basic intellectual competency but also such factors as the extent and the spatial structure of the area to be mapped, the child's familiarity with the area, and the mode of transport.

The Organization of Way-finding Strategies

When one moves through the environment, one often needs to track a current location; sometimes one needs to plan movements that ensure arrival at a destination. In other words, one needs to note where places are and how one place is related to another. Whether sequential or simultaneous display strategies will be used, we believe, depends upon the impact of such factors as the method of locomotion, the legibility or structure of the environment, the extent of the journey, previous familiarity with the environment, and the various purposes of the traveler. It also seems likely that within each type of strategy there are different degrees of locational and planning specificity that can be achieved. Some of the variety of strategies available are described below, along with accounts of the particular circumstances and locational and directional purposes that could call them into action.

SEQUENTIAL STRATEGIES

Ordinal Mapping. Beginning with sequential maps, we first propose order of encounter as a possible mapping strategy. This is likely to be a sequential strategy of remembering the place or event to be encountered next. In this case, a person would only anticipate what she will encounter in a more or less correct sequence, but she would not know the relative distances and times that separate the markers.

Some work by Berzok (1980) with children on their school bus journey suggests that this may be a strategy employed by young children in mapping the school bus trip, a journey they experience passively and in a special multidimensional context. Apparent age related differences were found, suggesting that a social base preceded a physical base as a location-establishing strategy.

Interval Mapping. More sophisticated sequential mapping includes knowledge of the relative distances or times separating the markers. Environmental information is again organized in a linear way, but distance and time separate the markers in memory. Directional information, however, is not tracked. It is useful to consider the experience of riding a bus or a commuter train. Until the final destination is reached, there is no need for directional information. It is possible, though, that passengers often track the succession of locations, giving them some sense of how much of the journey has been accomplished. Such interval mapping may also be adequate for way-finding if very clear markers are visible. We suspect this is a common mapping strategy.

Accurate Route Map.   Once directional information is inserted into the mapping strategy, one has a sophisticated route map. Not only are markers and the relative distances between them used, but also the directions to be utilized are available. In this case, enough information is available for very reliable way-finding.

Before leaving the subject of route mapping, we must admit that this discussion heavily emphasizes the planning aspect of way-finding. Much way-finding, we suspect, does not rely upon such planned and completed strategies. More likely it is a mixture of very simple ordinal or interval route mapping, with the reading of the vista that lies immediately before us: a more in-context sensorimotoric kind of way-finding. This is the kind of knowledge that leads to one of the most frustrating kinds of directional information: "I could tell you which way to go if I were there."

SURVEY STRATEGIES

For many situations, configurational mapping is more suitable. This is particularly the case when one needs to know alternative ways of reaching a number of places or to reevaluate one's position.

Loose Topological Mappings.   A rudimentary strategy would be one that is organized on a loose topological basis. Locations are mapped as "near to," "around," "behind," etc. (cf. Bremner, Chapter 4, this volume). This "regional" kind of mapping strategy allows gross locational judgments to be made. For instance, the strategy one utilizes in finding a car in a shopping center parking lot is probably of this "next to" type: "The car is in the lot next to the perfume counter door." This strategy only enables one to get to an approximate place and might therefore require additional strategies for completing the task—retracing one's steps or returning to a reliance on sensorimotor behavior may be a necessary next step, for example.

Accurate Simultaneous Display Strategy.   The other possibility for mapping is a spatially coordinated map in which various paths and locations are organized into an integrated configuration (i.e., with the relative positions of features correctly shown: a positional map). This mapping strategy allows directional choice, that is, planning of alternative routes to locations even at great distances from one another.

SOME ILLUSTRATIVE FIELD DATA

Two primary sources are used in this section to provide examples of the kinds of strategies just outlined. The first source is a description of the spatial representation of the landscape by children aged 4–11, part of a larger descriptive study by Hart (1979) of the development of children's place experience in the New England town of Inavale. In this investigation, a methodology was developed using a wide diversity of scale model elements that maximized children's ability to recreate their town in miniature. The second source is an unpublished pilot investigation, by Berzok, of children's cognitive mapping of their school bus journey.

Four- and Five-Year-Olds Construct Configurational Maps of Very Familiar Areas

Figure 7.2 is a schematic rendering of a "landscape model" produced by Christopher, 4:6, of "all of the places he knows, beginning with his home." (See Figure 7.3 for a key to maps of children's landscape models.) He located most of the features in a spatially arbitrary manner, choosing instead to place things near to each other that usually "go together" in experience. Piaget calls this "conceptual proximity." Surprisingly, however, Christopher's house, favorite tree, and road were carefully and correctly placed in their positional relationship to each other; that is, they were mapped configurationally. This example does not contradict Piaget's research, it merely reminds one to look more closely at it. Piaget noted that a preoperational level child (i.e., a child lacking truly logical powers) could be expected to recall in a simple manner not only familiar journeys but also vantage points. Our impression was that Christopher was recreating from memory a very familiar scene, mentally placing himself inside his home or on the stoop while he did so. It should be noted that this strategy can only be applied to sets of places or objects that have been perceived simultaneously by a child. It is of no utility for spatially linking features that are out of sight from one another. This example, however, does disprove the simple notion that "route mapping" is a necessary prior stage to "survey mapping" (Hart & Moore, 1971; Siegel & White, 1975). These three items mapped by Christopher are coincident with his very limited "free range" of movement. From the landscape models of other children under 8 years of age, it appears that most of their positional or survey clusters of places and objects lie within their "free range" of movement around their own homes. This leads us to add to Piaget's account the hypothesis that the ability of a child to decenter (i.e., abandon the egocentric reference system) and take another perspective will vary according to the object, place, or person to which the child

HOUSE THAT BURNT

HIDO'S HOUSE (THE HORSE)

HS

YOUR HOUSE

OUR TREE

MY TEACHER'S HOUSE

STREET

STREET

ES

Symbol	Description	Symbol	Description
★	Home	BG	School bus garage
▲ (filled, outlined)	Relative's home	V	Vacant building
▲ (circle)	Elementary School child's home	TG	The Town Garage
▲	House - named	ES	Elementary School
△	House - unnamed	HS	High School
●	Garage	DR	Doctor's office
†	Church	SKI	Ski lodge
⊞	Church School	◆	Barn, shed, sugarhouse
▮	Gas station	---	Wall
■	Store	‡	Fence
P	Police station	M	Manufacturing
F	Firehouse	▦	Lake
S	Supermarket	▒	Woods or bushes
R	Restaurant	♡	Trees
I	Inn, lodge, motel	▨	Grass
B	Bank	≋	River
BA	Bowling Alley	□	Other keyed individually on each map
O	Office	▨	Snow
PO	Post Office	⬡	Sign
L	Library		

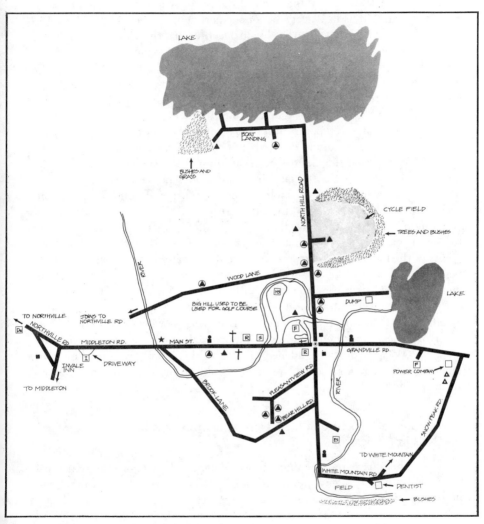

Figure 7.4. Casey's landscape model.

Figure 7.2. (*Top of facing page*) Christopher's landscape model.

Figure 7.3. (*Bottom of facing page*) Legend for maps of children's landscape models: Figures 7.2, 7.4, 7.5(a), and 7.5(b). (From Hart, 1979.)

is decentering. If it is a doll that a child has never seen before, standing at the side of a mountain as in Piaget's famous experiment (Piaget & Inhelder, 1947/ 1967), the child is less likely to be able to decenter to it than if it were a model of his own home, from which he has looked out many times. Christopher was most likely using this view-from-the-home method; we might call it a *domocentric* strategy. Within this positionally mapped cluster we may expect Christopher to be more competent in certain aspects of his spatial behavior than when outside of it. To go beyond the area of his frequent view from the doorstep, Christopher must rely upon a different strategy for representing the spatial relationships. At a minimum he needs to be able to remember the sequence of places visited. From the extensive research on the development of children's logical abilities, we suspect that Christopher does not have the logical operations necessary to turn sets of such sequential information into configurational types of mapping. When asked to lead Hart to his favorite play places not only could he not describe their location, other than "across the street," he needed his 9-year-old brother, Casey, to lead the way. Casey's own representational ability of the town was so comprehensive and accurate that he insisted on using pencil and crayon alone to "map" his town, rather than using the moveable models and road strips that the other children used (Figure 7.4). Christopher's cognitively mapped spatial world is undoubtedly expanding in a similar manner at this time. It is a mistake to conclude from this, however, that the development of logical abilities is the most important factor in children's cognitive mapping. We will provide evidence to suggest that there are numerous other factors affecting a child's cognitive mapping ability and the particular mapping strategies used.

Environmental Structure Influences Which Strategy Will Be Used in Navigating

Research into spatial cognition in developmental psychology has not paid attention to the role of environmental "legibility" or "imageability" in the selection of spatial strategies. This is because research has largely focused upon child development and theoretical questions concerning the ability to logically manipulate spatial relations in any environment. The field has probably overlooked an important contextual factor that has a strong influence on the particular cognitive mapping strategy a person will use. Urban designers, planners, and geographers have, in contrast, focused their interests on environmental issues. These studies were built upon the pioneering study by Kevin Lynch in *The Image of the City* (1960) and were designed to provide information regarding the imageability of certain areas as guidance for city planning practice.

Consider the differences in your own cognitive maps of the different room organizations found in a hotel or hospital. First, think of a pair of rooms in a convention building whose rectangular form is easily grasped. Paths in this building appear as linear or perpendicularly intersecting hallways. Contrast this with a second pair of rooms located in buildings composed of a number of annexes added over the years (as in main hospital complexes). Often in these cases, the relation of one wing to another is not symmetrical. If you were to walk from one room to another and then were asked to "map" the trip, in the first case, you would probably generate a fairly accurate map. In the second case, where the path system is not so easily grasped, it would be more difficult to establish the spatial relations.

In Hart's data (1979, Chapter 6), some environmental influences were so obvious that they could not be ignored, though this was not the focus of his study. In creating their landscape models the older children, in particular, commonly used the town's roads and the angles they formed with each other as a reference system. The remarkable point is that the crossroads in one part of town were so imageable that even the 6-year-old children were able to use them as a reference (e.g., Figures 7.5(a) and 7.5(b): Margaret, 6:3). There are a number of possible reasons why these crossroads were found to be more imageable than other places. Perhaps it was because symmetry such as theirs (the roads form a cross) is more readily perceived and represented. Alternatively, or additionally, there are traffic lights at the crossroads, so their parents' cars and the school bus stop there frequently, allowing the children time to observe the scene. A third possibility is that candy stores on three of the four corners make this one of the most highly valued places in town! This example suggests that imageability may be more than just a question of "good" spatial form.

As Lynch 1960 demonstrated long ago with the highly imageable Commons in downtown Boston, too great a reliance upon such a strong image can result in great distortion in a person's representation of the environment. Boston Commons is often represented as a four-sided figure rather than as the five-sided figure it is. Many children in Inavale imaged the crossroads so well that they modeled them even though they were unable to recall correctly how these crossroads related to other parts of the environment with which they were more intimately familiar. The result was maps with severe disjunctions or distortions.

These observations suggest the structure of an environment may have a considerable influence upon the ability of children to represent it. More specifically, some forms may be so much more imageable than others that they enable a child to use them as a reference for other places or objects in the environment.

It is likely that the cognitive mappings of encounters with the environment of adults take many forms. They vary with the demands and opportunities presented by the structure of the environment (Hart, 1979; Lynch, 1960;

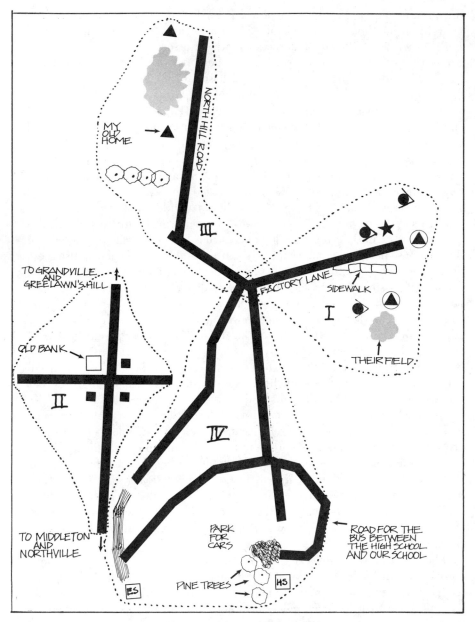

Figure 7.5(a). Margaret's landscape model. (From Hart, 1979.)

160

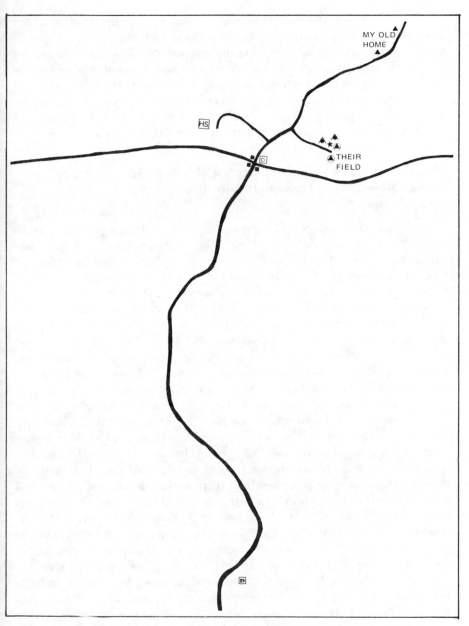

Figure 7.5(b). Location of features in Margaret's landscape model. (From Hart, 1979.)

Wood, 1973). When the environment is legible, adults easily form positional maps that correctly relate landmarks and paths to each other. When the environment is not legible, as in moving through streets whose patterns are not clear, we may maintain interval maps in which simple sequential or contiguous directions along a route may be encoded (i.e., left turn and right turn). In such cases, nonadjacent locations or routes are not encoded in relation to one another, and so it is not possible to integrate separate route maps. People may therefore rarely develop positional maps of these difficult environments.

Primary Factor Influencing Children's Ability to Make Configurational Maps: Freedom to Direct Their Own Movements

It was found in the New England study that positional type clusters lie within the children's "free range" or "range-with-permission" rather than within their much larger, but less "active," "ranges-with-older-children." Since it has been previously hypothesized that the representation of space begins with the internalization of action in space, we might anticipate that walking or cycling would be most important to a child's formation of topographical representations and that more passive modes of travel would not serve the same purpose (Appleyard, 1970; Hart, 1979; Lee, 1963). This point had been suggested by Lee as a result of his fascinating study of the effects of busing on children in an area of Devon, England where all children had previously walked to school. He found that those young children (6 and 7 years) who were bused to school suffered serious problems of social and emotional adjustment during their first year of busing, unlike their peers who walked to school each day. He hypothesized that the bus journey took them beyond their known (representable) world into a space that they had had no opportunity to articulate through their own bodily locomotion through the environment. As a result, he claims, the children felt separated from their mother, home, and the physical expressions of security, thus producing anxiety during difficult periods in school. Lee's notion that bodily movement is necessary to "articulate the schema" is in line with Piaget's theory of the development of the operational basis of knowing. Some recent experiments have investigated the importance of bodily locomotion in learning to cognitively map an environment (Acredolo, 1977; Herman & Siegel, 1978). Bodily locomotion alone, however, is probably not the most relevant factor. The parts of children's spatial ranges of movement that were limited to travel under the care of other persons were not mapped in the same sophisticated manner as the children's "free ranges." It is not possible to determine from this field study how much this might be due to the probably greater frequency of travel within the free ranges than in the ranges with others. However, it raises an interesting hypothesis for experimental investigation. What improvements in the cognitive mapping might we find if we allowed children to freely and purposefully direct their own movements in an experimental setting?

Figure 7.5(a) is a rendering of a landscape model by Margaret, 6:3. She had only moved into her new house 3 months previously but was already able to map accurately the elements around it, for this is the area of her "free range" of play (Cluster I). A second cluster of elements was also mapped positionally even though Margaret had never walked to this town center area alone and had usually been driven there by her mother. As has already been suggested, its spatial articulation is probably a reflection of the highly imageable properties of the crossroads. Cluster III shows the route Margaret has frequently taken on foot between her old and new homes and vice versa; no doubt such two-way trips help considerably with the common problem found in young children's mapping of not being able to logically reverse one's thinking in order to retrace a route. The fourth cluster is an expression of Margaret's daily journey to the high school where she waits for the school bus, sometimes going on foot with her brother, sometimes going in her mother's car. She made no attempt to express the turns and different stretches of road on this trip, probably because she had traveled it so few times on foot and never alone. Cluster IV also shows the journeys to and from school. Margaret chose first to express the route via the mode she knows best, her mother's car, in which she traveled from school throughout her preschool year. She then drew the route the bus takes coming from the elementary school to the high school, which is identical except for the last few hundred yards to her home. Except for the short strip of Factory Lane on which her new home stands, she expresses these two vehicle journeys as completely separate from each other. In fact, she even crosses her bus journey road over her car journey road. This, and all the other very similar maps found in Inavale, suggested that the children mentally represented very little of the physical qualities of their bus journey. Lee's suggestion was that such a journey may be through a nowhere land for the children. This conclusion, however, is drawn from the use of only one method. In recognition of our arguments that children use different strategies for different problems, Berzok (1980) is devising more suitable methods for allowing children to reveal what they are cognizing during such a seemingly "passive journey."

Children Transported through the Environment with No Control over the Locomotion May Construct Different Cognitive Mapping Strategies

The bus ride takes children from home to school, from one location in the environment to another. Lee's hypothesis that children younger than 8 cannot (with rare exceptions) incorporate a school bus journey into spatial representations of the environment seems to be confirmed by field research in the town of Inavale (Hart, 1979). The school was usually placed arbitrarily at the end of a section of road, with the beginning of the road representing the location where the child catches the bus and with few, if any, landmarks between origin and destination. It is possible, though, that children do mentally organize this trip, but in ways that are not tapped by conventional mapping research methods.

The bus journey to school has special characteristics. Not only is it a physical experience, but it is a social experience. Regularities may be expected to occur in both the physical and the social dimensions. Physical elements suitable as spatial landmarks would obviously include the buildings, houses, stores, traffic lights, etc. found along the route. But the same children get on the bus each day in fairly constant order. This could provide a system of recurrent social markers. A child could also mark the journey by noting other recurrent events (what the bus driver does when stopping and starting the bus, or stopping for a police officer directing traffic at an otherwise unmarked location).

If we order these domains on the basis of reliability for establishing location, it is clear that physical landmarks give the most reliable information for location. Buildings do not move. So if a child is tracking a particular position—anticipating where he gets off the bus by searching for a particular building—he will have reliable information about where he is. A social order map, one that marks and segments the journey according to the order by which children get on the bus, is less reliable. For example, if the child marks the journey by anticipating when Suzie gets on the bus, if Suzie does not get on that day the child's sense of the order of the trip may be disrupted. An event map is the least reliable one for establishing location. Stopping for a police officer directing traffic each day can help order the journey by establishing a regularity, but it does not necessarily establish a reliable place. It may be that the very young child cannot track or order the physical landmark regularities of the bus trip, so instead she organizes her trip schema around recurrent events.

There are, then, three domains from which a child may abstract regularities. He may form a physical map, a social map, or an event map. Some of the information may be more relevant to a "where am I map" while other information may be more relevant to a "what happens next map." Either kind of map, though, may serve as a regularity marking device, as the basis for structuring the route.

A PILOT STUDY: HOW YOUNG CHILDREN MARK AND ORGANIZE THE SCHOOL BUS TRIP

In a pilot study, we hoped to find out what about the school bus trip was salient for the children, and to gain some insight into the strategies they might be using to mentally structure and locate themselves in the trip.

The pilot work was conducted at a day camp in suburban New Jersey. The children ranged in age from 5:2 to 9:10 years. They were picked up at their homes each day and taken to the camp in minibuses. The trips were approximately 13 miles long and took the children at least partially through towns and routes they did not normally travel through.

The children were asked first to describe the bus trip to camp and the trip home. Later they were guided to reconstruct the trip by being asked to pretend they were telling someone else how to drive to the camp. The children's responses suggested a developmental sequence for segmenting and mapping this experience through the very large scale environment. Their responses to the trip-description question fell into two basic categories depending on age. Children from 5–7 years of age would generally describe the bus trip using social information; older children tended to use physical information to describe the trip.

A typical response to the question, "Tell me everything you remember. What happens when you go on the bus when you come to camp?" is offered by George (5:2).

"We pick up kids."

When prompted to tell more about the kids by being asked where the bus went after he got on, he said

"Uh Sam's house. He went on vacation—so now Allison. But then if Sam comes back, he's going to go after me again."

After that, the next thing that happens, according to George, is that they go to camp. When he was asked to tell how he knew he was almost at camp, his answer was

"Because we pick up these last two kids and then we go to camp."

It does appear that the social events are important trip marking and segmenting devices for George. He anticipates where the bus is going by noting the appearance of children. He marks the approach to camp by noting "these last two kids" get on.

Eve, who was 9 months older than George, answered similarly to the question, "Eve, do you remember the trip when you go home?"

"It's always the same thing. . . .
First I don't go home first. I go home after Ned . . . Barbara goes home before him."

It is interesting to note how she enters the social ordering list. Although at another time she answered a direct question about social order by starting at the beginning and listing five names of the list, here she answered the going-home question by telling me about the children who immediately precede her arrival at home.

These anecdotes begin to suggest that the early segmenting and marking system children use for the bus trip to school has a social basis. Only that part of the bus trip when children got on might be articulated in a "cognitive map" for the young child. A child may code successive locations but not anchor the intervening route into any sort of physical spatial system.

In contrast to the younger children, older children tended to respond to a request to describe the trip to camp with information about the physical environment. They would enumerate the buildings they pass with varying degrees of specificity or they would name streets.

> "Well, the first thing I know is we cross Crossbrook Bridge. And pass a flower shop... and make nine stops." (Larry, 9:10)

> "I come down my street, and then I go down Ardsley Drive and then I go up Pond Street until I come to Grace Street... I go left until I get onto Main Street." (Stephen, 9:2)

The children were also asked to pretend they were going to tell someone else how to drive to the camp. The intent here was to try to get some other indication of what children understand about negotiating the large scale environment. There were interesting differences in the patterns of response between the younger and older children. The youngest children gave two types of answers. Two early 5-year-olds just said they would not know how to tell someone else how to drive to the camp. The children between 6 and 7 years old seemed to have some grasp of the problem.

> "The other driver should look for the roads, the names of the roads."

> "Follow me."

> "I'll tell you where to turn." (Eve, 5:11)

> "First... look for Susan's house, then that lady's house, then Jack's house." (Kristin, 6:11)

These children between 6 and 7 years old seem to know something about how to convey information about an extensive route. What is not clear at this time is the relation of the form of the report to the child's actual knowledge. What is anchoring the trip information—a social system or a physical system? Is there a disjunction between the two, or is there a mapping of all or some of one kind of information onto the other? It is possible that the younger children have the physical information but cannot access it out of the context of an actual trip so that another strategy is used—imitating a partially understood adult routine or using social order information.

The older children answered the question with a strategy we might call "starting from a known place."

"If you were by Tech . . . ." (Jack, 8:11)
"If you knew where Cross Brook Bridge is." (Larry, 9:10)

In summary, it appears that older children are able to utilize a cognitive mapping strategy based on an analysis of physical elements. They show memory for places and seem to understand what is required for conveying information to another person so that the spatial information has relevant organization. Younger children either do not have, or are unable to utilize, spatial information that is physically grounded in the environment. Minimally, they seem to interpret the extensive space socially. Further research is necessary, of course, to establish that there is a social basis for mapping this very large scale spatial experience and that the social map exists not only as an "out-of-context" phenomenon.

THE NEED FOR GROUNDED RESEARCH

Little experimental research has been cited in this chapter because of our belief that questions in our field of interest have not been sufficiently grounded in the context of their everyday occurrence. This is a call for more descriptive, more ecologically valid research.

Developmental psychologists have recently become aware of the very real problems of generalizing from laboratory research to children's cognition of geographic spaces (see particularly Acredolo, 1977, p. 8 and Herman & Seigel, 1978, conclusion) and have begun to design experiments to deal with the complex contextual problems of geographic space representation and orientation. In all of the experiments known to us what is probably the most important factor of all has been systematically excluded from the research design: The subject never has control over the decision making. This is probably the factor that most distinguishes the experimental approach in psychology. Experiments in the development of spatial cognition have become more and more ingenious and, as described above, have now reached the point of comparing children's "active" versus "passive" learning of a route. But movement per se is probably not the crucial variable. More important, we hypothesize, is the control over the decision to move or not to move in a certain direction. Surely this is the factor that makes it much easier for the driver of a car to learn a route than for a passenger sitting next to her with the same view. The implications of this for further research into the development of children's spatial knowledge of the environment are considerable.

The question of what children represent when they "map" the world for different reasons has also been ignored by developmental psychologists, whereas it has been the focus of those planners, geographers, and environmental psychologists who have investigated spatial cognition. This may be an omission that has theoretical importance for developmental psychologists. Recent experimental research, for example, has investigated the effect of familiarity with landmarks; yet one landmark was still taken to be as good as any other landmark (Acredolo, Pick, & Olsen, 1975). Different landmarks may have different affective potential and even these may be subject to developmental changes. This is the kind of question that is best explored naturalistically: What landmarks do children spontaneously use in the everyday environment? Answers to such questions can serve to improve experimental research designs.

It is time for us to develop a synthesis of our research designs; we need to investigate children's cognition in settings that are familiar to them with tasks that are known to be meaningful to them. The unit of study in such complex ecological research must of course be the individual child. We can ask the same children to represent the environment spatially for different purposes (see Wapner, Kaplan, & Cohen, 1977 for some attempts at this). We can also use a range of different media of representation to match the different purposes, environments, and ages of children. The influence of the media of representation is itself an important area for further research.

What we are suggesting is more exploration of the commonplace. If we ask children how they go places, how they know when goals are achieved, and if we use more situationally appropriate tasks, we may find a variety of place-accomplishing strategies that are appropriate to their purposes.

Finally, the important relationship of research to the practical aspects of orientation, navigation, and cognitive mapping skills should be noted. A very useful direction for future research in this area would be the development of experimental interventions designed to improve spatial cognition. In sum, the important questions we need to have answered are: What do children themselves want to learn about places? For what purposes? How do they learn outside the classroom and how can this be facilitated by education?

REFERENCES

Acredolo, L. P. Developmental changes in the ability to coordinate perspectives of a large-scale space. *Developmental Psychology*, 1977, *13*, 1–8.

Acredolo, L. P., Pick, H. L., & Olsen, M. G. Environmental differentiation and familiarity as determinants of children's memory for spatial location. *Developmental Psychology*, 1975, *11*, 495–501.

Appleyard, D. Why buildings are known. *Environment and Behavior*, 1970, *2*, 100–117.

Berzok, M. *Children's cognition of the school bus journey.* Unpublished doctoral dissertation proposal, City University of New York, 1980.

Downs, R. M. Maps and mappings as metaphors for spatial representations. In L. Liben, N. Newcombe, & A. Pattison (Eds.), *Spatial representation and behavior across the life span.* New York: Academic Press, 1981.

Feldman, D. H. *Beyond universals in cognitive development.* Norwood, N.J.: Ablex Publishing Corp., 1980.

Hart, R. A. *Children's experience of place.* N.Y.: Irvington Publishers, 1979. (See literature review of environmental cognition in appendix.)

Hart, R. A. Children's spatial representation of the landscape: Lessons and questions. In L. Liben, N. Newcombe, & A. Patterson (Eds.), *Spatial representation and behavior across the life span.* New York: Academic Press, 1981.

Hart, R. A., & Moore, G. T. *The development of spatial cognition of large-scale environments: A review.* Monograph of the Center for Human Environments, City University of New York Graduate School, 1976. (Originally published Clark University, 1971.)

Hazen, N. *Young children's knowledge and exploration of large-scale spaces.* Paper presented at the conference of the Society for Research in Child Development, San Francisco, March 1979.

Hazen, N. L., Lockman, J. J., & Pick, H. L. The development of children's representations of large-scale environments. *Child Development,* 1978, *48,* 623–636.

Herman, J. F., & Siegel, A. W. The development of cognitive mapping of the large-scale environment. *Journal of Experimental Child Psychology,* 1978, *26,* 389–401.

Lee, T. R. On the relation between the school journey and social and emotional adjustment in rural infant children. *British Journal of Educational Psychology,* 1963, *27,* 100.

Lynch, K. *The image of the city.* Cambridge: M.I.T. Press, 1960.

Moore, G. T. *Developmental variations between and within individuals in the cognitive representations of large scale spatial environments.* Unpublished master's thesis, Department of Psychology, Clark University, 1973.

Nelson, K. How children represent knowledge of their world in and out of language: A preliminary report. In R. S. Seigler (Ed.), *Children's thinking: What develops.* Hillsdale, N.J.: Lawrence Erlbaum Associates, 1978.

Nelson, K. *Children's long-term memory for routine events.* Paper presented at the American Psychological Association convention, New York, September 1979.

Nelson, K., & Greundel, J. *From personal episode to sound script: Two dimensions in the development of event knowledge.* Paper presented at the SRCD conference, San Francisco, March 1979.

Piaget, J., & Inhelder, B. *The child's conception of space.* New York: W. W. Norton, 1967. (Originally published, 1947).

Piaget, J., Inhelder, B., & Szeminska, A. *The child's conception of geometry.* New York: Basic Books, 1970. (Originally published, 1948).

Pick, H. L., Jr., & Lockman, J. J. *The development of spatial cognition in children.* Paper presented at the Mind, Child and Architecture Conference, New Jersey School of Architecture, Newark, N.J., October 1979.

Shemyakin, F. N. Orientation in space. In B. G. Ananyev (Eds.), *Psychological sciences in the U.S.S.R.* (Vol. 1). (Rep. 62-11083) Washington, D.C.: Office of Technical Services, 1962. Pp. 186–255.

Siegel, A. W., & White, S. H. The development of spatial representations of large scale environments. In H. W. Reese (Ed.), *Advances in child development and behavior* (Vol. 10). New York: Academic Press, 1975.

Tolman, E. C. Cognitive maps in rats and men. *Psychological Review,* 1948, *55,* 89–208.

Wapner, S., Kaplan, B., & Cohen, S. B. An organismic developmental perspective for understanding transactions of men and environments. *Environment and Behavior,* 1973, *5,* 255–287.

Wapner, S., Kaplan, B., & Cohen, S. *Experiencing the environment.* New York: Plenum Press, 1977.

Wood, D. *I don't want to but I will.* Doctoral dissertation and monograph, Graduate School of Geography, Clark University, 1973 (available from Clark University Graduate School of Geography).

SPATIAL ABILITIES OF ADULTS:
THE INFLUENCE OF
HEREDITY AND GENDER

Mental Rotation:
Anatomy of a Paradigm[1]

> Then, sir, you will turn it over once more in what you are pleased to call your mind.
> —Lord Westbury (1800–1873)

Mental rotation is not new to psychology. Prior to the 1970s, however, it was studied primarily in the contexts of psychometrics and of cognitive development in children. For instance, mental rotation is the basis of the Space Test in Thurstone's (1938) Primary Mental Abilities, and is a critical component of two of the tests (Block Design and Visualization Memory) developed by Guilford and his colleagues within the "structure-of-intellect" model of human intelligence (e.g., Guilford & Hoepfner, 1971). In their studies of the development of cognition, Piaget and Inhelder have made use of tasks requiring mental rotations, whether in imagining scenes from different perspectives (Piaget & Inhelder, 1956) or in describing the appearance of a toppling stick (Piaget & Inhelder, 1971). They have concluded that children are unable to perform mental rotations until around 7 or 8 years of age, when they reach the "concrete

[1]The research described herein that was carried out by the author and his various colleagues was supported by grants from the National Research Council of Canada, the F.C.A.C. program of the Quebec Ministry of Education, the Faculty of Graduate Studies and Research at McGill University, the Vision Research Fund of the New Zealand Optometrical Association, and the University of Auckland.

173

SPATIAL ABILITIES
Development and Physiological Foundations

operational" stage of development. Subsequent research has shown that mental rotation continues to develop beyond this stage, however, since the *rate* of mental rotation approximately doubles between middle childhood and adulthood (Kail, Pellegrino, & Carter, 1980; Marmor, 1975).

Mental rotation was introduced to experimental cognitive psychology by R. N. Shepard and his colleagues in the early 1970s (e.g., Cooper & Shepard, 1973; Metzler & Shepard, 1974; Shepard & Metzler, 1971). Their main innovation was to measure the reaction time taken to make decisions requiring mental rotation of a shape, under varying experimental conditions. This technique has proven both sensitive and versatile in revealing basic cognitive and perceptual mechanisms, and indeed has become one of the flourishing *paradigms* of modern experimental psychology. Mental rotation is therefore one of the very few concepts to bridge successfully the chasm between what Cronbach (1957) called "the two disciplines of scientific psychology," the psychometric and the experimental; in its links with the work of Piaget and Inhelder it has spanned the American and European traditions in psychology.

Newell (1973) has criticized experimental psychology for being "phenomenon-driven"; that is, for relying too heavily on particular experimental paradigms. Echoed by Neisser (1974) and Allport (1975), among others, he argued that the study of relatively isolated paradigms has led to fragmentation, so that there is little sense of accumulated knowledge. Yet in many ways experimental psychology has been well served by its paradigms, and the apparent lack of progress may reflect in part the great complexity of human cognition and the unrealistically high levels of aspiration among those who study it (cf. Baddeley, 1976). Moreover experimental paradigms do not serve purely, or even primarily, as sources of facts that accumulate to provide ever more complete representations of the human mind. Rather, they serve often as heuristics, as foci for the illumination of theoretical issues (if not always for their resolution). I think that there *has* been an accumulation of knowledge in psychology that is not manifest in the stockpiling of facts so much as in the increased subtlety and sophistication of our theories. One need only recall the naive associationistic theories of the 1950s to appreciate that this is so. The advances in our understanding of human cognition surely owe at least something to such paradigms as the dichotic-listening and shadowing experiments of the 1950s and 1960s, or the various paradigms based on reaction time that have been prominent since the mid 1960s.

The mark of a good paradigm is that it can serve as a vehicle for studying much wider questions than are implicit in the paradigm itself. Mental rotation has already proven its value in just that respect. It has served not only as a source of new information, but also as a constructive focus for the broad issues that confronted cognitive psychology in the 1970s. These issues include the role of imagery in human cognition, the nature of mental representation and mental process, and the appropriateness of models based on analogy with the digital computer—issues that will no doubt continue to haunt us into the 1980s. These issues have not been resolved by the study of mental rotation, but they have been greatly clarified and refined, and we have learned new facts and gained new insights along the way.

In this review, I shall focus mainly on experimental work based on the reaction-time paradigm for studying mental rotation, for this work has lent a precision and depth to the study of mental imagery that was previously lacking. I shall also touch upon broader issues where it seems appropriate to do so, and especially where it bears upon themes discussed elsewhere in this book.

THE NATURE OF MENTAL ROTATION

The Reaction-Time Paradigm

There are two variants of the reaction-time paradigm for the study of mental rotation, both introduced by Shepard and his colleagues. One involves the matching of pairs of patterns or objects depicted in different orientations. For example, Shepard and Metzler (1971) presented pictures of pairs of solid forms that were either the same or mirror images of one another, and timed subjects as they judged whether they were the same or different. The reaction times to make "same" judgments increased linearly as a function of the angle between the forms, suggesting that the subjects mentally rotated one form into congruence with the other before making the judgment. This interpretation was consistent with the subjects' introspective reports.

The second variant of the reaction-time paradigm involves timed judgments about patterns presented singly. For example, Cooper and Shepard (1973) had subjects decide whether letters or digits, presented in different angular orientations, were normal or backward (i.e., mirror reversed). The reaction time to make the decision increased monotonically with the angular departure of the character from the upright. Examples of this function are shown later in Figure 8.1 and Figure 8.2. Cooper and Shepard interpreted the function to mean that the subjects mentally rotated each character to its normal upright position before making their decisions, even though the function in their experiment departed slightly from linearity (see Figure 8.2), and again this interpretation was verified by most of the subjects' introspections. With this particular procedure, the rate of mental rotation as estimated from the slope of the function is about 400 degrees per second, which is about six times the rate estimated from the function derived from the matching procedure used by Shepard and Metzler (1971). The reason for this discrepancy is not entirely clear, and will be discussed later in this chapter.

The Representational Space Underlying Mental Rotation

Mental rotation takes place in a representational space that is not confined to the retina or its internal equivalent. For instance, Shepard and Metzler's experiments revealed that two-dimensional pictures of three-dimensional objects can be mentally rotated in depth as well as in the plane of the picture (Metzler & Shepard, 1974; Shepard & Metzler, 1971). Even when subjects mentally rotate two-dimensional patterns they do so within a reference frame that is fundamentally perceptual rather than retinal. For instance, Corballis, Zbrodoff, and Roldan (1976) had subjects judge disoriented letters normal or backward, and found that the function relating reaction time to the orientation of the letters was essentially unaltered when the subjects tilted their heads 66 degrees to the left or right (see Figure 8.1). That is, the subjects evidently mentally rotated the letters to the true, gravitationally defined upright rather than to the retinal upright. In a subsequent series of experiments Corballis, Nagourney, Shetzer, and Stefanatos (1978) found that the subjective upright in mental rotation does not always coincide with the gravitational vertical, but may be influenced by the physical surround, the degree of head tilt, and to some extent by verbal instructions. However it never coincided with the retinal vertical, and usually lay closer to the gravitational than to the retinal vertical, at least for head tilts up to 90 degrees. Mental rotation therefore takes place within a frame of reference that is subjective, or perceptual, and that need coincide neither with gravitational axes nor with retinal axes.

Mental rotation need not even be visual. Marmor and Zaback (1976) had blind and blindfolded-sighted subjects judge whether pairs of tactile forms oriented at different angles to one another were the same or different. Again, the function relating decision time to the angular disparity between forms, as well as subjects' introspective reports, suggested that most subjects mentally rotated one form into congruence with the other before making their judgments. This was true of the congenitally blind, who were presumably devoid of visual imagery, as well as of the adventitiously blind and of the sighted subjects, although the results did suggest that mental rotation was easier and faster among those with access to visual representations. Carpenter and Eisenberg (1978) obtained similar results with blind subjects in a task requiring haptic judgments of whether disoriented letters were normal or backward (cf. Hermelin & O'Connor, Chapter 2, this volume).

The representational space in which mental rotation and other imaginary spatial transformations occur is therefore spatial in a general sense, rather than tied to any specific modality. We can refer to this space through several dif-

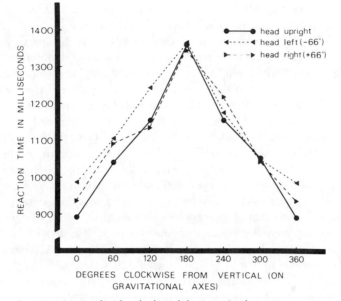

Figure 8.1. Reaction time to decide whether alphanumeric characters were normal or backward as a function of stimulus orientation and of 66 degree tilt of the head. These data show that mental rotation took place within gravitational rather than retinal coordinates. (From Corballis et al., 1976, and reprinted with permission of the Psychonomic Society.)

ferent imagined senses; we can imagine objects visually located in space or visually rotating in it, just as we can imagine touching objects or feeling them rotate. In the larger context, we can imagine ourselves moving through space, whether walking, running, or driving an automobile.

Experiments on mental rotation, especially those in which the reaction-time paradigm has been used, have generally involved only that limited region of representational space that immediately contains the patterns to be rotated. However representational space has a multilevel, embedded structure (cf. Neisser, 1976) in that regions can be considered to be embedded in larger regions. For instance Attneave and Farrar (1977) had subjects observe a set of objects on a shelf and then turn around so that the objects were behind them; the subjects were then able to make rapid decisions about the locations and orientations of the objects as though seeing them through eyes at the backs of their heads. As Attneave and Farrar put it, "the *mind's* eye has a cycloramic, 360-degree field [p. 561, emphasis in original]." But of course representational space does not stop there; our immediate surroundings are also embedded conceptually in more distant surroundings, such as a house, a street, a city, a country. In these larger contexts, we speak of the representation as a *cognitive map* (Tolman, 1948), although there are no rigid boundaries between one level and another (cf. Hart & Berzok, Chapter 7, this volume).

At some level, too, one can say that the representational space ceases to be merely a space and becomes a potential *environment*—a distinction drawn by Gibson (1972)—in which one can imagine oneself moving about. With respect to mental rotation, one can then constrast mental rotation of the environment with mental rotation of the self. (The issue probably does not seriously arise in, say, the mental rotation of letters, as in the Cooper and Shepard task, although I fancy I *can* perform the task by imagining myself tiny, the letters huge, and mentally rotating myself instead of the letters.) In imagining a scene from a different point of view, is it easier to mentally rotate the environment, or to imagine oneself appropriately translated and rotated into the new position?

The evidence suggests that children can better assess the locations of objects from some imagined vantage point when they mentally rotate the environment than when they mentally relocate themselves (e.g., Hardwick, McIntyre, & Pick 1976; Huttenlocher & Presson, 1979). Rather perversely, however, the children in the experiment by Hardwick *et al.* (1976) insisted that they found it easier to mentally relocate themselves than to mentally rotate the room, even though their performance was worse with mental relocation; perhaps they were influenced by the fact that it was *actually* more feasible to move about the room than to *actually* rotate the room itself! Why, then, should it prove more difficult to imagine oneself relocated, even when this is physically the more practicable? Perhaps this is because self-relocation involves the more complex series of mental operations; one must imagine a translation as well as a rotation, and one must imagine moving one's self and changing one's own orientation relative to the larger environment. By contrast, imagining the environment rotate involves just one rotation. Thus Pick and Reiser, in their chapter in this book, note that it is much easier for subjects to imagine a scene from a different location if they actually walk to that location, even if they are blindfolded and wear earphones to prevent them from seeing or hearing the environment. Actual locomotion reorientation may eliminate the need to imagine these components, and so make it easier to focus on the environment itself from the new perspective.

It may be worth noting that images of the self often seem to possess an "out of body" character that may not be conducive to imagining oneself viewing something from a different perspective. This disembodied experience is some-times associated with hallucinations or deathbed visions (e.g., Moody, 1975), but may well be a normal property of self-imagery. For instance ,Siegel (1980) suggests a simple demonstration.

> Recall the last time you went swimming in the ocean. Now ask yourself if this
> memory includes a picture of yourself running along the beach or moving about in
> the water. Such a picture is obviously fictitious, since you could not have been

looking at yourself, but images in the memory often include fleeting pictures of this kind [p. 924].

This objective quality to the self-image may make it difficult to establish the subjective perspective that would be created if one was in a different location.

Interface between Perception and Imagination

Although representational space goes beyond the immediately visible or tangible, it nevertheless underlies perception as well as imagination. Things perceived can be compared with things imagined, or interpreted in a wider spatial context than is immediately accessible to the senses. Some of the interfaces between perception and imagination have been nicely demonstrated in experiments using the reaction-time paradigm.

In their epic study Cooper and Shepard (1973) were especially concerned with the interactions between perceived and imagined *position*. As already explained, if a person must judge a disoriented letter or digit normal or backward, he or she typically mentally rotates it to the upright before making the judgment. But if the subject is told in advance what the character is going to be, and what its orientation will be, the mental-rotation function flattens out, as shown with the labels "B-1000" and "C" in Figure 8.2. The interpretation of this result is that the subject can prepare a mentally rotated representation of the character in advance, and so compare the presented character directly with that rotated representation. Mental rotation is therefore no longer necessary. This result illustrates what is probably one of the most important natural functions of imagery, which is to serve in readiness for an anticipated stimulus (cf. Neisser, 1976).

When Cooper and Shepard's subjects were informed in advance of the identity of each character but not its orientation, or of the orientation but not the identity, the mental-rotation functions did not flatten out. Rather they retained the characteristic peaked shape (labeled "I" and "O," respectively, in Figure 8.2) indicating that the subjects still mentally rotated the characters to the upright to judge whether they were normal or backward. It is of some interest that advance knowledge of the orientation of a character does not eliminate the necessity to mentally rotate it unless one also knows its identity. This means that subjects apparently cannot mentally rotate an abstract frame or references. Yet, as noted earlier, they apparently can adjust their frames of reference to compensate for head tilt (Corballis *et al.*, 1976; Corballis, Nagourney, Shetzer, & Stefanatos, 1978). It seems likely that the frame of reference in visual imagery is determined primarily by automatic processes, and that there is little voluntary control over it.

The interface between imagined and perceived patterns was even more

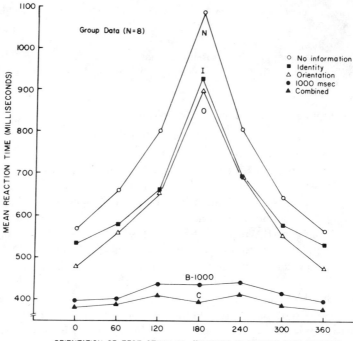

Figure 8.2.   Reaction time to decide whether alphanumeric characters are normal or backward as a function of angular orientation. The different plots represent different kinds of advance information about the identity and/or the orientation of the test character. With no information (N), information about identity only (I), or about orientation only (O), the data show the characteristic mental-rotation function. With advance information about both identity and orientation (B-1000 and C), mental rotation is no longer required. (From Cooper & Shepard, 1973.)

strikingly demonstrated in a further experiment in which Cooper and Shepard instructed subjects to imagine a letter in successively rotated orientations. Periodic instructions gave the required orientations in 60 degree steps: "up," "tip," "tip," "down," "tip," "tip," "up," and so on. The character was then actually presented coincident in time with one of the imagined orientations, and the time taken to judge it normal or backward was a function of the angular deviation of the actual character from the *imagined* one. That is, the imagined stimulus now served as the standard against which the actual character was matched. Because of the striking interaction between perception of an actual pattern and an imagined pattern, Shepard (1978) has inferred that there is a "second-order isomorphism" between image and percept. At some level, image and percept must share common representational elements; indeed, Finke (1980) has argued that mental imagery can result in activation of perceptual mechanisms in the brain even at the very lowest stages of perceptual processing.

These elegant experiments nicely demonstrate an isomorphism between the perceived and imagined *positions* of letters and digits. However they do not necessarily imply any interface between perceived and imagined *movement* per se; Cooper and Shepard (1973) concede that in the experiment just described the subjects may have generated the successive positions of the image in discrete steps. Yet it is often claimed that mental rotation is smooth, that the image passes through all intermediate stages between one imagined position and the next (e.g., Attneave, 1974; Shepard, 1978), although this claim is essentially no more than an inference based on the smooth (but not always linear) increase in reaction time with increasing angle of rotation.

Two graduate students, Sharon Cullen and Rachael McLaren, and I have tried to demonstrate a more direct interface between perceived and imagined *rotation*. Our approach was to induce a rotation aftereffect on the pattern to be mentally rotated to determine whether the perceived rotation due to the aftereffect would influence the mental-rotation function. The rotation aftereffect provides precisely the perceptual impression we sought, since earlier evidence shows that a pattern subject to a motion aftereffect is seen paradoxically to move without changing its position (Wohlgemuth, 1911). Any influence on mental rotation should therefore be due to the perception of motion *per se* and not to perceived changes in orientation. We used the mental-rotation technique described by Cooper and Shepard (1973); that is, subjects were shown letters (*F, G,* or *R*) in different angular orientations and timed as they decided whether each one was normal or backward (i.e., mirror-reversed). Before each letter appeared, the subjects watched a textured disk rotating at 25 rpm for 9.5 sec. After practice trials, the subjects had 72 trials with the disk rotating in one direction, and then another 72 with the disk rotating in the other direction. They watched the disk for a minute or two before the trials began, and more or less continuously between trials. Twelve subjects were tested with 2-sec exposure of each letter, and 12 further subjects saw each letter exposed for 90 msec followed by a dark field for 1.91 sec.

The mean decision times are plotted as a function of angular orientation for each direction of aftereffect in Figure 8.3. Notice that the function was affected by the aftereffect, and indeed was generally displaced in the direction of the aftereffect. We have thought of three possible hypotheses to explain this. First, the aftereffect may summate with the mental rotation so that mental rotation is faster when it is in the same direction as the aftereffect and slower when it is in the opposite direction. Thus decision times would be increased on one arm of the mental-rotation function and decreased on the other. Second, the rate of mental rotation might remain constant, but subjects might be induced to mentally rotate through the larger angle to the upright rather than

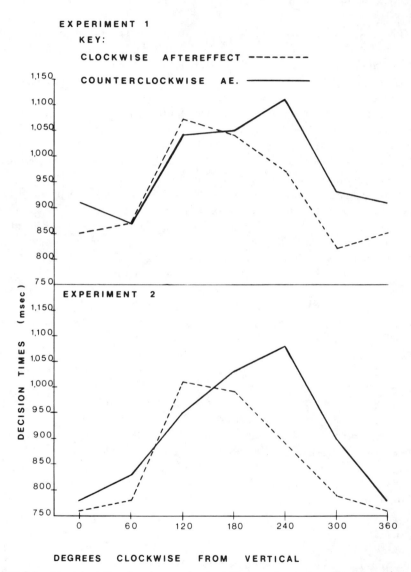

Figure 8.3. Mental-rotation functions under the influence of a motion aftereffect on the test letter. In Experiment 1, the stimulus duration was 2 sec, whereas in Experiment 2 it was 90 msec.

through the smaller angle when the aftereffect is in the direction of the larger angle. Discussions with subjects suggested that this was likely only when the letter was oriented at 120 degrees from the upright in the same direction as the aftereffect, so that they had a tendency to rotate through the 240 degree angle. According to this hypothesis, the mental-rotation functions should in influenced at only one point, the 120 degree point, where decision times should be lengthened.

The third hypothesis is that subjects actually perceived the characters as shifted in orientation in the direction of the aftereffect, implying a lateral shift in the whole mental-rotation function. Although this hypothesis runs counter to the doctrine that the motion aftereffect is a pure motion effect, leaving perceived position unaltered, there seems no other obvious explanation for why, when the letters were exposed for 2 sec (top panel of Figure 8.3), the *minima* as well as the maxima of the mental-rotation functions were shifted 60 degrees against the aftereffect. When the duration was cut back to 90 msec (bottom panel of Figure 8.3), however, the minima remained at 0 degrees, whereas the maxima were shifted. Reducing the duration may have attenuated or eliminated any changes in perceived orientation, so that the effect was then due primarily to the perceived motion. Overall, the second hypothesis fitted most closely to the data: Subjects quite often mentally rotated the characters at 120 degrees though the larger 240 degree angle back to the upright if the aftereffect is in that direction. Some subjects spontaneously suggested that this is what they did. But whatever the precise nature of the effect, it does suggest that there is an interface between perceived and imagined rotation, an interface that is not attributable to changes in perceived position.

In summary, then, experiments show that mental rotation does appear to take place in real time, that imagining a shape in a different orientation is in some respects like seeing it in that orientation, and that mental rotation interacts with perceived rotation. By and large, these empirical results confirm the subjective impression that mental rotation is a smooth, holistic process, in some respects analogous to physical rotation. Indeed, a good deal of the fascination with the mental-rotation paradigm has been generated by the claim that mental rotation is fundamentally an analogue process rather then, say, a propositional or a digital one (e.g., Attneave, 1974; Cooper & Shepard, 1973; Kosslyn & Pomerantz, 1977; Paivio, 1976). This claim is highly contentious, if only because it seems to threaten the widely held belief that human cognition can be modeled in terms of the digital computer. I do not think that this is a real threat, although close study of the phenomena of mental rotation and other properties of mental imagery may force changes in the kinds of models we construct; in particular they may reveal the limitations of the network theories of memory and cognition that were prominent in the 1970s (e.g., Anderson & Bower, 1973; Norman, Rumelhart, & LNR Research Group, 1975).

There are really two parts to the claim that mental rotation is analogous. One part is that images themselves are in some sense analogues of the shapes or objects they represent, more like pictures, say, than sets of propositions or descriptions (Anderson, 1978). The second part is that the process of mental rotation is itself an analogue of physical rotation, in that it is smooth and takes place in real time. I now considered each of these issues in turn.

Are Images Analogue or Propositional?

In specifying what is meant by analogue and propositional representations, it is easier to begin with the latter. A *propositional* representation consists of one or more propositions, and there are in turn three characteristics that define a proposition (Anderson, 1978). First, it is abstract in that it does not depend on the specific features of its instantiation. For instance a proposition may be denoted by a sentence, but remains the same whether the sentence is spoken, written, translated into different languages, or paraphrased. A proposition is thus more abstract than a sentence. Second, it has a truth value in that it makes sense to ask whether a proposition is true or false. Third, it has rules of formation that determine whether or not it is well formed. These rules are reflected in the rules that determine whether a sentence is properly constructed, or in the rules that govern the formation of a well-formed formula in mathematics or logic. In terms of these criteria, then, we can say that there is an implicit propositional system underlying natural language, or mathematic expressions, or computer programs, or even network theories of meaning such as that of Quillian (1969).

An *analogue* representation[2] does not possess these characteristics. It implies instead an isomophism, or one-to-one mapping, between the representation and the thing represented, as a map represents a terrain or a photograph represents a visual scene. The representation is thus not abstract and there is no sense of truth value, at least in terms of the dichotomy between true and false, although a given representation might be more or less faithful or detailed. An analogue representation need not imply a lack of processing. For instance objects may be represented in terms of outlines or line drawings, implying selective extraction of information at boundaries or at other discontinuities in the distribution of light over the surface. An analogue representation might be selective in more subtle ways; for instance, a caricature might be considered an analogue representation, even though it is a simplified and distorted version of the original.

---

[2]Some authors have eschewed the term *analogue*. For instance, Anderson (1978) writes of "pictorial" representations and Kosslyn and Pomerantz (1977) of "imagery" representations. For present purposes, there are no important distinctions among these terms.

An analogue representation also implies a holistic, templatelike character. The different properties of the thing represented are not listed separately, as in a representation consisting of a set of propositions, but are represented in global fashion. Some authors have taken this to mean that operations on an analogue representation would also take place in a global rather than a piecemeal manner. Although this is not necessarily the case, it has provided one of the bases of attempts to distinguish empirically between analogue and propositional representations (e.g., Cooper & Podgorny, 1976; Pylyshyn, 1979a). In particular, it has been suggested that if a representation is propositional, then the greater the complexity of the pattern to be rotated the slower the mental rotation, since there would be more propositions to adjust in the act of mental rotation. However, if a representation is analogue, there is no reason to expect the rate of mental rotation to depend on complexity, since the pattern would be rotated as a whole.

Cooper and Podgorny (1976) found that the rate at which subjects mentally rotated random polygons was not influenced by the number of points in the polygons, which ranged from 6 to 24. This result seems to rule out one kind of propositional theory: Suppose the representation of a polygon includes a listing of the coordinates of its points each of which must be transformed separately and serially in the act of mental rotation. One would then expect the rate of mental rotation to increase with the number of points. That it did not might be taken as evidence for a more holistic process of mental rotation, and thus a more holistic representation.

Cooper and Podgorny's result does not rule out all propositional theories, however. For instance, the different propositions required to represent a pattern might be rotated in parallel. Alternatively, it is possible, as Anderson (1978) has shown, to represent a pattern as a set of propositions only one of which represents the angular orientation of the pattern. No matter how many propositions are required, then, only one needs to be transformed in the process of mental rotation. Indeed, one might say that a propositional theory of this kind is more plausible than an analogue theory, which would require that some spatial representation in the brain itself be subject to a global transformation representing rotation. It is not easy to conceptualize in neurophysiological terms how this might be done. Computationally, and in terms of what is known of coding in the visual system, it seems much more likely that orientation is extracted as a single proposition and subject to a simple transformation in perceived or imagined rotation.

In any event, the rate of mental rotation does not appear to be entirely free of variations in the patterns to be rotated, and there is some evidence that it is not always holistic. For instance, when Shepard and Metzler (1971) timed subjects as they mentally rotated one angular, boxlike shape to match it to another, the estimated rate of mental rotation was only about 60 degrees per second. In marked contrast, studies of the mental rotation of letters and digits typically yield average estimates of about 400 degrees per se (e.g., Cooper & Shepard, 1973; Corballis *et al.,* 1976). It is not entirely clear what explains this striking difference, although it seems likely from studies of eye movements during the mental rotation of the sorts of shapes used by Shepard and Metzler that the subjects mentally rotated the shapes in parts rather than holistically (Just & Carpenter, 1976). Kail *et al.* (1980) have suggested that children might tend to mentally rotate letters or digits in parts, which would explain why their rate of mental rotation is slower than that of adults. These authors also found that unfamiliar shapes were mentally rotated more slowly than letters or digits, perhaps again because there is a greater tendency to mentally rotate unfamiliar shapes in piecemeal fashion.

Pylyshyn (1979a) has shown that the rate of mental rotation is influenced by properties of the pattern to be rotated, as well as by practice and by the difficulty of the task. In one task he required subjects to mentally rotate part of a pattern to match it to a whole pattern, and found that mental rotation was faster when the part was a "good" rather than a "poor" subfigure of the whole. Other observations suggest that the physical properties of the objects to be imaged, for example, size, inertia, viscosity of the surroundings, do not affect the rate of mental rotation. In particular, Robyn Brannigan (unpublished) has shown that the projected size of a letter does not influence the rate at which it is mentally rotated, although projected size has been shown to be important in other aspects of image-based performance (Hayes, 1973). On these grounds, Pylyshyn (1979a) has argued that the evidence is more consistent with a propositional than with an analogue theory of imagery. In particular, some of the data suggest that images may be manipulated in piecemeal rather than holistic fashion, even though we still do not have a very coherent notion of the rules governing the decomposition of images, and thus of their propositional format.

Yet it is one thing to show that images may be decomposed, but quite another to prove that they are fundamentally propositional in form. Anderson (1978) has argued that analogue and propositional theories cannot be distinguished on the basis of behavioral evidence. For any analogue interpretation of a given psychological result, one can always construct a propositional account that is equally fitting, and vice versa. For instance, demonstrations that the rate of mental rotation is subject to psychological manipulation, or even that mental rotation can be piecemeal rather than holistic, can be attributed to properties of the mental-rotation *process* rather than of the image itself. That is, the process itself might operate on parts of a holistic image, and rotate each part in turn (cf. Kosslyn & Shwartz, 1977). In general, one can never finally ease apart the properties of a representation from the properties of the process that acts upon the representation, at least with psychological techniques, because any given psychological result must involve both representation and process.

For all that, Anderson argues that we should not abandon theory-building, whether based on analogue or propositional principles. It is not uncommon in science to find quite different models with equal explanatory power, as in the wave and particle models of light, yet the fundamental ambiguity inherent in this state of affairs need not impede the development of knowledge within *either* framework. Pylyshyn (1979a, 1979b) has argued that analogue models are "computationally opaque," or inaccessible to further articulation, like the wired-in functions of a computer. In this he appears to neglect the considerable advances that have been made in the articulation of analogue models (e.g., Kosslyn & Shwartz, 1977). The point is that further elaboration of the properties of imagery will no doubt be possible within either a propositional or analogue framework, and there may be no reason to suppose that either is in any sense more fundamental than the other.

Is Mental Rotation an Analogue Process?

Shepard (1978) appears to have accepted that experiments on mental imagery do not resolve the issue of whether images are analogue or propositional in representational format, but he argues that the mental rotation *process* is analogue, at least in the sense that the representation at each point in time during the process should correspond to the image of that object at each intermediate orientation. Shepard contrasts this with a computational process of rotation in which intermediate states would be psychologically meaningless. For instance, a pattern might be represented as a set of coordinates that could then be rotated by multiplying by a matrix, but stopping that multiplication at an intermediate stage would not in general yield a representation that is meaningful.

The evidence that mental rotation represents the smooth, continuous transition from one orientation to another remains somewhat indirect. Certainly, the *impression* is one of continuous motion. For instance, Shepard and Judd (1976) presented perspective views of the same three-dimensional object in two orientations, presented in alternation. This produced an illusion of rigid rotation, suggesting that the subjects "filled in" the interval between successive presentations with mental rotation. Moreover the minimum duration between alternations for the illusion to occur increased linearly with the angular difference between the orientations, and this linear increase possessed the same slope when the orientations differed in depth as when they differed in the plane of the pictures. These properties also characterize mental rotation as measured by the reaction-time paradigm (Shepard & Metzler, 1971), and reinforce the interpretation that the rotation illusion is mediated by mental rotation.

The very fact that mental rotation takes place in real time is further evidence that it is an analogue process. Just as it takes longer to actually rotate a

pattern the larger the angle of rotation, so it takes longer to *mentally* rotate a
pattern the larger the angle. There is no reason to expect this correspondence if
mental rotation from one orientation to another is achieved by a single matrix
operation. At best, one might argue that mental rotation is necessarily achieved
by a series of small rotations, each requiring a separate matrix operation.

To assert that mental rotation is an analogue process in the sense specified
by Shepard is not to assert that something is actually turning around in the
brain. One could easily achieve the property of smooth or fine-grained rotation
by reducing an image to a set of propositions, extracting a single parameter to
represent the orientation of the image, and incrementing or decrementing that
parameter in a more or less continuous manner. Thus even if one begins with a
propositional theory of the image, it is not necessary to assume that mental
rotation requires a complex operation such as multiplication by a matrix. By
describing a pattern in polar rather than Cartesian coordinates, the act of
mental rotation can be modeled in quite simple terms, and the time-bound
nature of mental rotation is easily encompassed by assuming that changes in the
orientation parameter can only be accomplished incrementally.

One might ask why mental rotation is time bound, whereas in other respects
imagery can escape the constraints of reality. The answer is that imagination
can never be entirely free, since it must serve an anticipatory role in perception;
the freedom to anticipate absolutely anything would serve no useful purpose,
since it would not be informative. In particular, the nature of time imposes
some absolute constraints on physical movement, and because of their very
absoluteness these constraints are also modeled in imagined movement. Lord
Macaulay once said of John Dryden, English poet and dramatist, "His imagination resembled the wings of an ostrich. It enabled him to run but not to soar."
This was not intended as a compliment, but it may serve to remind us that
imagination is only useful as long as it keeps our feet on the ground.

Mental Rotation and Hemispheric Specialization

In assessing the nature of mental rotation, it is of interest to inquire whether
it is mediated primarily by the left or by the right cerebral hemispheric. Given
that mental rotation seems to be a fundamental component of spatial ability
(e.g., Thurstone, 1938), one might expect it to be fundamentally right hemispheric, given the right hemisphere's well documented superiority in the mediation of visuospatial skills (e.g., De Renzi, 1978). Indeed most, but not all, of
the evidence supports this expectation (cf. Benton, Chapter 11, this volume,
Newcombe, Chapter 10, this volume, and Ratcliff, Chapter 13, this volume).

Cohen (1975) found that right handers judged disoriented letters to be
normal or backward more rapidly when the letters were flashed in the left than
in the right visual half field. Since there was no evidence for a difference

between half fields in the rate of mental rotation, Cohen concluded that mental rotation is always carried out "in the right hemisphere." Letters presented to the left hemisphere must be relayed across the corpus callosum to the right, causing a delay in overall response. In our own laboratory, however, we have been unable to detect *any* systematic differences between fields in the mental rotation of letters. Nevertheless Ratcliff (1979) reported that patients with posterior lesions of the right hemisphere were significantly worse than other groups (including those with left-hemispheric lesions) in judging whether inverted human figures were depicted with black discs over their left hands, or right hands. Since these patients performed normally when the figures were upright, Ratcliff inferred that the deficit was one of mental rotation. Findings of Cohen and of Ratcliff are consistent with other evidence that spatial transformations are subserved primarily by the right cerebral hemisphere in most people (cf. Butters, Barton, & Brody, 1970; Franco & Sperry, 1977).

Surprisingly, however, Ornstein, Johnstone, Herron, and Swencionis (1980) reported greater EEG activation over the *left* than over the right cerebral hemisphere when subjects performed a mental-rotation task, whereas activation was greater over the right hemisphere during performance of other visuospatial tasks. This curious reversal may have had something to do with the fact that Ornstein *et al.* used the mental-rotation task devised by Shepard and Metzler (1971), in which the subjects judged pairs of three-dimensional shapes to be the same or different. Recall that mental rotation in this task is considerably slower and presumably more difficult than in tasks involving the mental rotation of flat patterns such as letters or digits. De Renzi (1978) has summarized other evidence that left-hemisphere involvement in spatial tasks is greater the more complex or sophisticated the task. Moreover, as noted earlier, there is reason to believe that mental rotation in Shepard and Metzler's paradigm may be accomplished in piecemeal rather than holistic fashion, which would impose a sequential aspect on the task, perhaps calling upon the specialized sequencing skills of the left hemisphere (cf. Kimura, 1976; Zangwill, 1976). Mental rotation is in any event both temporal and spatial, and the relative demands of either aspect may call upon the respective resources of left or right hemisphere. It may therefore be misleading to locate mental rotation exclusively or even predominantly in either hemisphere.

As Anderson (1978) has pointed out, studies of the role of hemispheric specialization in tasks involving imagery probably do not bear very directly on the question of whether images are propositional or analogous, despite the popular suggestion that the left hemisphere is fundamentally propositional and the right hemisphere is appositional (Bogen, 1969). For one thing, any right-hemispheric advantage in mental rotation can be attributed to that hemisphere's specialization for spatial representation, just as any left-hemispheric advantage might be attributed to a specialization for sequencing, and there is no need to appeal to any deeper dichotomy (cf. Corballis, 1980). Indeed, Basso, Bisiach, and Luzzatti (1980) have reported a rather striking loss of visual imagery in a right-handed patient with a vascular legion in the *left* occipital and temporal lobes. This patient seemed to have access to propositional knowledge about objects and scenes, but to be unable to picture them, and indeed reported that he no longer dreamed. This result may of course be exceptional, but it does warn against associating analogue representations with the right hemisphere and propositional representations with the left. However, Basso *et al.* do admit the possibility that the patient may have been capable of visual imagery, but may have been unable to describe it due to a functional disconnection between visual and verbal centers; a similar denial of right-hemispheric imagery occurs in patients with section of the corpus callosum (Gazzaniga, 1970).

FUNCTIONS OF MENTAL ROTATION

Mental images probably serve a role in perception by preparing the observer for possible events (cf. Neisser, 1976). The flexibility of imagery is the brain's way of coping with the infinite variety and occasional unexpectedness of real-world scenes and events. Indeed, *every* input is unique, since we never receive exactly the same pattern of stimulation on separate occasions. It is because of this fundamental element of unpredictability that imagery must escape at least some of the constraints of the physical world, and it is precisely this quality that lends imagery its potential for creativity (cf. Shepard, 1978).

Recognizing Disoriented Shapes

We may now raise the question of whether mental rotation serves a specific function in the perception or recognition of disoriented shapes. In his monograph *Perception and Orientation,* Rock (1973) suggested that mental rotation does indeed do so. According to Rock's account, shapes are first interpreted according to their retinal coordinates, but are then corrected for any tilt of the shape relative to the retina, whether due to the tilt of the shape itself or to tilt of the head. This correction, in Rock's view, is accomplished by mental rotation.

There is a logical difficulty here, however. One must presumably identify a shape before mentally rotating it to its upright or "normal" position, for otherwise one would not know what its upright position is. Presumably, one must also know its orientation in order to know which way and how far to mentally rotate it. Remember that Cooper and Shepard (1973) did not measure the time it took subjects to identify disoriented alphanumeric characters; they measured the time it took the subjects to judge the characters to be normal or backward.

It seems logical to conclude that their subjects knew the identity and the orientation of each character before they mentally rotated it.

Our own experiments seem to confirm this conclusion. Latencies to simply identify disoriented characters, or to judge their orientations, do depend to some extent on orientation, but the function does not resemble the steep, single-peaked function indicative of mental rotation (Corballis, Zbrodoff, Shetzer, & Butler, 1978; White, 1980). Even in classifying disoriented characters as letters or digits, subjects do not seem to resort to mental rotation (Corballis & Nagourney, 1978); the function is shown in Figure 8.4. Corballis, Zbrodoff, Shetzer, and Butler (1978) concluded from their experiments that subjects first identify a character, then determine its orientation, and only then would they mentally rotate it to the upright if required to judge whether it is normal or backward. The results also revealed something of a paradox, however, which is illustrated in Figure 8.4: The subjects responded significantly more rapidly to normal than to backward versions regardless of angular orientation. This implies that the difference between normal and backward versions is registered, to some degree at least, prior to mental rotation. Presumably the difference is registered too weakly to serve as a reliable basis for judgment or is for some other reason unaccessible to the judgmental process. In most of our experiments on mental rotation there has been a small minority of subjects who do not seem to mentally rotate the characters in making their judgments, and these subjects may well rely on this earlier, pre-rotational information.

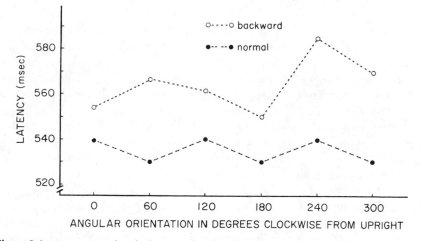

Figure 8.4. Latency to classify disoriented normal and backward alphanumeric characters as letters or digits. Note that the function in no way resembles a mental-rotation function, but paradoxically the function does distinguish normal from backward letters. (From Corballis & Nagourney, 1978. Copyright (1978), Canadian Psychological Association. Reprinted by permission.)

In a limited sense, however, Rock's theory is correct, since one might say that an alphanumeric character is not completely identified until one has determined whether it is normal or backward. In most subjects, at least, this does require mental rotation. Similarly, one might recognize a band as a hand, but it generally requires a mental rotation to determine whether it is a left or a right hand (Cooper & Shepard, 1975). In the natural environment it is not normally important to discriminate shapes from their mirror images; indeed, the same object may often appear in either of two mirror-image profiles, as when a face, for example, may be seen in left or right profile, so that recognition may actually be enhanced if left–right orientation is ignored (see Corballis & Beale, 1976 for more extensive discussion). Left–right orientation is only important in the artificial, asymmetrical world of human beings, and it may therefore be only in that restricted context that mental rotation plays a part in the recognition of shapes. Recognizing words and letters as normal rather than reversed, or a hand as a left or right hand, or an automobile as a left- or right-hand drive— these are activities that normally have significance only to humans.

Mental Rotation and Mirror-Image Discrimination

Nearly all studies of mental rotation have involved mirror images. Why should mental rotation be so critically allied to the discrimination of mirror images? The answer, I think, has to do with the fact (sic) that a bilaterally symmetrical organism cannot discriminate mirror images (Corballis & Beale, 1976), and indeed cannot perform the majority of mental-rotation tasks so far devised no matter how sophisticated its capacity for mental transformation. There is no way, for instance, a bilaterally symmetrical organism can label a letter such as R as distinct from its mirror image, Я . Despite our handedness and cerebral asymmetry, we human beings remain strikingly symmetrical, and the tendency toward mirror-image equivalence in the recognition of patterns is strong. When it comes to actually discriminating mirror images, then, we do so by aligning each pattern with our own bodily coordinates to refer opposite senses of the mirror-image relation to our own left and right sides. For instance, we can only understand which is the left or right hand of another person by mapping that person's bodily coordinates onto our own. Similarly, in order to determine whether a letter, such as $R,$ is normal or backward, the letter must be aligned with some internal representation of a normal, upright $R,$ and the left–right orientation of that internal representation must itself be dependent upon reference to an internal sense of which is left and which is right. That is, left and right are fundamentally egocentric, and can only be understood with reference to the left and right sides of one's own body. Mirror-image discrimination thus depends on alignment with one's own left–right axis, and on our ability to use this axis as a basis for classification. The task used by

Shepard and Metzler (1971) does not require true mirror-image discrimination, but because the patterns are in fact either the same or mirror images of one another, the problem is solved by aligning one pattern with the other and directly judging sameness or differentness.

These processes of alignment, whether the pattern is aligned with one's own bodily coordinates or with another pattern, can be accomplished mentally. I have already discussed evidence that the act of imagining an object in a different orientation is equivalent at some level to actually seeing it in that orientation. It is also pertinent to note here a remarkable report by Bisiach and Luzzatti (1978) showing that unilateral neglect following right-hemispheric brain injury applies to imagined scenes as well as to perceived ones. Two patients asked to imagine the Piazza del Duomo in Milan and to describe it showed more or less complete neglect of details to the left of their imagined vantage point. When asked to imagine the square from the opposite end, they again neglected details on the left, thus omitting those they had previously described and describing those they had previously omitted. This result seems to confirm that an imagined representation may be physically aligned with the axes of the brain. This need not imply that the representation is completely analogue or picturelike; rather, it suggests that the left–right dimension, however encoded, may depend in direct fashion on the left–right axis of the brain itself.

The critical role of mental rotation in the discrimination of mirror images may create ambiguity in the interpretation of tests designed to assess the ability to tell left from right. Such tests typically require the subject to identify the left and right sides of his or her own body and also of other people or objects. Since identification of the left or right side of another person or object typically requires a mental rotation to align its axes with one's own, performance may depend on the ability to perform mental rotations as well as on the ability to tell left from right. Several studies have suggested, for instance, that most children can discriminate between the left and right parts of their own bodies by the age of about 6 years, but that they continue to improve on items related to the left and right sides of other people or objects up until about 10 years of age (Belmont & Birch, 1963; Piaget, 1928; Swanson & Benton, 1955). This later improvement probably depends primarily on the development of mental rotation, which does not seem to emerge until the age of 7 or 8 (Piaget & Inhelder, 1971), rather than on any development in the basic ability to distinguish left from right.

There is even evidence that these two components have different neural substrates. Following lesions of the left hemisphere, left–right discrimination may be impaired (e.g., Benton, Chapter 11, this volume) without additional problems in mental rotation (Ratcliff, Chapter 13, this volume). Conversely, patients with Turner's syndrome (McGee, Chapter 9, this volume; Newcombe, Chapter 10, this volume) may have problems in mental rotation, as witnessed by their impaired performance on standardized Road-Map Tests of Directional Sense (Money, Alexander, & Walker, 1965), but have no difficulty identifying the left and right sides of their own body or in drawing a line to the left or right according to verbal instructions.

CONCLUSIONS

This review of studies of mental rotation, although by no means exhaustive, has nevertheless taken us into some of the most fundamental topics of cognitive psychology, including the interface between perception and imagery, the nature of internal representations, the mechanisms of shape recognition, and the neurological bases of spatial cognition. I know of no experimental paradigm in mental imagery that has had broader implications or is more compelling.

Yet the vivid subjective quality of mental rotation has also proven to be something of an impediment to understanding, since it has caused some observers to mistake the phenomenon itself for its explanation. Because mental rotation is subjectively smooth and holistic, there is a natural tendency to suppose that it depends on a mechanism that also operates in a smooth, holistic fashion, like a picture turning on a wheel. However, to be privy to internal events is not to be privy to their explanation. Pylyshyn (1973, 1978, 1979a, 1979b) seems to have perceived the error more clearly than most, and to have asserted the legitimacy of a propositional account. At the same time, however, those who have insisted on an analogue account (e.g., Kosslyn & Pomerantz, 1977; Kosslyn & Shwartz, 1977) have developed it to a level of sophistication beyond naive realism, so it is now perhaps debatable which account has the greater utility.

For my own part, I find it difficult to avoid the conclusion that imagery has both propositional and analogous elements. Clearly, there is much about perception itself that is propositional; studies of the neurophysiology of vision, for instance, increasingly emphasize the extraction of propositionlike features encoding shape, orientations, movement, spatial frequency, and color, rather than the formation of picturelike representations in the brain. Yet despite the intricate mechanism for the extraction of local features in the visual system there is still a gross level at which topographic representation of the visual world is maintained. Some of the evidence I have reviewed suggests that even images may be mapped onto brain structures in such a way that the topography of the image matches the topography of the brain itself, at least with respect to the left–right axis. The critical role of mental rotation in the discrimination of mirror images, and to a lesser extent in the detection of symmetry, implies an analogue quality to mapping of left and right, and the phenomenon of hemifield neglect even in the visual image (Bisiach & Luzzatti, 1978) seems to provide even more direct evidence. The neurophysiological evidence of O'Keefe and Nadel (1978) suggests that, in rats at least, the "cognitive map" may be mapped onto the hippocampus in a way that is invariant with respect to the animal's orientation to the environment. This evidence is still controversial (Solomon, 1979), but it is precisely this kind of mapping that is implied by an analogue model of mental rotation.

One way to partially reconcile analogue and propositional accounts of imagery is to suppose that images are *stored* in propositional fashion but "displayed" in analogue format. When we imagine something, the generated image may involve activation of many of the neural elements that would be activated in actual *perception* of the object or event, perhaps at a quite peripheral perceptual level (Finke, 1980). Operations on the image, such as mental rotation, may occur at the level of the analogue display rather than at the level of the propositions used to generate the display (cf. Kosslyn & Shwartz, 1977). Such an account helps explain why the omissions and distortions in images of remembered scenes seem to follow propositional rules (Pylyshyn, 1973), even though the display itself seems in some respects to be mapped in analogue fashion onto the underlying brain structures.

Turning now to questions of function, mental rotation may play an important anticipatory role in perception: Mental rotation has more to do with the "where" than with the "what" of perception; it is involved in such spatial tasks as map reading, giving and receiving spatial instructions, and planning to rearrange furniture, build a house, or landscape a garden. These are all activities that are better described by maps or plans than by words, suggesting that they might be better modeled by an analogue device than by a digital or propositional device. As with the image on a television screen, however, the information that generates these maplike structures may be fundamentally propostional in character.

REFERENCES

Allport, D. A. The state of cognitive psychology. *Quarterly Journal of Experimental Psychology*, 1975, 27, 141–152.

Anderson, J. R. Arguments concerning representations for mental imagery. *Psychological Review*, 1978, 85, 249–277.

Anderson, J. R., & Bower, G. H. *Human associative memory*. Washington, D.C.: Hemisphere Press, 1973.

Attneave, F. How do you know? *American Psychologist*, 1974, 29, 493–499.

Attneave, F., & Farrar, P. The visual world behind the head. *American Journal of Psychology*, 1977, 90, 549–563.

Baddeley, A. D. *The psychology of memory*. New York: Basic Books, 1976.

Basso, A., Bisiach, E., & Luzzatti, C. Loss of mental imagery: A case study. *Neuropsychologia*, 1980, 18, 435–442.

Belmont, L., & Birch, H. G. Lateral dominance and right–left awareness in normal children. *Child Development,* 1963, *34,* 257–270.

Bisiach, E., & Luzzatti, C. Unilateral neglect of representational space. *Cortex,* 1978, *14,* 129–133.

Bogen, J. E. The other side of the brain. II: An appositional mind. *Bulletin of the Los Angeles Neurological Society,* 1969, *34,* 135–162.

Butters, N., Barton, M., & Brody, B. A. Role of the right parietal lobe in the mediation of cross-model associations and reversible operations in space. *Cortex,* 1970, *6,* 174–190.

Carpenter, P. A., & Eisenberg, P. Mental rotation and frame of reference in blind and sighted individuals. *Perception and Psychophysics,* 1978, *23,* 117–124.

Cohen, G. Hemispheric differences in the utilization of advance information. In P. M. A. Rabbitt & S. Dornic (Eds.), *Attention and performance. V.* London: Academic Press, 1975.

Cooper, L. A. Mental transformation of random two-dimensional shapes. *Cognitive Psychology,* 1975, *7,* 20–43.

Cooper, L. A., & Podgorny, P. Mental transformations and visual comparison processes: Effects of complexity and similarity. *Journal of Experimental Psychology: Human Perception and Performance,* 1976, *2,* 503–514.

Cooper, L. A., & Shepard, R. N. Chronometric studies of the rotation of mental images. In W. G. Chase (Ed.), *Visual information processing.* New York: Academic Press, 1973.

Cooper, L. A., & Shepard, R. N. Mental transformation in the identification of left and right hands. *Journal of Experimental Psychology: Human Perception and Performance,* 1975, *1,* 48–56.

Corballis, M. C. Laterality and myth. *American Psychologist,* 1980, *35,* 284–295.

Corballis, M. C., & Beale, I. L. *The psychology of left and right.* Hillsdale, New Jersey: Lawrence Erlbaum Associates, 1976.

Corballis, M. C., & Nagourney, B. A. Latency to categorize disoriented alphanumeric characters as letters or digits. *Canadian Journal of Psychology,* 1978, *23,* 186–188.

Corballis, M. C., Nagourney, B., Shetzer, L. I., & Stefanatos, G. Mental rotation under head tilt: Factors influencing the location of subjective reference frame. *Perception & Psychophysics,* 1978, *24,* 263–273.

Corballis, M. C., Zbrodoff, N. J., & Roldan, C. E. What's up in mental rotation? *Perception & Psychophysics,* 1976, *19,* 525–530.

Corballis, M. C., Zbrodoff, N. J., Shetzer, L. I., & Butler, P. B. Decisions about identity and orientation of rotated letters and digits. *Memory and Cognition,* 1978, *6,* 98–107.

Cornbach, L. J. The two disciplines of scientific psychology. *American Psychologist,* 1957, *12,* 671–684.

De Renzi, E. Hemispheric asymmetry as evidenced by spatial disorders. In M. Kinsbourne (Ed.), *Asymmetrical function of the human brain.* New York: Academic Press, 1978.

Finke, R. A. Levels of equivalence in imagery and perception. *Psychological Review,* 1980, *87,* 113–132.

Franco, L., & Sperry, R. W. Hemisphere lateralization for cognitive processing of geometry. *Neuropsychologia,* 1977, *15,* 107–114.

Gazzaniga, M. S. *The bisected brain.* New York: Appleton–Century–Crofts, 1970.

Gibson, J. J. A theory of direct visual perception. In J. R. Royce & W. W. Rozeboom (Eds.), *Psychology of knowing,* New York: Gordon & Breach, 1972.

Guilford, J. P., & Hoepfner, R. *The analysis of intelligence.* New York: McGraw–Hill, 1971.

Hardwick, D. A., McIntyre, C. W., & Pick, H. Content of manipulation of cognitive maps in children and adults. *Monographs of the Society for Research in Child Development,* 1976, *41*(3, Serial No. 166).

Hayes, J. R. On the function of visual imagery in elementary mathematics. In W. G. Chase (Ed.), *Visual information processing.* New York: Academic Press, 1973.

Huttenlocher, J., & Presson, C. C. The coding and transformation of spatial information. *Cognitive Psychology*, 1979, *11*, 375–394.

Just, M. A., & Carpenter, P. A. Eye fixations and cognitive processes. *Cognitive Psychology*, 1976, *8*, 441–480.

Kail, R., Pellegrino, J., & Carter, P. Developmental changes in mental rotation. *Journal of Experimental Child Psychology*, 1980, *29*, 102–116.

Kimura, D. The neural basis of language qua gesture. In H. Whitaker & H. A. Whitaker (Eds.), *Studies in neurolinguistics* (Vol. 2). New York: Academic Press, 1976.

Kosslyn, S. M., & Pomerantz, J. R. Imagery, propositions, and the form of internal representations. *Cognitive Psychology*, 1977, *9*, 52–76.

Kosslyn, S. M., & Shwartz, S. P. A data-driven simulation of mental imagery. *Cognitive Science*, 1977, *1*, 265–296.

Marmor, G. S. Development of kinetic images: When does the child first represent movement in mental images? *Cognitive Psychology*, 1975, *7*, 548–559.

Marmor, G. S., & Zaback, L. A. Mental rotation by the blind: Does mental rotation depend on visual imagery? *Journal of Experimental Psychology: Human Perception and Performance*, 1976, *2*, 515–521.

Metzler, J., & Shepard, R. N. Transformational studies of the internal representations of three-dimensional objects. In R. L. Solso (Ed.), *Theories of cognitive psychology: The Loyola Symposium*. Hillsdale, New Jersey: Lawrence Erlbaum Associates, 1974.

Money, J., Alexander, D., & Walker, H. T. *A standardized road-map test of directional sense*. Baltimore: Johns Hopkins Press, 1965.

Moody, R. *Life after death*. New York: Bantam/Mockinbird, 1975.

Neisser, U. Cognitive psychology. (Review of *Visual information processing* W. G. Chase, Ed.). *Science*, 1974, *183*, 402–403.

Neisser, U. *Cognition and reality*. San Francisco: Freeman, 1976.

Newell, A. You can't play 20 questions with nature and win. In W. G. Chase (Ed.), *Visual information processing*. New York: Academic Press, 1973.

Norman, D. A., Rumelhart, D. E., & the LNR Research Group. *Explorations in cognition*. San Francisco: Freeman, 1975.

O'Keefe, J., & Nadel, L. *The hippocampus as a cognitive map*. Oxford: Clarendon Press, 1978.

Ornstein, R., Johnstone, J., Herron, J., & Swencionis, C. Differential right hemisphere engagement in visuospatial tasks. *Neurophychologia*, 1980, *18*, 49–64.

Paivio, A. Images, propositions, and knowledge. In J. M. Nicholas (Ed.), *Images, perception, and knowledge*. Dordecht, The Netherlands: Reidel, 1976. (Part of The Western Ontario Series in the Philosophy of Science.)

Piaget, J. *Judgment and reasoning in the child*. London: Routledge & Kegan Paul, 1928.

Piaget, J., & Inhelder, B. *The child's conception of space*. New York: Basic Books, 1956.

Piaget, J., & Inhelder, B. *Mental imagery in the child*. New York: Basic Books, 1971.

Pylyshyn, Z. W. What the mind's eye tells the mind's brain: A critique of mental imagery. *Psychological Bulletin*, 1973, *80*, 1–24.

Pylyshyn, Z. W. Imagery and artificial intelligence. In W. Savage (Ed.), *Perception and cognition: Issues in the foundation of psychology* (Vol. 9 of *The Minnesota studies in the philosophy of science*). Minneapolis: The University Press, 1978.

Pylyshyn, Z. W. The rate of "mental rotation" of images: A test of a holistic analogue hypothesis. *Memory and Cognition*, 1979, *7*, 19–28. (a)

Pylyshyn, Z. W. Validating computational models: A critique of Anderson's indetermincy of representation claim. *Psychological Review*, 1979, *86*, 383–394. (b)

Quillian, M. R. The teachable language comprehender. *Communications of the ACM*, 1969, *12*, 459–476.

Ratcliff, G. Spatial thought, mental rotation, and the right cerebral hemisphere. *Neuropsychologia,* 1979, *17,* 49–54.

Rock, I. *Perception and orientation.* New York: Academic Press, 1973.

Shepard, R. N. The visual image. *American Psychologist,* 1978, *33,* 125–137.

Shepard, R. N., & Judd, S. Perceptual illusion of rotation of 3-dimensional objects. *Science,* 1976, *191,* 952–954.

Shepard, R. N., & Metzler, J. Mental rotation of three-dimensional objects. *Science,* 1971, *171,* 701–703.

Siegel, R. K. The psychology of life after death. *American Psychologist,* 1980, *35,* 911–931.

Solomon, P. R. Temporal versus spatial information processing theories of hippocampal function. *Psychological Bulletin,* 1979, *86,* 1272–1279.

Swanson, R., & Benton, A. L. Some aspects of the genetic development of right–left discrimination. *Child Development,* 1955, *26,* 123–133.

Thurstone, L. L. Primary mental abilities. *Psychometric Monographs,* 1938, *1.*

Tolman, E. C. Cognitive maps in rats and men. *Psychological Review,* 1948, *55,* 189–208.

White, M. J. Naming and categorization of tilted alphanumeric characters do not require mental rotation. *Bulletion of Psychonomic Society,* 1980, *15,* 153–156.

Wohlgemuth, A. On the after-effect of seen movement. *British Journal of Psychology, Monograph Supplement,* 1911, *1,* 1–117.

Zangwill, O. L. Thought and the brain. *British Journal of Psychology,* 1976, *67,* 301–314.

MARK G. McGEE 9

Spatial Abilities:
The Influence of Genetic Factors[1]

SEX DIFFERENCES ON SPATIAL TESTS

In this chapter the empirical evidence for sex differences in spatial abilities is reviewed and evidence for the heritability of spatial abilities is presented. The X-linked recessive gene hypothesis that has served as a tentative explanation for sex differences on spatial tests and for the mode of genetic transmission is critically examined.

Factor analytic studies conducted during the past five decades (for review, see McGee, 1979a) demonstrate the existence of at least two Spatial factors, Visualization and Orientation (cf. Ratcliff, Chapter 13, this volume). Spatial visualization is an ability to manipulate or rotate two- and three-dimensional pictorally presented stimulus objects (cf. Corballis, Chapter 8, this volume), whereas spatial orientation is an ability to remain unconfused by the changing orientation in which a spatial configuration may be presented (McGee, 1979a). Visualization ability enters into abstract reasoning required in solving mathematical problems. Orientation ability enters into tasks requiring geographical sense of direction such as map reading and piloting an airplane through three-dimensional space. A sex difference on tasks requiring these

[1]The preparation of this manuscript was supported in part by PHS grant MH 15442 and by NIMH, NRSA 1-F32-MH08680-D1.

199

SPATIAL ABILITIES
Development and Physiological Foundations

abilities is among the most persistent of individual differences in all the abilities literature (Anastasi, 1958; Bouchard & McGee, 1977; Buffery & Gray, 1972; Garai & Scheinfeld, 1968; Harris, 1978; Maccoby & Jacklin, 1974; McGee, 1977, 1978a, 1979b; McGee, Cozad, & Pate, 1980, 1982; O'Connor, 1943; Sherman, 1967; Smith, 1964; Wittig & Petersen, 1979), contrary to reports by Sherman (1978) and Fairweather (1976).

Consider sex differences on three illustrative spatial tests: Thurstone's Primary Mental Abilities (PMA) Space Test (Thurstone & Thurstone, 1941), the Differential Aptitude Space Relations Test (Bennett, Seashore, & Wesman, 1974), and the Mental Rotations Test (Vandenberg & Kuse, 1978). Sex differences on Thurstone's Space Test and the Kuhlman–Anderson IQ Test have been examined by Hobson (1947) for ninth-grade boys ($N = 222$) and girls ($N = 250$). The most salient result was the margin by which boys exceeded the girls in the Space factor despite the higher IQ of the girls. This difference was redressed in other areas with girls exceeding the boys most consistently in the verbal area, most notably in Word Fluency, but also in Inductive Reasoning and Memory. Herzberg and Lepkin (1954) administered the intermediate form of the PMA to 1049 Pittsburg high school seniors. Numerous sex differences at

TABLE 9.1
Means and Standard Deviations for Men and Women on Two Parts of Four Spatial Tests ($N = 312$)[a]

Variable	Men ($N = 115$)		Women ($N = 197$)		t	p
	M	SD	M	SD		
Mental rotations						
Part 1	9.92	4.42	7.02	4.09	5.70	.001
Part 2	9.14	4.18	6.15	3.45	6.37	.001
Spatial relations						
Part 1	19.00	5.51	15.45	5.66	5.30	.001
Part 2	10.70	5.10	9.68	4.02	1.80	
Chair–window						
Part 1	9.16	2.78	7.02	3.61	5.48	.001
Part 2	7.68	3.42	6.40	3.52	2.97	.01
Identical blocks						
Part 1	6.07	3.40	3.42	2.79	6.97	.001
Part 2	5.73	3.11	4.09	2.65	4.56	.001

[a]Reprinted with permission of authors and publisher from: S. G. Vandenberg and A. R. Kuse, Mental rotations, a group test of three-dimensional spatial visualization. *Perceptual and Motor Skills*, 1978, 47, Table 3, p. 602.

16, 17, and 18 years of age on the primary factors were reported. Differences favoring females in the younger ninth-grade sample, particularly in Verbal Meaning and Inductive Reasoning, disappeared in the later ages. However, girls at various age levels tended to do better than boys in Word Fluency. The most highly significant and consistent sex difference on the PMA appeared in the Space factor, with the difference at each age level favoring boys significant at the .01 level.

Boys and girls have been found to differ in the abilities measured by the Differential Aptitude Tests (DAT). For instance, boys tend to score higher than girls on Space Relations and Mechanical Reasoning, whereas girls score higher than boys on Clerical Speed and Accuracy, Spelling, and Language Usage. As noted by Wesman (1949): "To the extent that competition in any curriculum or vocation comes preponderantly from a single sex, separate sex norms are needed for test score interpretation [p. 227]." Thus, norms for the DAT subtests have been provided (Bennett *et al.*, 1974) for both boys and girls.

A recently developed paper-and-pencil test of spatial visualization, the Mental Rotations Test (Vandenberg & Kuse, 1978), was constructed from the figures used in the chronometric study of Shepard and Metzler (1971). In large samples in which this test has been used (Bouchard & McGee, 1977; DeFries, Ashton, Johnson, Kuse, McClearn, Mi, Rashad, Vandenberg, & Wilson, 1976; DeFries, Johnson, Kuse, McClearn, Polovina, Vandenberg, & Wilson, 1979; Kuse, 1977; McGee *et al.*, 1980; Park, Johnson, DeFries, Ashton, McClearn,

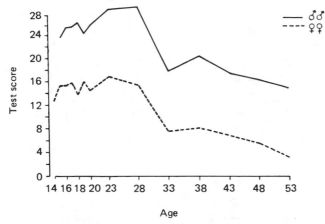

Figure 9.1. Male and female age curves for the Mental Rotations Test. (Reprinted with permission of authors and publisher from: S. G. Vandenberg & A. R. Kuse, Mental rotations, a group test of three-dimensional spatial visualization. *Perceptual and Motor Skills*, 1978, 47, Figure 2, p. 603.)

Mi, Rashad, Vandenberg, & Wilson, 1978; Spuhler, 1976; Vandenberg, 1975; Wilson, DeFries, McClearn, Vandenberg, Johnson, & Rashad, 1975; Yen, 1975) there have been clear indications of substantial internal consistency and test–retest reliability, and consistent sex differences in favor of males over the entire age ranges investigated. Means and standard deviations for the Mental Rotations Test, on which the maximum score is 20, along with those from three additional measures of spatial abilities in a sample of 115 men and 197 women reported by Vandenberg and Kuse (1978) are indicated in Table 9.1. The sex difference in performance on the Mental Rotations Test is consistent over a wide age range in the general population. This is demonstrated by the results shown in Figure 9.1. An analysis of Mental Rotations Test items has been tabulated (McGee, 1977) to determine the extent to which male and female item profiles differ. These results are presented in Figure 9.2. The male–female score profiles based on the percentage of males and females passing each Mental Rotations Test item indicated: (*a*) a consistent sex difference in favor of males for each of the 20 items on the test; and (*b*) a strong positive correlation ($r = .96$) between the rank ordered item difficulties for the males and the

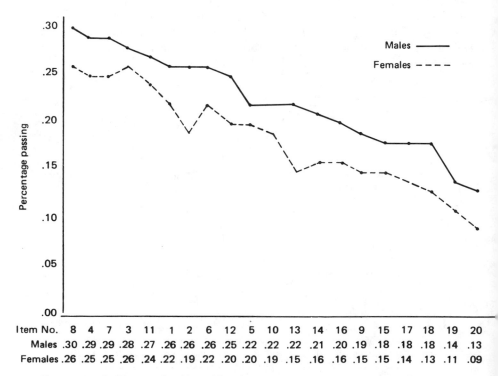

Item No.	8	4	7	3	11	1	2	6	12	5	10	13	14	16	9	15	17	18	19	20
Males	.30	.29	.29	.28	.27	.26	.26	.26	.25	.22	.22	.22	.21	.20	.19	.18	.18	.18	.14	.13
Females	.26	.25	.25	.26	.24	.22	.19	.22	.20	.20	.19	.15	.16	.16	.15	.15	.14	.13	.11	.09

Figure 9.2. Percentage of males and females passing each Mental Rotations Test item. (Compiled by the author, McGee, 1977.)

females. The results would seem to dispel any explanation of the observed sex difference that relies solely on motivational differences between the sexes.

HERITABILITY OF SPATIAL ABILITIES

Accumulated evidence shows that spatial abilities are as heritable as or more heritable than verbal abilities (Ashton, Polovina, & Vandenberg, 1979; Blewett, 1954; Block, 1968; Bock, 1973; DeFries, Vandenberg, McClearn, Kuse, Wilson, Ashton, & Johnson, 1974; DeFries et al., 1976, 1979; McGee, 1978a; Osborne & Gregor, 1968; Park et al., 1978; Vandenberg, 1962, 1967, 1968, 1969, 1971; Vandenberg & Kuse, 1979; Vandenberg, Stafford, & Brown, 1968; Williams, 1975; Wittig, 1979) and much less correlated with traditional measures of environmental quality such as level of education and socioeconomic status (Bock & Vandenberg, 1968; Marjoribanks, 1972, McGee, 1977; Vandenberg, 1971). As an illustration, heritability estimates from four twin studies for six primary mental abilities are shown in Table 9.2. A summary review of heritability results reported by various investigators led Yen (1975) to conclude that the genetic influence appears more direct in orientation than in visualization tests and that test-item dimensionality (i.e., three-dimensional versus two-dimensional items) does not appear to affect heritability.

Evidence has accumulated, more recently, concerning the mode of genetic inheritance and the presence of major-gene effects on spatial abilities. A number of studies have suggested that spatial abilities may be enhanced by an X-linked, recessive gene (Bock & Kolakowski, 1973; Goodenough, Gandini, Olkin, Pizzamiglio, Thayer, & Witkin, 1977; Guttman, 1974; Hartlage, 1970; Stafford, 1961; Walker, Krasnoff, & Peaco, 1981; Yen, 1975). This hypothesis has served as a tentative explanation for the sex difference on spatial tests *and* for the mode of genetic transmission.

TABLE 9.2
Heritability Estimates for Six Abilities from Four Twin Studies[a]

Ability	Blewett (1954)	Thurstone (1955)	Vandenberg (1962)	Vandenberg (1966)
Spatial	.51	.76	.59	.72
Verbal	.68	.64	.62	.43
Word fluency	.64	.59	.61	.55
Number	.07	.34	.61	.56
Reasoning	.64	.26	.28	.09
Memory	—[b]	.39	.20	—

[a] From Vandenberg, 1971. Reprinted with permission.
[b] Not available.

THE POSSIBLE INFLUENCE OF AN X-LINKED MAJOR GENE

The X-Linked Model of Inheritance

Predictions of X-linked inheritance for human characteristics have been presented elsewhere recently (McGee, 1981). Here, a brief review of the basic features of the X-linked inheritance model will be presented and available evidence from family studies that bears on the X-linked recessive hypothesis for human spatial visualizing ability will be critically examined.

Traits, in human beings, affected by the transmission of a single gene on the X chromosome are said to be X-linked and are determined to be either dominant or recessive based upon the relative frequency of affected males and females in the population. If the X-linked trait is recessive more males than females will be affected, whereas if the X-linked trait is dominant more females than males will be affected. This is true because in a population at equilibrium one-third of the X-linked genes are carried by males and two-thirds are carried by females, since females inherit two X chromosomes (one from each parent) and males (XY) inherit only one.

We can designate the recessive enhancing allele as a, and the dominant, nonenhancing allele as A. Assuming complete dominance of A over a, the genotypic values and population proportions of the X-linked recessive gene for males and females will be as shown in Table 9.3.

A recessive, X-linked trait will be expressed in hemizygous recessive males $(X^a Y)$ and homozygous recessive females $(X^a X^a)$, but not in hemizygous dominant males $(X^A Y)$ nor in heterozygous $(X^A X^a)$ or homozygous dominant $(X^A X^A)$ females. Females with the double recessive genotype $(X^a X^a)$ would be expected

TABLE 9.3
Theoretical Population Proportions and Genotypic Values for a Recessive, X-Linked Major Gene[a]

Sex	Genotype	Proportion	Genotypic value	Percentage enhanced $(q = .5)$
Male	X^A	p	0^b	.50
	X^a	q	1^c	.50
Female	$X^A X^A$	p^2	0	.25
	$X^A X^a$	$2pq$	0	.50
	$X^a X^a$	q^2	1	.25

[a] Adapted from Jensen (1975). Reprinted with permission of author and publisher from: A. R. Jensen, a theoretical note on sex linkage and race differences in spatial visualization ability. *Behavior Genetics*, 1975, 5, Table 1, p. 156.
[b] 0 = nonenhancement.
[c] 1 = enhancement.

to occur in the population with a frequency q^2—the square of the frequency q of males (X^aY) carrying the single recessive allele. If the gene frequency of the enhancing allele q were .5 then the proportion of enhanced males to females will be as shown in the last column of Table 9.3. As the value q departs from .5, the absolute sex difference $(q-q^2)$ will decrease (Jensen, 1975). Where the frequency of the recessive enhancing allele q equals either zero or 1.00, the difference between males and females will be zero. With a gene frequency of q = .5, the difference between males and females will be at a maximum with the ratio of enhanced females to males being 1:2. Thus, the observation made by O'Connor (1943) that only one-fourth of all females score above the male median on tests of spatial abilities, a finding replicated by numerous investigators (e.g., Bock & Kolakowski, 1973; Bouchard & McGee, 1977; Loehlin, Sharan, & Jacoby, 1978; Yen, 1975), is in accordance with the X-linked, recessive model.

Since many inherited human traits occur with greater frequency in males than females (e.g., color blindness, Laurence–Moon–Biedl syndrome, allergy, baldness, harelip, and cleft palate), this finding alone is insufficient evidence for X-linked inheritance. It is difficult, in fact, to distinguish X-linked recessive inheritance from sex linked, autosomal dominant inheritance. The pedigree pattern of a family with testicular feminization syndrome (a male-limited, autosomal dominant trait) where reproduction by affected males cannot occur, is explained equally well by either mode of inheritance (McKusick, 1964). Another step in detecting X-linkage is to determine the proportion of enhanced males and females among the children of different types of matings (Csik & Mather, 1938; McKusick, 1964). Table 9.4 shows the theoretical proportion of mating types in a randomly ascertained sample of marital couples and the distribution of phenotypes in their children (based on a gene frequency .5 of the recessive allele which can be derived from the phenotypic frequencies in the male). Specifically, two kinds of sibships would be expected: one having enhanced boys *and* girls, and one having only enhanced boys.

The method of testing this expectation has been applied by Csik and Mather (1938) and is derived in the following manner. First, if the father transmits the spatial enhancing X-linked recessive gene a then both his sons and his daughters could show the trait if the mother was heterozygous (X^AX^a) or homozygous recessive (X^aX^a). The frequency of enhanced daughters, in this case, should be the same as that of enhanced sons since a homozygous (X^aX^a) recessive female and a hemizygous (X^aY) recessive male will have X^aX^a daughters and X^AY sons in equal proportions. Second, if the father transmits the nonenhancing gene A, then all his daughters would be nonenhanced but his sons could be enhanced if the mother was either homozygous recessive (X^aX^a) or heterozygous (X^AX^a). Thus if the trait is X-linked, the expectation is an equal number of enhanced girls and boys among the sibs of enhanced girls, but more enhanced boys than girls among the sibs of enhanced boys.

TABLE 9.4
Expected Distribution of Mating Types in a Randomly Ascertained Series of Couples and the Distribution of Phenotypes in their Children[a,b]

Parents		Proportion of types	Sons		Daughters	
Father	Mother		Dominant (X^AY)	Recessive (X^aY)	Dominant $(X^AX^A$ or $X^AX^a)$	Recessive (X^aX^a)
Dominant phenotype $\times$ (X^AY) p	Dominant phenotype $(X^AX^A$ or $X^AX^a)$ $p^2 + 2pq$	$p^3 + 2p^2q$	$1\text{-}pq$ (75%)	pq (25%)	1 (100%)	0
Dominant phenotype $\times$ (X^AY) p	Recessive phenotype (X^aX^a) q^2	pq^2	0	1 (100%)	1 (100%)	0
Recessive phenotype $\times$ (X^aY) q	Dominant phenotype $(X^AX^A$ or $X^AX^a)$ $p^2 + 2pq$	$p^2q + 2pq^2$	$1\text{-}pq$ (75%)	pq (25%)	$1\text{-}pq$ (75%)	pq (25%)
Recessive phenotype $\times$ (X^aY) q	Recessive phenotype (X^aX^a) q^2	q^3	0	1 (100%)	0	1 (100%)

[a]Adapted from McKusick (1968).
[b]p and q are frequencies of dominant A and recessive a alleles, respectively. Proportions are based upon $q = .5$.

Mendelian analysis of pedigree data, however, has not been attempted. Unreliability of the tests and errors in classifying genotypes makes such an attempt unproductive. Methods are available to correct for errors resulting from the arbitrary identification of thresholds (Elston & Campbell, 1970) but an equivalent quantitative test is provided by examining the fit of the X-linked model to observed intrafamilial phenotypic correlations.

Evidence from Family Studies

The characteristic pattern of intrafamilial correlations for an X-linked trait is distinct from that expected for an autosomal, polygenic trait (Mather & Jinks, 1963). The X-linked recessive model predicts a higher father–daughter than father–son correlation and a higher mother–son than mother–daughter correlation (Hogben, 1932). Opposite sex pairs of siblings will tend to show less similarity than pairs of sisters, while the correlation among pairs of brothers should be an intermediate value. Sisters will be most similar because for brothers the one X chromosome may be either of the mother's two, while for sisters the paternal X chromosome is identical (McKusick, 1964).

Under random mating and a recessive gene frequency of $q = .5$ (the frequency that best explains the mean sex difference and the shape of the male and female score distributions), the expected order of family correlations is as follows:

$$rSis\text{—}Sis \ > \ rMo\text{—}Son \ = \ rFa\text{—}Dau \ > \ rBro\text{—}Bro \ > \ rMo\text{—}Dau \ >$$
$$rBro\text{—}Sis \ > \ rFa\text{—}Son.$$

The theoretically expected correlation, based on X-linkage, of $r = .00$ between fathers and sons is not easily predicted from environmental hypotheses in which modeling effects and shared experiences would ordinarily be expected to lead to higher same-sex than opposite-sex parent–child correlations. Consequently, the original studies supporting the X-linkage hypothesis (Bock & Kolakowski, 1973; Hartlage, 1970; Stafford, 1961) generated considerable interest, despite the fact that they were based on rather small samples and failed to provide correlations among siblings. Age-corrected familial correlations reported in these and several more recent family studies for a variety of spatial tests are summarized in Table 9.5.

Only three studies in the literature (Bouchard & McGee, 1977; Corley, DeFries, Kuse, & Vandenberg, 1980; Loehlin et al., 1978) reported the complete array of family correlations consisting of both parent–child and sibling correlations. In each of these studies, the father–son correlation equals or exceeds the mother–son correlation and therefore it does not conform to the expected pattern. Furthermore, in Bouchard and McGee's (1977) study, the difference between the brother–brother and sister–sister correlation is highly significant ($p < .005$) in the direction opposite that predicted by the X-linkage model. The expected pattern of parent–child correlations (with unlike-sex parent–child correlations highest, the mother–daughter correlation intermediate, and the father–son correlation lowest), although obtained in the initial studies supporting the X-linkage hypothesis, has not been obtained for a variety of spatial tests used in recent, larger family studies. Considered together the results presented in Table 9.5 provide little evidence for the spatial-enhancing effect of the X-linked recessive gene postulated by O'Connor (1943). A similar conclusion has been reached by other reviewers of this literature (c.f. Boles, 1980; Vandenberg & Kuse, 1979).

Research Directions

Failure to find the expected pattern of intrafamilial correlations does not of itself rule out the possibility that spatial abilities are enhanced by an X-linked

TABLE 9.5

Age-Corrected Familial Correlations for a Variety of Spatial Tests[a]

Study	Test	Father–mother	Father–son	Mother–daughter	Mother–son	Father–daughter	Brother–sister	Brother–brother	Sister–sister
Stafford (1961)	Identical Blocks Test	.03(99)	.02(51)	.14(64)	.31(50)	.31(63)	—	—	—
Corah (1965)	Embedded Figures	.14(60)	.18(30)	.02(30)	.31(30)	.28(30)	—	—	—
Hwang (1969)	Progressive Matrices	—	.52(90)	.41(68)	.43(90)	.55(68)	—	—	—
Bock (1970)	Embedded Figures	.18(60)	-.05(25)	.36(26)	.18(26)	.49(22)	—	—	—
Hartlage (1970)	DAT Space Relations	—	.18(25)	.25(25)	.39(25)	.34(25)	—	—	—
Bock and Kolakowski (1973)	Guilford–Zimmerman Spatial Visualization	.26(121)	.15(99)	.12(97)	.20(115)	.25(84)	—	—	—
Guttman (1974)	Progressive Matrices	.26(100)	.36(89)	.39(119)	.24(89)	.23(119)	—	—	—
Williams (1975)	WAIS Block Design	.19(55)	.28(55)	—	.05(55)	—	—	—	—
Yen (1975)	Paper Folding	—	—	—	—	—	.25(192)	.35(86)	.48(102)
	Paper Form Board	—	—	—	—	—	.29(187)	.51(84)	.68(100)
	PMA Space Relations	—	—	—	—	—	.30(197)	.07(83)	.46(104)
	Mental Rotations	—	—	—	—	—	.27(191)	.32(84)	.41(103)
Spuhler (1976)[b]	Mental Rotations	-.07(104)	.25(81)	.04(81)	.10(81)	.32(81)	—	—	—
	Paper Form Board	.08(104)	.37(81)	.24(81)	.12(81)	.35(81)	—	—	—
	Hidden Patterns	.15(104)	.29(81)	.35(81)	.29(81)	.34(81)	—	—	—
	Card Rotations	.02(104)	.25(81)	.03(81)	.16(81)	.15(81)	—	—	—
	Progressive Matrices	.22(104)	.22(81)	.47(81)	.14(81)	.26(81)	—	—	—
Bouchard & McGee (1977)	Mental Rotations	.06(144)	.23(185)	.16(196)	.20(204)	.17(172)	.33(249)	.50(132)	.21(112)
Carter–Saltzman (1977)	WAIS Block Design	.15(96)	.33(95)	.17(119)	.37(97)	.38(117)	—	—	—
Kuse (1977)	WAIS Block Design	.07(118)	.26(80)	.18(81)	.24(80)	.06(81)	—	—	—
Loehlin et al. (1978)	Cube Comparisons	.01(192)	.16(183)	.19(201)	.04(183)	.17(201)	.32(99)	.43(42)	.14(51)
	Card Rotations	.28(192)	.27(183)	.40(201)	.27(183)	.32(201)	.24(99)	.44(42)	.52(51)
	Hidden Patterns	.21(192)	.40(183)	.22(201)	.44(183)	.38(201)	.39(99)	.76(42)	.55(51)
	Paper Folding	.09(192)	.27(183)	.21(201)	.24(183)	.30(201)	.17(99)	.44(42)	.44(51)
McGee (1978)	Hidden Patterns	.26(144)	.07(185)	.35(196)	.23(204)	.39(172)	.11(249)	-.03(132)	.20(112)
Park et al. (1978)	Mental Rotations	—	.22(99–103)	.46(113–121)	.26(100–105)	.41(107–117)	—	—	—

208

Study / Test									
DeFries et al. (1979)[c]									
Paper Form Board		—	—	.59(99–103)	.57(113–121)	.63(100–105)	.53(107–117)	—	—
Card Rotations		—	—	.12(99–103)	.61(113–121)	.36(100–105)	.54(107–117)	—	—
Progressive Matrices		—	—	.33(99–103)	.51(113–121)	.25(100–105)	.39(107–117)	—	—
Hidden Patterns		—	—	.51(99–103)	.65(113–121)	.58(100–105)	.56(107–117)	—	—
Mental Rotations	AEA	—	—	.20(672)	.30(692)	.13(666)	.20(685)	—	—
	AJA	—	—	.20(241)	.11(248)	.17(244)	.24(237)	—	—
Card Rotations	AEA	—	—	.26(672)	.34(692)	.21(666)	.23(685)	—	—
	AJA	—	—	.24(241)	.17(248)	.10(244)	.11(237)	—	—
Hidden Patterns	AEA	—	—	.24(672)	.32(692)	.24(666)	.27(685)	—	—
	AJA	—	—	.09(241)	.19(248)	.13(244)	.22(237)	—	—
Paper Form Board	AEA	—	—	.28(672)	.33(692)	.29(666)	.35(685)	—	—
	AJA	—	—	.21(241)	.29(248)	.20(244)	.27(237)	—	—
Progressive Matrices	AEA	—	—	.23(672)	.25(692)	.32(666)	.25(685)	—	—
	AJA	—	—	.09(241)	.25(248)	.24(244)	.20(237)	—	—
Guttman and Shoham (1979)									
Hidden Patterns		.25(261)	.24(—)	.18(—)	.17(—)	.19(—)	—	—	—
Hidden Blocks		.07(261)	.28(—)	.12(—)	.17(—)	.13(—)	—	—	—
Identical Blocks		.27(261)	.24(—)	.25(—)	.22(—)	.25(—)	—	—	—
Mental Rotations		.00(261)	.32(—)	.21(—)	.12(—)	.08(—)	—	—	—
PMA Space Test		.29(261)	.36(—)	.26(—)	.27(—)	.17(—)	—	—	—
DAT Space Relations		.20(261)	.34(—)	.22(—)	.23(—)	.25(—)	—	—	—
Corley et al. (1980)									
Identical Blocks Test	AEA	.01(171)	.35(126)	.15(122)	.14(128)	.13(121)	.07(77)	.32(50)	.11(43)
	AJA	.11(98)	.25(69)	.28(60)	.10(69)	−.03(60)	.44(30)	.01(18)	.15(14)
Walker et al. (1981)									
Identical Blocks Test		—	−.02(30–34)	.19(30–34)	.29(30–34)	.29(30–34)	—	—	—
DAT: Spatial Relations Test		—	.07(30–34)	.16(30–34)	.45(30–34)	.23(30–34)	—	—	—
Expected test correlations with X-linked recessive gene frequency of .5		.00	.00	.33	.58	.58	.13	.50	.67

[a] Correlations are corrected for age effects. The number of pairs entering into each correlation is shown in parentheses.

[b] Reported by Vandenberg and Kuse (1979).

[c] Results reported separately for Americans of European ancestry (AEA) and Americans of Japanese ancestry (AJA).

major gene. In addition to examining intrafamilial correlations, a quantitative analysis of score distributions for spatial tests enables us to estimate the magnitude of influence, if any, of the postulated major gene. Segregation analysis permits a comparison between observed within-sex phenotypic score distributions and those expected from either a single-gene or polygene model; thus, estimates of the proportion of individuals within each genotype as well as within genotype means and variances may be calculated. The assumption made in calculating these parameters requires that the within-sex score distributions be normally distributed with unequal means and equal variances. Research into the genetics of spatial abilities has turned in this direction. A few studies reporting segregation analyses are available.

Bock and Kolakowski (1973) estimated scaled spatial ability scores for males ($N = 380$) and females ($N = 347$) using the normal-ogive, latent trait model proposed by Lord and Novick (1968). A metric analysis of these distributions was made using the maximum likelihood method of Day (1969). The sex difference was highly significant ($p < .001$). Additionally, scores for males were distributed bimodally, as predicted by the model, with an antimode near the fiftieth percentile. The observed proportion of enhanced females, .20, was found to be close to that expected, $q^2 = .25$ (Bock & Kolakowski, 1973).

The recessive gene frequency estimate has been found to vary, however, depending upon the type of spatial test. For example, Yen (1975) derived maximum likelihood estimates from male and female score distributions for four spatial tests. Although sex differences on all four tests were statistically significant ($p < .001$), the gene and genotype frequency estimates varied considerably across tests. The resolution of the score distribution for the Mental Rotations Test was in closest agreement with the X-linked model, assuming a gene frequency of $q = .5$. The observed proportion of high-ability females (.31), although higher than that expected ($q^2 = .25$), was explained by Yen (1975) as resulting from incomplete dominance in heterozygous females.

More recently, Ashton *et al.* (1979), using the method described by MacLean, Morton, and Lew (1974), reported distributions for the Mental Rotations Test very similar to those of Bock and Kolakowski (1973). However, their evidence suggested that Mental Rotations Test performance was controlled by a dominant gene in both males and females, not a recessive gene. Detection of a major gene effect was found for two tests in addition to the Mental Rotations Test (the Hidden Patterns Test and the Progressive Matrices Test), although not for three other tests (the Paper Form Board, the Card Rotations, and the Lines and Dots Tests) which also contributed significantly to the Spatial factor (Ashton *et al.*, 1979). Discouraged by failure to replicate findings that support the X-linked recessive hypothesis, some researchers have suggested alternative explanations for the observed sex-related difference in spatial abilities.

POSSIBLE INFLUENCES OTHER THAN
A MAJOR GENE, A SEX-LINKED GENE, OR BOTH

Hormonal Influences

There are three lines of evidence suggesting the influence of hormones on sex differences in spatial abilities: (1) developmental studies; (2) studies of persons with chromosomal–hormonal abnormalities; and (3) studies investigating the relationship between physical–somatic androgenicity and spatial abilities.

A sex difference in spatial abilities that is observed so reliably in adolescents and adults fails to appear reliably in prepubescent children (Drew, 1944; Emmett, 1949; Fruchter, 1954; Gardner, Jackson, & Messick, 1960; Maccoby & Jacklin, 1974). Also, the widely documented sex difference in perceptual-cognitive tasks, such as field dependence, does not reliably appear until age 9 or 10 (Witkin, Lewis, Hertzman, Machover, Meissner, & Wapner, 1954). The onset of observed sex-related cognitive differences thus appears to coincide with hormonal changes that occur during puberty (Rifkind, Kulin, Rayford, Cargile, & Ross, 1970; Tanner, 1969).

The hypothesis that hormone balance may be implicated insofar as any explanation for sex differences in spatial abilities is concerned derives mainly from studies of individuals with various chromosomal abnormalities, particularly patients with Turner Syndrome. This syndrome is associated with abnormalities of the sex chromosome pair. About half of all females with Turner Syndrome have a single X chromosome (XO), rather than the normal pair (XX) for females (Mittwoch, 1967). One reason for interest in persons with Turner Syndrome is that, although such persons are of normal intelligence and distributed throughout the intellectual range (Garron, 1977), they appear to have a characteristic pattern of abilities. Verbal abilities are normal, whereas spatial abilities are impaired (Alexander, Ehrhardt, & Money, 1966; Garron & Vander Stoep, 1969; Pennington, Bender, Puck, Salbenblatt, & Robinson, in press). This impairment was first described by Shaffer (1962) and is revealed by particularly low scores on numerous measures including the Mental Rotations Test (Rovet & Netley, 1980), the Block Design subtest of the Wechsler Adult Intelligence Scale (Cohen, 1957, 1959), the Spatial subtest of the Primary Mental Abilities Test (Money, Alexander, & Walker, 1965), the Road and Map Test of Direction Sense (Alexander, Walker, & Money, 1964), and tests requiring imaginary locomotions, rotation, and direction of the body in space (Alexander & Money, 1966). The spatial deficit in Turner Syndrome women is paradoxical and contradicts the hypothesis that spatial abilities are transmitted by an X-linked recessive gene. The expression of X-linked recessive gene traits in Turner Syndrome women ought to be similar to the expression of these traits in normal males, who also have only one X chromosome. However, as noted by Garron (1970), the impaired spatial abilities of the former do not necessarily invalidate the X-linkage hypothesis, since it is possible that this specific impairment of abilities is secondary, resulting from hormonal abnormalities associated with the syndrome. It may be, for instance, as suggested by Garron (1970), that a minimum level of androgen, specifically testosterone, is required for normal expression of spatial abilities.

Studies investigating the relationship between physical–somatic androgenicity and spatial abilities in normal subjects have been reviewed elsewhere (e.g., McGee, 1979b; Petersen, 1979; Reinisch, Gandelman, & Spiegel, 1979). High body androgenization seems to be associated with *low* spatial test scores among males (Klaiber, Broverman, & Kobayashi, 1967; Maccoby, 1966; Petersen, 1976), and with *high* spatial scores among females (Broverman, Klaiber, Kobayashi, & Vogel, 1968; Petersen, 1976). It might be that spatial abilities are facilitated not by any absolute level of androgen, but rather by an optimal estrogen–androgen balance. It follows from results found in the few studies done in this area that the estrogen–androgen balance would be optimal, and consequently spatial abilities would be highest for males low in androgen and for females high in androgen. Thus, for females, the more androgen one has the better as far as spatial abilities are concerned. This hypothesis if supported in future studies would explain why individuals with Turner Syndrome (phenotypic females, the majority of whom have a single X chromosome and no gonadal hormones) demonstrate poorer spatial abilities (Alexander *et al.,* 1966; Alexander & Money, 1966; Alexander *et al.,* 1964; Garron, 1970, 1977; Money, 1963; Money & Granoff, 1965; Serra, Pizzamiglio, Boari, & Spera, 1978; Silbert, Wolff, & Lilienthal, 1977), poorer direction sense (Alexander *et al.,* 1964), and greater field dependence (Serra *et al.,* 1978) than males who also have a single X chromosome and than genetically normal females.

There is need for further research to clarify the relationship between somatic androgenization, hormone balance, and the evidence reviewed above for an X-linked, major gene influence on spatial skill enhancement. Measures of body androgenization used by Petersen (1976) and by Broverman *et al.* (1968), obtained by rating photographs of physical characteristics including muscle development, body shape, genital or breast size, and pubic hair distribution, are crude and imprecise indices of sexual differentiation controlled by estrogen–androgen balance. More direct methods of hormonal assay employed on larger samples are encouraged.

A second strategy for examining the effects of hormones on the development of spatial abilities is to make experimental hormonal manipulations and then observe the subsequent effects on behavioral sex differences (Reinisch *et al.,* 1979). Since such research with human subjects is unacceptable for obvious ethical and other reasons, we are likely to see an increase in studies involving experiments in nature, that is, involving individuals who suffer from sex chromosome anomalies known to be associated with alterations in gonadal hormones. Such attempts to delineate the effects of hormones on within-sex variation in spatial skills are complicated, however, by the fact that androgen levels are related to activity levels in children, with abnormally elevated levels of prenatal androgen associated with higher activity during childhood (Ehrhardt & Baker, 1974) and abnormally low levels of androgen associated with less childhood activity (Money, Ehrhardt, & Masica, 1968). Activity level will likely influence the kinds of social experiences a child will have and "differential experience may, in turn, influence the development of certain skills measured on tests of cognitive ability [Reinisch et al., 1979, p. 226]."

In sum, although available evidence suggests that hormone levels have an effect on sex-related differences in cognitive functioning, the precise nature of the relationship between spatial abilities and hormone balance remains an open question.

Neurological Influences

An abundance of psychological literature is now available that suggests a difference between males and females in precisely those areas of cognitive functioning that are believed to be differentially represented by the two cerebral hemispheres. For example, males tend to show performance advantages over females on various (right hemisphere) tasks requiring spatial abilities, whereas females tend to show performance advantages over males on various (left hemisphere) tasks requiring verbal abilities. Neurological hypotheses tested in a rapidly increasing number of clinical, behavioral, and electrophysiological studies, none of which is conclusive, suggest that observed sex-related differences in spatial as well as in verbal functions may be due to fundamental differences between the sexes in the functional organization of the brain (for review, see Newcombe, Chapter 10, this volume). In this section a critique of this literature is offered.

Findings from clinical studies tend to support the general conclusions that among right-handed adults spatial and nonverbal information processing is more a function of areas of the right cerebral hemisphere than the left and that males have relatively greater hemispheric specialization than females. Females are believed to be more bilateral (less specialized) than males in their cerebral representation of both spatial and verbal functions (c.f., McGlone, 1980). Unfortunately, the vast majority of clinical studies reported in the neuropsychological literature have involved relatively small samples where sex-related differences were ignored or were nearly impossible to detect. A recent report by Inglis and Lawson (1981), however, has provided convincing evidence for a sex-related difference in the functional asymmetry of the damaged human brain. A sex difference was reflected by a test-specific laterality effect in male but not in female patients. This evidence was used to explain inconsistencies in the clinical literature concerning the effects of unilateral brain damage on cognitive functioning in studies in which the influence of sex was overlooked.

Behavioral studies with adults have provided inconsistent results. It is not clear whether tasks and measures used to date, such as dichotic listening tasks and visual half-field presentations of various stimuli, are reliable indicators of cerebral specialization, let alone sensitive enough to be able to distinguish patterns and degrees of cerebral specialization. Also there has been a general lack of consistency among behavioral studies in controlling for subject and task variables that are now believed to influence measures of cerebral specialization.

Results from behavioral studies with children suggest that sex-related differences in cerebal functional specialization emerge reliably only after puberty, a hypothesis that finds support in the abilities literature where the widely documented sex differences on psychometric tests of spatial abilities, as well as on numerous tasks requiring these abilities, do not reliably appear until puberty, as noted earlier in this chapter. Although anatomical data suggest that cerebral *structural* asymmetries are present at birth (Wada, Clark, & Hamm, 1975; Witelson & Pallie, 1973), it may be that cerebral *functional* asymmetries do not become reliably observed until later in development. However, a few electrophysiological studies conducted to date that circumvent some of the methodological problems that characterize behavioral and clinical studies of cerebral functional specialization have found sex-related hemispheric asymmetries in evoked potentials in infants (Davis & Wada, 1977; Molfese, Freeman, & Palermo, 1975; Shucard, Shucard & Cummins, 1981) as well as in adults (e.g., Eason, Groves, White, & Oden, 1967).

If cerebral functional asymmetries are present at birth, we might reasonably expect infants and young children to show a pattern of hemispheric specialization similar to adults. If the hemispheres gradually become more specialized so that in adults males are more specialized and females are more bilateralized in their cerebral representation for specific cognitive functions, then we might reasonably expect to observe signs of this in behavioral studies using dichotic listening and tachistoscopic procedures or in neurophysiological studies using electrophysiological techniques. In fact, we would expect to find that girls shift to the expected adult pattern of hemispheric specialization earlier than boys, in light of the general maturational advantage in physical and cognitive development enjoyed by girls (Garai & Scheinfeld, 1968; Maccoby & Jacklin, 1974; Money & Ehrhardt, 1972). The picture would be complicated, however, in that our recent work and that of other investigators suggests that sex-related differences in cognitive functioning and cerebral specialization may interact with a number of subject variables including age (Bryden & Allard, 1978), handedness (Hardyck & Petrinovich, 1977; McGee, 1976, 1978b), the existence of a familial history of left-handedness (Hardyck, 1977; Hardyck & Petrinovich, 1977; Lake & Bryden, 1976; McGee *et al.*, 1980), physical somatic androgenicity (Petersen, 1979; Reinisch *et al.*, 1979), rate of maturation (Waber, 1976, 1977, 1979), and a variety of developmental factors associated with sex role differentiation and socialization present during adolescence (e.g., Conger, 1977; Sherman, 1967, 1978).

If through further research we find that patterns of performance in specific cognitive abilities are related to patterns of brain function, then we are faced with answering the question, Why? There are at least three possible explanations, as noted by Bryden (1979). First, there may be a fundamental biological neurologic difference in cerebral organization between males and females, so that cognitive information processing is more likely to be bilaterally represented in females than males. Second, the observed differences may arise from the test procedures; that is, females may use different strategies to perform the behavioral tests (e.g., dichotic listening and tachistoscopic procedures) that are used to measure hemispheric specialization. That problem-solving strategies may account for sex-related differences in mental rotation task performance has been suggested elsewhere (McGee, 1978c). The notion that observed sex-related differences on tachistoscopic and dichotic listening measures of hemispheric specialization may reflect strategy differences rather than neurophysiological differences in hemispheric functioning gains some support from Bryden's (1978) failure to find sex-related differences in dichotic listening when such factors were controlled. Third, it is possible that whatever sex-related differences are observed result from an interaction of strategy effects with cerebral organization. Males and females may use different strategies for solving different kinds of problems because their cerebral organization is different. In view of available anatomical and clinical evidence, this hypothesis seems quite reasonable.

SUMMARY AND CONCLUSIONS

Evidence that variation in spatial test scores is to some degree heritable remains positive. Applying general causal models to human spatial visualizing ability, Rice, Cloninger, and Reich (1980) have estimated the proportion of variance that is accounted for by transmissible factors to be 45%. However, the X-linked recessive gene hypothesis that has served as a tentative explanation for the widely documented sex difference in spatial test performance and the mode of genetic transmission is not supported in recent family studies. Results from segregation analyses lead to the conclusion that segregation of a major gene may influence performance on some but not all spatial tests. Whether this major gene is recessive or dominant has yet to be determined. As suggested elsewhere (McGee, 1979a), somewhat different assessments of spatial abilities (e.g., two-dimensional versus three-dimensional, rotation versus transformation of spatial objects, analytic versus gestalt processing, visualization versus orientation) most likely have somewhat different genetic structures. It is not at all clear from the few attempts to address these issues (Ashton *et al.*, 1979; Bock & Perline, 1979; Guttman, 1974; Loehlin *et al.*, 1978; Yen, 1975) just what kinds of spatial test performance the major gene should be expected to influence most.

Further research into the genetic and environmental sources of variance in spatial test performance is encouraged. Particularly important is the development of scientific knowledge concerning cerebral specialization of function, and genetic and hormonal mechanisms, either of which might affect the development of the brain. As far as spatial abilities are concerned, it seems reasonable to conclude that areas of the right cerebral hemisphere are specialized for processing spatial, nonverbal stimuli and that males have relatively greater right hemisphere specialization than females. An emphasis in future studies on individual differences and multivariate analyses of the data obtained should lead to results that will help to clarify and extend our understanding of the complex relationships among measures of cerebral functional specialization, sex, age, handedness, and specific cognitive abilities. Such information may have significant applied implications. The study of cerebral specialization for spatial as well as verbal functions may contribute to our further understanding, diagnosis, and successful treatment of a variety of learning and language disorders such as developmental dyslexia, a specific disability in reading found predominantly among males and believed to be related to hemispheric dysfunction (Ansara, Geschwind, Galaburda, Albert, & Gartrell, 1981; Ferry, Culbertson, Fitzgibbons, & Netsky, 1979). Scientific knowledge of cerebral specialization of function may enable diagnosticians to better diagnose and educators to better educate the increasing numbers of children seen for evaluation of learning, language, and reading disabilities. Ultimately such information will be useful in suggesting ways to tailor instructional programs to the full range of individual students' needs and abilities.

REFERENCES

Alexander, D., Ehrhardt, A. A., & Money, J. Defective figure drawing, geometric and human, in Turner's syndrome. *Journal of Nervous and Mental Disease,* 1966, *142,* 161–167.

Alexander, D., & Money, J. Turner's syndrome and Gerstmann's syndrome: Neuropsychologic comparisons. *Neuropsychologia,* 1966, *4,* 265–273.

Alexander, D., Walker, H. T., Jr., & Money, J. Studies in direction sense I: Turner's syndrome. *Archives of General Psychiatry,* 1964, *10,* 337–339.

Anastasi, A. *Differential psychology: Individual and group differences in behavior* (3rd ed.). New York: Macmillan, 1958.

Ansara, A., Geschwind, N., Galaburda, A., Albert, M., & Gartrell, N. (Eds.), *Sex differences in dyslexia*. Towson, Maryland: The Orton Dyslexia Society, 1981.

Ashton, G. C., Polvina, J. J., & Vandenberg, S. G. Segregation analysis of family data for 15 tests of cognitive ability. *Behavior Genetics*, 1979, *9*, 329–347.

Bennett, G. K., Seashore, H. G., & Wesman, A. G. *Manual for the differential aptitude tests: Forms S and T* (5th ed.). New York: The Psychological Corporation, 1974.

Blewett, D. B. An experimental study of the inheritance of intelligence. *Journal of Mental Science*, 1954, *100*, 922–933.

Block, J. B. Hereditary components in the performance of twins on the WAIS. In S. G. Vandenberg (Ed.), *Progress in human behavior genetics*. Baltimore: Johns Hopkins University Press, 1968.

Bock, R. D. *A study of familial effects in certain cognitive and perceptual variables*. Final Report of the Illinois Psychiatric Training and Research Grant No. 17-137 to the University of Chicago, 1970.

Bock, R. D. Word and image: Sources of the verbal and spatial factors in mental test scores. *Psychometrika*, 1973, *38*, 437–457.

Bock, R. D., & Kolakowski, D. Further evidence of sex-linked major-gene influence on human spatial ability. *American Journal of Human Genetics*, 1973, *25*, 1–14.

Bock, R. D., & Perline, R. A lod score method for detecting linkage on the X chromosome between a marker locus and a major locus for a quantitative character. *Behavior Genetics*, 1979, *9*, 139–149.

Bock, R. D., & Vandenberg, S. G. Components of heritable variation in mental test scores. In S. G. Vandenberg (Ed.), *Progress in human behavior genetics*, Baltimore: Johns Hopkins University Press, 1968.

Boles, D. B. X-Linkage of spatial ability: A critical review. *Child Development*, 1980, *51*, 625–635.

Bouchard, T. J., Jr., & McGee, M. G. Sex differences in human spatial ability: Not an X-linked recessive gene effect. *Social Biology*, 1977, *24*, 332–335.

Broverman, D. M., Klaiber, E. L., Kobayashi, Y., & Vogel, W. Roles of activation and inhibition in sex differences in cognitive abilities. *Psychological Review*, 1968, *75*, 23–50.

Bryden, M. P. Strategy effects in the assessment of hemispheric asymmetry. In G. Underwood (Ed.), *Strategies of information processing*. London: Academic Press, 1978.

Bryden, M. P. Evidence for sex-related differences in cerebral organization. In M. A. Wittig & A. C. Petersen (Eds.), *Sex-related differences in cognitive functioning*. New York: Academic Press, 1979. Pp. 121–143.

Bryden, M. P., & Allard, F. Dichotic listening and the development of linguistic processes. In M. Kinsbourne (Ed.), *Hemispheric asymmetries of function*. Cambridge, England: Cambridge University Press, 1978.

Buffery, A. W. H., & Gray, J. H. Sex differences in the development of spatial and linguistic skills. In C. Quinstead & D. C. Taylor (Eds.), *Gender differences: Their ontogeny and significance*. London: Churchill, 1972.

Carter–Saltzman, L. Patterns of cognitive abilities and lateralization in adoptive and biological families. *Behavior Genetics*, 1977, *7*, 48. (Abstract)

Cohen, J. A factor-analytically based rationale for the Wechsler Adult Intelligence Scale. *Journal of Consulting Psychology*, 1957, *21*, 451–457.

Cohen, J. The factorial structure of the WISC at ages 7–6, 10-6, and 13-6. *Journal of Consulting Psychology*, 1959, *23*, 285–289.

Conger, J. J. *Adolescence and youth: Psychological development in a changing world*. New York: Harper & Row, 1977.

Corah, N. L. Differentiation in children and their parents. *Journal of Personality,* 1965, *33,* 300–308.

Corley, R. P., DeFries, J. C., Kuse, A. R., & Vandenberg, S. G. Familial resemblance for the Identical Blocks Test of spatial ability: No evidence for X linkage. *Behavior Genetics,* 1980, *10,* 211–216.

Csik, L., & Mather, K. The sex incidence of certain hereditary traits in man. *Annals of Eugenics Quarterly,* 1938, *8,* 126–145.

Davis, A. E., & Wada, J. A. Hemispheric asymmetries in human infants: Spectral analysis of flash and click evoked potentials. *Brain and Language,* 1977, *4,* 23–31.

Day, N. E. Estimating the components of a mixture of normal distributions. *Biometrika,* 1969, *56,* 463–474.

DeFries, J. C., Vandenberg, S. G., McClearn, G. E., Kuse, A. R., Wilson, J. R., Ashton, G. C., & Johnson, R. C. Near identity of cognitive structure in two ethnic groups. *Science,* 1974, *183,* 338–339.

DeFries, J. C., Ashton, G. C., Johnson, R. C., Kuse, A. R., McClearn, G. E., Mi, M. P., Rashad, M. N., Vandenberg, S. G., & Wilson, J. R. Parent-offspring resemblance for specific cognitive abilities in two ethnic groups. *Nature,* 1976, *261,* 121–135.

DeFries, J. C., Johnson, R. C., Kuse, A. R., McClearn, G. E., Polovina, J., Vandenberg, S. G., & Wilson, J. R. Familial resemblance for specific cognitive abilities. *Behavior Genetics,* 1979, *9,* 23–43.

Drew, L. J. An investigation into the measurement of technical ability. *Occupational Psychology,* 1944, *21,* 34–48.

Eason, R. G., Groves, P., White, C. T., & Oden, D. Evoked cortical potentials: Relation to visual field and handedness. *Science,* 1967, *156,* 1643–1646.

Ehrhardt, A. A., & Baker, S. W. Fetal androgens, human central nervous system differentiation, and behavior sex differences. In R. C. Friedman, R. M. Richart, & R. L. Vande Wiele (Eds.), *Sex differences in behavior.* New York: Wiley, 1974 Pp. 33–51.

Elston, R. C., & Campbell, M. A. Schizophrenia: Evidence for the major gene hypothesis. *Behavior Genetics,* 1970, *1,* 3–10.

Emmett, W. G. Evidence of a space factor at 11 plus and earlier. *British Journal of Psychology,* 1949, *2,* 6–16.

Fairweather, H. Sex differences in cognition. *Cognition,* 1976, *4,* 231–280.

Ferry, P. C., Culbertson, J. L., Fitzgibbons, P. M., & Netsky, M. G. Brain function and language disabilities. *International Journal of Pediatric Otorhinolaryngology,* 1979, *1,* 13–24.

Fruchter, B. Measurement of spatial abilities: History and background. *Educational and Psychological Measurement,* 1954, *14,* 337–395.

Garai, J. E., & Scheinfeld, A. Sex differences in mental and behavioral traits. *Genetic Psychology Monographs,* 1968, *77,* 169–299.

Gardner, R. W., Jackson, D. N., & Messick, S. J. Personality organization in cognitive controls and intellectual abilities. *Psychological Issues,* 1960, *2* (Whole No. 8).

Garron, D. C. Sex-linked recessive inheritance of spatial and numerical abilities and Turner's syndrome. *Psychological Review,* 1970, *77,* 147–152.

Garron, D. C. Intelligence among persons with Turner's syndrome. *Behavior Genetics,* 1977, *7,* 105–127.

Garron, D. C., & Vander Stoep, L. P. Personality and intelligence in Turner's syndrome: A critical review. *Archives of General Psychiatry,* 1969, *21,* 339–346.

Goodenough, D. R., Gandini, E., Olkih, I., Pizzamiglio, L., Thayer, D., & Witken, H. A. A study of X chromosome linkage with field dependence and spatial visualization. *Behavior Genetics,* 1977, *7,* 373–387.

Guttman, R. Genetic analysis of analytical spatial ability: Raven's progressive matrices. *Behavior Genetics*, 1974, *4*, 273–284.

Guttman, R., & Shoham, I. Intrafamilial invariance and parent–offspring resemblance in spatial abilities. *Behavior Genetics*, 1979, *9*, 367–378.

Hardyck, C. A model of individual differences in hemispheric functioning. In H. Avakaian–Whitaker & H. A. Whitaker (Eds.), *Studies in neurolinguistics* (Vol. 3). New York: Academic Press, 1977.

Hardyck, C., & Petrinovich, L. F. Left-handedness. *Psychological Bulletin*, 1977, *84*, 385–404.

Harris, L. J. Sex differences in spatial ability: Possible environmental, genetic, and neurological factors. In M. Kinsbourne (Ed.), *Asymmetrical function of the brain*. New York: Cambridge University Press, 1978.

Hartlage, L. C. Sex-linked inheritance of spatial ability. *Perceptual and Motor Skills*, 1970, *31*, 610.

Herzberg, F., & Lepkin, M. A study of sex differences in the PMA Test. *Educational and Psychological Measurement*, 1954, *14*, 687–689.

Hobson, J. R. Sex differences in primary mental abilities. *Journal of Educational Psychology*, 1947, *41*, 126–132.

Hogben, L. Filial and fraternal correlations in sex-linked inheritance. *Proceedings Royal Society of Edinburgh*, 1932, *52*, 331–336.

Hwang, C. Parent–child resemblance in psychological characteristics. *Psychology and Education*, 1969, *3*, 29–36. (Taiwan)

Inglis, J., & Lawson, J. S. Sex differences in the effects of unilateral brain damage on intelligence. *Science*, 1981, *212*, 693–695.

Jensen, A. R. A theoretical note on sex linkage and race differences in spatial visualization ability. *Behavior Genetics*, 1975, *5*, 151–164.

Klaiber, E. L., Broverman, D. M., & Kobayashi, Y. The automatization cognitive style, androgens, and monoamine oxidase (MAO). *Psycholopharmacologia*, 1967, *11*, 320–336.

Kuse, A. R. *Familial resemblances for cognitive abilities estimated from two test batteries in Hawaii.* Unpublished doctoral dissertation, University of Colorado, 1977.

Lake, D. A., & Bryden, M. P. Handedness and sex differences in hemispheric asymmetry. *Brain and Language*, 1976, *3*, 266–282.

Loehlin, J. C., Sharan, S., & Jacoby, R. In pursuit of the "spatial gene": A family study. *Behavior Genetics*, 1978, *8*, 27–41.

Lord, F. M., & Novick, M. R. *Statistical theories of mental test scores*. Reading, Massachusetts: Addison–Wesley, 1968.

Maccoby, E. E. Sex differences in intellectual functioning. In E. E. Maccoby (Ed.), *The development of sex differences*. Stanford: Stanford University Press, 1966.

Maccoby, E., & Jacklin, C. N. *The psychology of sex differences*. Stanford: Stanford University Press, 1974.

MacLean, C. J., Morton, N. E., & Lew, R. C. Analysis of family resemblance, IV: Operational characteristics of segregation analysis. *American Journal of Human Genetics*, 1974, *27*, 365–384.

Marjoribanks, K. Environment, social class, and mental abilities. *Journal of Educational Psychology*, 1972, *63*, 103–109.

Mather, K., & Jinks, J. L. Correlation between relatives arising from sex-linked genes. *Nature*, 1963, *198*, 314–315.

McGee, M. G. Laterality, hand preference, and human spatial ability. *Perceptual and Motor Skills*, 1976, *42*, 781–782.

McGee, M. G. A family study of human spatial abilities (Doctoral dissertation, University of

Minnesota, 1976). *Dissertation Abstracts International*, 1977, 37, 6396. (University Microfilms No. 77-12836.)

McGee, M. G. Intrafamilial correlations and heritability estimates for spatial ability in a Minnesota sample. *Behavior Genetics*, 1978, 8, 77–80. (a)

McGee, M. G. Handedness and mental rotation. *Perceptual and Motor Skills*, 1978, 47, 641–642. (b)

McGee, M. G. Effect of two problems solving strategies on Mental Rotation Test scores. *The Journal of Psychology*, 1978, 100, 83–85. (c)

McGee, M. G. Human spatial abilities: Psychometric studies and environmental, genetic, hormonal, and neurological influences. *Psychological Bulletin*, 1979, 86, 889–918. (a)

McGee, M. G. *Human spatial abilities: Sources of sex differences*. New York: Praeger, 1979. (b)

McGee, M. G. The effect of brain asymmetry on cognitive functions depends upon what ability, for which sex, at what point in development. *The Behavioral and Brain Sciences*, 1980, 3, 233–234.

McGee, M. G. Predictions of X-linked inheritance for human characteristics. *Developmental Review*, 1981, 1, 289–295.

McGee, M. G., Cozad, T. W., & Pate, J. L. Cognitive deficits and human hand preference. Paper presented at the 88th annual meeting of the American Psychological Association, Montreal, September, 1980.

McGee, M. G., Cozad, T. W., & Pate, J. L. Spatial abilities in normal adolescents. Paper presented at the 90th annual meeting of the American Psychological Association, Washington, D.C., August, 1982.

McGlone, J. Sex differences in human brain asymmetry. *Brain and Behavioral Sciences*, 1980, 3, 215–264.

McKusick, V. A. *On the X chromosome of man*. Washington, D. C.: American Institute of Biological Sciences, 1964.

McKusick, V. A. *Mendelian inheritance in man: Catalogs of autosomal dominant, autosomal recessive, and X-linked phenotypes* (2nd ed.). Baltimore: Johns Hopkins Press, 1968.

Mittwoch, U. *Sex chromosomes*. New York: Academic Press, 1967.

Molfese, D. L., Freeman, R. B., & Palermo, D. S. The ontogeny of brain lateralization for speech and nonspeech stimuli. *Brain and Language*, 1975, 2, 356–368.

Money, J. Cytogenetic and psychosexual incongruities with a note on space-form blindness. *American Journal of Psychiatry*, 1963, 119, 820–827.

Money, J., Alexander, D., & Walker, H. T., Jr. *A standardized test of direction sense*. Baltimore: Johns Hopkins University Press, 1965.

Money, J., & Ehrhardt, A. A. *Man and woman: Boy and girl*. Baltimore: Johns Hopkins University Press, 1972.

Money, J., Ehrhardt, A. A., & Masica, D. N. Fetal feminization induced by androgen insensitivity in the testicular feminization syndrome: Effect on marriage and maternalism. *Johns Hopkins Medical Journal*, 1968, 123, 105–114.

Money, J., & Grandoff, D. IQ and the somatic stigmata of Turner's syndrome. *American Journal of Mental Deficiency*, 1965, 70, 69–77.

O'Connor, J. *Structural visualization*. Boston: Human Engineering Laboratory, 1943.

Osborne, R. T., & Gregor, A. J. Racial differences in heritability estimates for tests of spatial ability. *Perceptual and Motor Skills*, 1968, 27, 735–739.

Park, J., Johnson, R. C., DeFries, J. C., Ashton, G. C., McClearn, G. E., Mi, M. P., Rashad, M. N., Vandenberg, S. G., & Wilson, J. R. Parent–offspring resemblance for specific cognitive abilities in Korea. *Behavior Genetics*, 1978, 8, 43–52.

Pennington, B. F., Bender, B., Puck, M., Salbenblatt, J., & Robinson, A. Learning disabilities in children with sex chromosome anomalies. *Child Development*, in press.

Petersen, A. C. Physical androgny and cognitive functioning in adolescence. *Developmental Psychology*, 1976, *12*, 524–533.

Petersen, A. C. Hormones and cognitive functioning in normal development. In M. A. Wittig & A. C. Petersen (Eds.), *Sex-related differences in cognitive functioning*. New York: Academic Press, 1979. Pp. 189–214.

Reinisch, J. M., Gandelman, R., & Spiegel, F. S. Prenatal influences on cognitive abilities: Data from experimental animals and human genetic and endocrine syndromes. In M. A. Wittig & A. C. Petersen (Eds.), *Sex-related differences in cognitive functioning*. New York: Academic Press, 1979. Pp. 215–239.

Rice, J., Cloninger, C. R., & Reich, T. General causal models for sex differences in the familial transmission of multifactorial traits: An application to human spatial visualizing ability. *Social Biology*, 1980, *27*, 36–47.

Rifkind, A. B., Kulin, H. E., Rayford, P. L., Cargile, C. M., & Ross, G. T. 24-hour urinary lutenizing hormone (LH) and follicle stimulating hormone (FSH) excretion in normal children. *Journal of Clinical Endocrinology and Metabolism*, 1970, *31*, 517–525.

Rovet, J., & Netley, C. The mental rotation task performance of Turner Syndrome subjects. *Behavior Genetics*, 1980, *10*, 437–443.

Serra, A., Pizzamiglio, L., Boari, A., & Spera, S. A comparative study of cognitive traits in human sex chromosome aneuploids and sterile and fertile euploids. *Behavior Genetics*, 1978, *8*, 143–154.

Shaffer, J. W. A specific cognitive deficit observed in gonadal aplasia (Turner's syndrome). *Journal of Clinical Psychology*, 1962, *18*, 403–406.

Shepard, R. N., & Metzler, J. Mental rotation of three-dimensional objects. *Science*, 1971, *171*, 701–703.

Sherman, J. A. Problem of sex differences in space perception and aspects of intellectual functioning. *Psychological Review*, 1967, *74*, 290–299.

Sherman, J. A. *Sex-related cognitive differences*. Springfield: Charles C Thomas, 1978.

Shucard, J. L., Shucard, D. W., & Cummins, K. R. Auditory evoked potentials and sex related differences in brain development. *Brain and Language*, 1981, *13*, 91–102.

Silbert, A., Wolff, P. H., & Lilienthal, J. Spatial and temporal processing in patients with Turner's syndrome. *Behavior Genetics*, 1977, *7*, 11–21.

Smith, I. M. *Spatial ability: Its educational and social significance*. London: University of London, 1964.

Spuhler, K. P. *Family resemblance for cognitive performance: An assessment of genetic and environmental contributions to variation*. Unpublished doctoral dissertation, University of Colorado, 1976.

Stafford, R. E. Sex differences in spatial visualization as evidence of sex-linked inheritance. *Perceptual and Motor Skills*, 1961, *13*, 428.

Tanner, J. M. Growth and endocrinology of the adolescent. In L. I. Gardner (Ed.), *Endocrine and genetic diseases of childhood*. Philadelphia: Saunders, 1969.

Thurstone, L. L., & Thurstone, T. G. *The Primary Mental Abilities Tests*. Chicago: Science Research Associates, 1941.

Vandenberg, S. G. Twin data in support of the Lyon hypothesis. *Nature*, 1962, *194*, 505–506.

Vandenberg, S. G. Hereditary factors in psychological variables in man with special emphasis on cognition. In J. N. Spuhler (Ed.), *Genetic diversity and behavior*. Chicago: Aldine, 1967.

Vandenberg, S. G. The nature and nurture of intelligence. In D. C. Glass (Ed.), *Genetics*. New York: Rockefeller, 1968.

Vandenberg, S. G. A twin study of spatial ability. *Multivariate Behavioral Research*, 1969, *4*, 273–294.

Vandenberg, S. G. The genetics of intelligence. In L. C. Deighton (Ed.), *Encyclopedia of education*. New York: Macmillan, 1971.

Vandenberg, S. G. Sources of variance in performance on spatial tests. In J. Eliot & N. J. Salkind (Eds.), *Children's spatial development*. Springfield: Charles C Thomas, 1975.

Vandenberg, S. G., & Kuse, A. R. Mental Rotations: A group test of three-dimensional spatial visualization. *Perceptual and Motor Skills*, 1978, *47*, 599–604.

Vandenberg, S. G., & Kuse, A. R. Spatial ability: A critical review of the sex-linked major gene hypothesis. In M. A. Wittig & A. C. Petersen (Eds.), *Sex-related differences in cognitive functioning*. New York: Academic Press, 1979. Pp. 67–95.

Vandenberg, S. G., Stafford, R. E., & Brown, A. M. The Louisville twin study. In S. G. Vandenberg (Ed.), *Progress in human behavior genetics*. Baltimore: Johns Hopkins University Press, 1968.

Waber, D. P. Sex differences in cognition: A function of maturation rate? *Science*, 1976, *192*, 572–574.

Waber, D. P. Sex differences in mental abilities, hemispheric lateralization and rate of physical growth at adolescence. *Developmental Psychology*, 1977, *13*, 29–38.

Waber, D. P. Cognitive abilities and sex-related variations in the maturation of cerebral cortical functions. In M. A. Wittig & A. C. Petersen, (Eds.), *Sex-related differences in cognitive functioning*. New York: Academic Press, 1979. Pp. 161–186.

Wada, J. A., Clark, R., & Hamm, A. Cerebral hemispheric asymmetry in humans. *Archives of Neurology*, 1975, *32*, 239–246.

Walker, J. T., Krasnoff, A. G., & Peaco, D. Visual spatial perception in adolescents and their parents: The X-linked recessive hypothesis. *Behavior Genetics*, 1981, *11*, 403–413.

Wesman, A. G. Separation of sex groups in test reporting. *Journal of Educational Psychology*, 1949, *40*, 223–229.

Williams, T. Family resemblance in abilities: The Wechsler scales. *Behavior Genetics*, 1975, *5*, 405–409.

Wilson, J. R., DeFries, J. C., McClearn, G. E., Vandenberg, S. G., Johnson, R. C., & Rashad, M. N. Cognitive abilities: Use of family data as a control to assess sex and age differences in two ethnic groups. *International Journal of Aging and Human Development*, 1975, *6*, 261–276.

Witelson, S. F., & Pallie, W. Left hemisphere specialization for language in the newborn: Neuroanatomical evidence of asymmetry. *Brain*, 1973, *96*, 641–646.

Witkin, H. A., Lewis, H. B., Hertzman, M., Machover, K., Meissner, P. B., & Wapner, S. *Personality through perception*. New York: Harper, 1954.

Wittig, M. A. Genetic influences on sex related differences in intellectual performance: Theoretical and methodological issues. In M. A. Wittig & A. C. Petersen (Eds.), *Sex-related differences in cognitive functioning*. New York: Academic, 1979. Pp. 21–65.

Wittig, M. A., & Petersen, A. C. (Eds.). *Sex-related differences in cognitive functioning*. New York: Academic Press, 1979.

Yen, W. M. Sex-linked major-gene influences on selected types of spatial performance. *Behavior Genetics*, 1975, *5*, 281–298.

Sex-Related Differences in Spatial Ability: Problems and Gaps in Current Approaches

The evidence that males perform better than females on tests of spatial ability, at least by adulthood, has been summarized by many authors (e.g., Harris, 1978, 1981; McGee, Chapter 9, this volume). There is general agreement that such differences exist. (For a contrasting view see Fairweather, 1976.) There is far less consensus, however, concerning three questions: When do such differences arise; do they increase in size with age; and, how can they be explained? Several possible explanations have been offered for these sex-related differences. McGee (Chapter 9, this volume) has reviewed the present status of genetic and hormonal theories. Other hypotheses involve the effects of sex-differentiated patterns of hemispheric specialization, and the consequences of sex-related differences in environment and experience with spatial activity.

The first two issues, the descriptive ones, are discussed in the first section of the chapter. Assumptions about when sex-related differences appear and whether they change in size with age often substantially influence the reasoning of different authors on the likely causes of sex-related differences in spatial ability. (See Harris's [1979] review of Wittig & Petersen [1979] for an example.) The next two sections deal, respectively, with lateralization and experiential explanations. Because there have been recent reviews of the existing research in each of these areas (Harris, 1981; McGlone, 1980; Nash, 1979) the focus here will be on the unsolved problems associated with each position and

223

SPATIAL ABILITIES
Development and Physiological Foundations

the kinds of evidence that need to be gathered either to support or refute these hypotheses. In the fourth section, the current status of Waber's (1976, 1977a, 1977b) maturation-rate hypothesis is discussed. The fifth and final section contains a brief discussion of recent interactionist approaches. The argument is made that there is need to design research that tests relatively specific (and opposing) models of the development of sex-related differences in spatial ability.

DEVELOPMENTAL ISSUES: DESCRIPTION

The age when sex-related differences in spatial ability first appear is of importance because it is often used in arguments concerning explanations of the difference. For instance, Maccoby and Jacklin's (1974) conclusion that sex-related differences in spatial ability first appear in early adolescence has been taken as support for an environmental explanation, under the assumption that sex roles become more salient and sex-related differences in activities and interests more pronounced at adolescence (Nash, 1979).

Unfortunately, arguments of this type are suggestive rather than truly decisive. If sex-related differences are apparent early in childhood, this may be because children's play is already sex typed, and boys get more experience with spatial activities, rather than because the differences are biological. Connor and Serbin (1977) and Serbin and Connor (1979) have gathered evidence that play with masculine, and presumably spatial, toys (e.g., Lincoln Logs, blocks) in preschool is associated with higher spatial ability. However, if sex-related differences in spatial ability first appear in adolescence this need not be due to socialization; the fact can also be taken as evidence for the role of pubertal hormones in creating the difference (Petersen, 1976; Waber, 1977a, 1977b).

Despite this indeterminacy, however, it would clearly be helpful to reach conclusions concerning the age of first appearance of sex-related differences. The Maccoby and Jacklin (1974) conclusion that differences first appear at adolescence was based on studies concerning two types of spatial ability. The first, termed *visual, nonanalytic* spatial ability, included a wide variety of spatial memory and reproduction tasks, as well as some standardized tests of spatial ability. Sex-related differences favoring males did not appear in the table summarizing studies in this area until age 13 or 14, the one exception being a study by Keogh (1971) which showed that 8- and 9-year-old boys excelled at copying designs by walking, in conditions where they were given added cues. This table did not, however, include several older studies summarized in Maccoby (1966) which showed sex-related differences favoring boys beginning at age 6 to 7 (Mellone, 1944) and present at various ages under 13 (Emmett, 1949; Havighurst & Janke, 1944; Lord, 1941; Porteus, 1918; Schiller, 1934). Thus, even in 1974, some evidence existed that boys excelled at spatial tasks prior to puberty.

The second Maccoby and Jacklin table summarized results on *visual, analytic* tasks, chiefly Wechsler block design tests and tests of field dependence–independence. (Field dependence–independence tests appear to be tests of spatial ability rather than cognitive style, as shown by Widiger, Knudson, & Rorer, 1980.) Several studies cited showed boys excelling between the ages of 7 and 13 (Bergan, McManis, & Melchert, 1971; Keogh & Ryan, 1971; Wapner, 1968; Witkin, Goodenough, & Karp, 1967), although other studies of preadolescents found no differences. The findings of male superiority appeared to become more consistent in adolescence and adulthood. Not included in the review were findings by Witkin, Oltman, Raskin, and Karp (1971) that boys did better than girls on the Children's Embedded Figures Test (CEFT) as early as age 7, or findings by Dreyer, Dreyer, and Nebelkopf (1971) that kindergarten boys did better on the CEFT and a Portable Rod-and-Frame Test.

Other studies showing better performance on spatial tasks for boys in childhood have appeared since 1974. Vandenberg and Kuse (1979) summarize evidence from large-sample WISC and WISC-R standardization studies showing better male performance on Block Design at age 6. Working with smaller samples, Connor, Schackman, and Serbin (1978) and Connor, Serbin, and Schackman (1977) have found marginally significant male superiority among preschoolers on the CEFT. Short-lived female superiorities on the Preschool Embedded Figures Test (PEFT) and WPPSI Block Design for girls aged 4 and 5 years (Coates, 1974) are possibly due to these tests being of a different nature from spatial tasks typically used with school-age children.

Recent interest in children's spatial representations and cognitive maps has led to a number of studies of spatial cognition involving memory for real-world spaces (Hart & Berzok, Chapter 7, this volume). Several of these have shown sex-related differences prior to adolescence. Boys are more accurate than girls in constructing models of their classrooms and towns in kindergarten (Hart, 1979; Siegel & Schadler, 1977) and in second to fifth grade (Hart, 1979; Herman & Siegel, 1978; but see Herman, 1980 and Siegel, Herman, Allen, & Kirasic, 1979 for failures to find a sex difference).

There may be some aspect of reconstruction, perhaps its far-reaching demand on subjects to create their own spatial framework and scale, that results in the detection of early sex-related differences. This speculation can be supported in two ways. First, studies of spatial representation in children not using reproduction tasks have not generally shown sex-related differences (Allen, Kirasic, Siegel, & Herman, 1979; Cohen & Weatherford, 1980; Cohen, Weatherford, Lomenick, & Koeller, 1979; Feldman & Acredolo, 1979), although differences did appear in the distance-estimation task of Cohen, Weatherford, and Byrd (1980). Second, some studies have shown directly that the existence of sex-related differences on spatial tasks is dependent on the demands of the task. Cohen (1981) has found that second-grade boys and girls perform equivalently when asked to estimate distances along a line parallel to the orientation of the original locations. In this case, no reorienting is required. But when lines were not parallel, girls' but not boys' errors increased directly with the amount of reorienting necessary. Thus, early sex-related differences may appear in studies of memory for real-world spaces when subjects are required to mentally manipulate or transform spatial information. Alternatively, or in addition, early sex differences may appear primarily when spatial tasks are difficult, irrespective of the need for mental manipulation.

In sum, male superiority on spatial tasks prior to adolescence has been shown in many studies, albeit less consistently than in studies of adolescents and adults. Sex-related differences may increase with age. So with less sensitive methodologies or with small samples they would be less detectable during childhood.

What evidence exists to directly support the idea that sex-related differences increase in size in childhood and adolescence? In large-sample studies, both Droege (1967) and Flanagan, Davis, Dailey, Shaycroft, Orr, Goldberg, and Neyman (1964) have reported increases in the size of sex-related differences in spatial ability from the first to the last year of high school. Wilson, DeFries, McClearn, Vandenberg, Johnson, and Rashad (1975), with a large sample of parents and children tested across a wide age range, have found that male scores (on a space factor defined by the Mental Rotations Test and other tests) increase more sharply than female ones from age 14 through the late twenties. Nash (1975), in a smaller study, found higher male than female scores on the DAT Space Relations Test at age 14 but not at age 11. By contrast, Vandenberg and Kuse (1979) report that studies of preadolescents' and children's performance on WISC Block Design have shown sex-related differences relatively stable in size. Witkin, Goodenough, and Karp (1967) found no evidence of age by sex interactions on tests of field dependence–independence over the age range of 8 years to young adulthood. Porteus (1965) reported that sex-related differences favoring males on his maze test were constant in size through the school years, in a wide variety of studies.

The evidence is presently consistent with the idea that small sex-related differences present in childhood increase somewhat in size with age. This conclusion must be based, however, in large part on the observation that positive findings are more consistently found in studies of adolescents and adults. The preferable kind of evidence, that derived from finding age by sex interactions within a single study, is very mixed.

The problem created by the great variety of spatial tests that have been used in the studies discussed above is a serious one. Fundamentally, we lack agreement concerning what is meant by *spatial ability*. To what extent are tests tapping a single ability dimension, or to what extent are different entities involved? A variety of distinctions have been proposed, among them Maccoby and Jacklin's analytic versus nonanalytic division mentioned earlier.

Although there appears to be some common variance among the tests, there have been many proposals, based primarily on factor-analytic work, to subdivide "spatial ability" into different types. The most commonly cited distinction is between factors called *spatial orientation* and *spatial visualization* (e.g., McGee, Chapter 9, this volume; but see Carroll & Maxwell, 1979). Unfortunately, these factors do not appear to be clearly the same for different investigators, although the same terms are used. Michael, Guilford, Fruchter, and Zimmerman (1957) suggest five bases for distinguishing the factors. Two appear to be widely used by researchers. One distinction is that *spatial orientation* refers to spatial manipulation in which the observer is a reference point "embedded within the configurations pictured in the stimulus," whereas *spatial visualization* refers to spatial processes in which the observer seems "removed from elements presented... [and] manipulates objects that are at a distance as if they were at his finger tips... [Michael *et al.,* 1957, p. 193]." A second basis for distinguishing the factors is that spatial orientation is said to involve imagining movement of an entire stimulus without movement among parts, whereas spatial visualization involves movement of parts within the configuration. However, different investigators place different amounts of stress upon these two kinds of distinctions, while at times invoking still other differences among tasks (McGee, 1979).

Consistent differential findings regarding which factors show sex differences at which ages have yet to emerge. Richmond (1980) has recently reported finding sex-related differences among 10-year-olds on tests measuring Thurstone's space factor S_1 which includes tasks involving mental rotation of rigid configurations, but not on factor C_2 (flexibility of closure) which includes an embedded figures test, as well as tests of paper-folding and copying. But this is not easily interpretable in light of data summarized earlier showing sex-related differences on embedded figures tests even among preschoolers, and at least by age 6. Thus factor analytic work does not appear to be very helpful in grouping the variety of studies presented above in a way such that generalizations could be reached concerning whether sex-related differences increase in size with age.

Given these problems with traditional psychometric approaches to analyzing the meaning of *spatial ability,* it would seem fruitful to explore alternative approaches to the problem, particularly those concerned with individual differences. Pellegrino and Glaser (1979) have reviewed analyses of performance on "mental rotation" (see Corballis, Chapter 8, this volume) and have suggested the relevance of this work to the study of individual differences. For instance, there is evidence of important but subtle differences in rotation of two- as opposed to three-dimensional shapes, an issue often raised by those working in a psychometric tradition (e.g., Yen, 1975) but not satisfactorily resolved by correlational approaches. As another example, individuals seem to differ in strategies used for processing identical tasks (Cooper, 1976; Packer, 1979). This indicates that one cannot assume, as psychometricians do, that tasks index a certain ability by forcing a particular type of processing. It may be better to study how individuals approach tasks and to relate their styles or strategies to overall efficiency in terms of speed and/or accuracy.

Such an approach may be helpful in understanding sex-related differences in spatial ability. Many spatial tasks require that spatial arrays be accurately represented in memory and then manipulated or transformed in some way. For instance, on the DAT Space Relations Tests, subjects must first form an accurate representation of a two-dimensional stimulus and then fold it mentally into a three-dimensional object. It has often been suggested that only this second, kinetic or dynamic, process is harder for females (Guilford, 1947; Harris, 1978).

Whether or not this is the case has been the subject of several recent studies. Kail and Siegel (1977) reported that females from third and sixth grades and college perform more poorly than males on a spatial memory task involving presentation of five or seven letters in a 4 × 4 (16-cell) grid. Females had better memory for letters, males had better memory for which positions had been occupied by leters. This suggests that males may excel in memory for spatial position as well as in mental transformation of spatial information. However, Kail, Pellegrino, and Carter (1980, footnote 2) found that males mentally rotated stimuli more quickly, but they did not appear to excel at the encoding and comparison phases of the task, as reflected in intercept values. Further work may well shed light on whether the static component of spatial tasks shows sex differences, or whether it is in fact primarily in mental manipulation of spatial information that the sexes differ. Another relevant dimension, occurring in locomotor–spatial tasks, may be the ability to benefit from visual framework and past-performance cues. The enhanced performance of boys under these conditions (Keogh, 1971) could be due to differences in gross motor skills, but appears more likely to be due to differences in the ability to integrate spatial information (Vasta, Regan, & Kerley, 1980).

The main point of the above examples is that cognitive and cognitive-developmental techniques could be used in the study of spatial cognition to establish in a more analytic fashion than is possible with factor analyses what specific mental processes are more difficult for females than males.

DEVELOPMENTAL ISSUES: EXPLANATION

Patterns of Hemispheric Specialization

McGlone (1980) has recently reviewed the evidence bearing on the existence of sex differences in patterns of hemispheric specialization. She concludes that there is at least modest support for the proposition that males may be more lateralized than females, at least among the right-handed. That is, males may have speech and verbal functions represented more exclusively in the left hemisphere and visuospatial functions represented more exclusively in the right hemisphere, with females having more bilateral representation. Fairweather (1980), Kinsbourne (1980), and other commentators on McGlone's article take issue with it. Bryden (1979), reviewing much of the same material, emphasizes the *lack* of striking sex differences and the possibility that those that have been demonstrated could be due to differences in subjects' approaches to the tasks.

Nevertheless, for the moment let us assume that males are, in fact, more lateralized than females. What is then of concern to those interested in spatial ability is whether sex differences in cerebral organization are the cause of sex differences in spatial ability. This area is bedeviled by a classic "third variable" problem: Sex might be associated with different lateralization patterns and also with different cognitive patterns, without different lateralization being causally related to different cognition. At least three lines of investigation can be brought to bear on this issue. First, it must be shown that certain patterns of lateralization are related to spatial ability within sex; that is, they are independent of the association of each with sex. Second, the association of lateralization with spatial ability should be shown to be not plausibly attributable to people of high spatial ability using certain strategies on tests of lateralization. That is, lateralization-ability relationships may be due to the use of strategies (e.g., holistic comparisons) that are likely to differentially engage the right hemisphere by people of high ability. Third, one should be able to show that sex-differentiated patterns of lateralization developmentally precede sex-related differences in spatial ability.

LATERALIZATION-ABILITY RELATIONS

One important way to support a theory attributing sex-related differences in spatial ability to sex differences in cerebral organization is to show that certain patterns of lateralization are related to spatial ability within sex. Unfortunately, research on this question has been impeded by confusion regarding whether lower spatial ability would be expected to be associated with *reversals* of usual patterns of hemispheric asymmetry (e.g., right rather than left visual field advantages on tachistoscopic dot localization tasks) or with *weak* specialization (e.g., no visual field advantages on the tachistoscopic dot localization task). If the proposal is that females are less lateralized than males and therefore worse on spatial tasks, then clearly the extent of lateralization rather than its direction is of interest.

A second problem is that investigators have chosen very different methods of assessing patterns of lateralization. Levy (1974) has proposed that bilateralization of *verbal* functions leads to lower spatial ability by invasion of areas of the right hemisphere normally reserved for spatial processes. On this hypothesis, the relevant data involve assessment of language lateralization through techniques such as dichotic or tachistoscopic studies using verbal material. However, if one believes that the important issue is whether spatial mental processes are themselves lateralized, one will want to concentrate on techniques to assess right-hemispheric specialization.

A third wrinkle is provided by Gur and Gur's (1980) suggestion to distinguish studies of hemispheric activation from studies of lateralized cognitive capacity. Hemispheric activation is measured by electroencephalographic (EEG) recordings of alpha activity, by measurement of blood flow to the left versus the right hemispheres, or by conjugate lateral eye movements (CLEM)—the tendency to shift gaze in a direction away from the activated hemisphere. By contrast, dichotic, tachistoscopic, or dichhaptic tasks are considered to reflect the lateralization of specific cognitive functions.

There are several studies of the relationship of hemispheric activation to spatial skill. In males with presumed or demonstrated high spatial ability, faster reaction times and higher performance scores on paper-folding or picture completion tasks were positively correlated with right hemisphere activation as measured by EEG alpha asymmetry (Furst, 1976; Ray, Newcombe, Semon, & Cole, 1981) or cerebral blood flow (Gur & Reivich, 1980). However, Ray et al. found that correlations for females of both high and low spatial ability were low and nonsignificant. For low ability males the correlation was reversed ($r = -.77$, $p < .05$). These men did better with greater *left-hemispheric* activation. The available strategies for solving the problems for this group may have been more verbal in nature.

It should be noted that in the Ray et al. study, as might be expected from the opposite signs of the correlations for high- and low-ability males, no differences were found in hemispheric activation when comparing high-ability males as a group with low-ability males. This same negative result has been reported by Willis, Wheatley, and Mitchell (1979).

In sum, differential right-hemispheric activation during performance of spatial tasks has been found to be related to spatial ability, but only in high-ability males. Perhaps, for individuals who habitually use a holistic approach to spatial tasks, right-hemispheric activation indexes the efficiency of doing the task.

Other studies have looked at lateral specialization of function rather than at patterns of activation. McGlone and Davidson (1973) administered Wechsler Block Design and PMA Spatial Relations Tests to right- and left-handed males and females. Dichotic listening was used as a measure of lateralization for speech and a tachistoscopic dot enumeration task was used as a measure of lateralization for nonverbal functions. For females, right-hemispheric speech, as inferred from higher left- than right-ear scores on dichotic listening, was associated with lower PMA Spatial Relations scores. But no significant relationships were found for men, or for either sex using the Block Design scores or the tachistoscopic task. Thus, although the McGlone and Davidson study is often cited as showing relationships of cerebral organization and spatial ability, the evidence is actually quite weak. Similarly, Birkett (1980) found a linear relationship between left visual field advantages on a tachistoscopic form recognition task and spatial ability as measured by the DAT, for right-handed females. But there were no relationships for right-handed males, or for the females for two other spatial ability tests.

Other attempts to demonstrate relationships of indices of lateralization and spatial ability have failed completely. Dichotic listening scores were unrelated to spatial ability in studies by Fennell, Satz, Van Den Abell, Bowers, and Thomas (1978), Herbst (1980), and Waber (1977b). Studies using visual (Fennell et al., 1978) and haptic (Newcombe & Bandura, in press) tests of lateralization also failed to find a correlation with spatial ability. However, these studies should be evaluated in the light of Berenbaum and Harshman's (1980) point that sample sizes much larger than those used in most investigations would be required to discover a true difference of the size envisaged by most theories, given the known variability of spatial tests and a reasonably stringent standard for minimization of Type II errors (failures to find true differences).

Studies examining the relationship of patterns of lateralization to field dependence–independence (EFT, Rod-and-Frame Test, Articulation of Body Concept) have had more positive results. Pizzamiglio (1974) found that groups of subjects selected on the basis of having large right-ear advantages on a dichotic listening test were more field independent on two of three tests than those with minimal asymmetries. Pizzamiglio did not report results separately by sex. Oltman, Ehrlichman, and Cox (1977) found significant correlations for both males and females between EFT performance and greater left-visual-field lateralization of face perception. Zoccolotti and Oltman (1978) found field independence in a male sample associated with faster reaction times in the right visual field on letter discrimination and in the left field on face discrimination.

Most of the surprisingly few positive findings on within-sex relationships between lateralization and spatial ability involve spatial ability as assessed by tests of field dependence–independence. In the future, theoretical predictions concerning which measures of lateralized function should be expected to relate to spatial ability should be contrasted, as well as predictions of whether lower spatial ability is to be expected in cases of reversed asymmetries or cases of nonsignificant asymmetries (bilateralization of function).

An indirect way of studying lateralization-ability relationships within sex is through research on the effects of handedness. This is an argument by triangulation. Left-handers are said to be (like females) more bilateralized and (like females) lower in spatial ability. Since the patterns are similar for females and left-handers, one is enabled to deemphasize the problems of third variables in each case: Presumably left- and right-handers' environmental experience does not differ as strongly or in the same ways as the experiences of males and females. The evidence on the relationship of handedness to spatial ability does not, however, show unequivocally that left-handers have lower spatial skills (see review by Carter–Saltzman, 1979). The complexities of the findings may in part reflect the fact that not all left-handers have bilateral representation of speech, the situation supposed by Levy (1974) to lead to lowered spatial ability. Opinions differ as to what proportion of left-handers have this form of cerebral organization (Satz, 1980) and how to identify them.

Levy and Reid (1978) argued that hand posture during writing could provide a convenient index. Their tachistoscopic evidence suggested that lateral differentiation in the brains of left-handers with noninverted writing posture was as strong as that for noninverted right-handers, although reversed. Left-handers with inverted writing posture showed weaker lateral differentiation as did females. Thus, crucial data for the relationship of handedness, cerebral organization, and spatial ability is whether left-handers with inverted writing posture and presumed bilateralization of function show low spatial skills. It should be noted, however, that Levy's conclusions have not been uniformly supported. Herron, Galin, Johnstone, and Ornstein (1979) and Moscovitch and Smith (1979) present evidence suggesting that inversion of writing posture may primarily index some aspect of visual processing or visuomotor interaction rather than bilateralization *per se*.

In summary, although research on the spatial ability of left-handers is intriguing, it is not yet clear enough to support the argument that sex-related and (putative) handedness-related differences in spatial ability can be explained by a common mechanism.

STRATEGIES AND EFFECTS OF EXPERIENCE

A second important issue is what causal relationship underlies any correlation between cerebral lateralization and sex-related differences in spatial ability. Although differential lateralization patterns could cause cognitive differences, cognitive differences could also cause lateralization patterns: this can be called the *strategy issue,* or the idea that people who have high ability use spatial-problem-solving strategies that are nonverbal, or holistic, or for some other reason likely to engage the right hemisphere. A third possibility is that another variable (such as sex-differentiated experience) could also cause both lateralization and cognitive differences independently of each other.

McGlone (1980) notes that it is often proposed that females are less lateralized on tests of nonverbal spatial functioning because they use verbal–analytic strategies rather than spatial–holistic ones. The former are presumably less efficient than the latter, resulting in lower performance as well as less lateralization. Although this would explain in terms of strategy why males are often more lateralized on spatial tests, it would not explain, McGlóne comments, why they are also more lateralized for language, as shown by dichotic listening studies and by sex differences in verbal deficits following left-hemisphere damage. Explanation of these facts would seem to require postulation of females using nonverbal strategies for language tasks and verbal ones for spatial tasks, truly a paradoxical situation.

Bryden (1979) has argued that the dichotic listening data may show sex differences not because females use nonverbal strategies for a verbal task, but because they use different strategies for allocation of attention, order of report, and decisions concerning whether to report items about which they are uncertain. When these factors are minimized, he reports, sex-related differences are not found. If true, this would leave McGlone's own lesion data (that males' verbal abilities are more affected than females' by left-hemisphere lesions and their spatial abilities are more affected by right-hemisphere lesions) as the major (admittedly impressive) support for sex differences in lateralization for speech. A second point, made by Ray and Newcombe (1980), is that sex-related strategy differences could account for some but not all of the lateralization differences, specifically for results related to spatial but not verbal processing. Lateralization of spatial functions seems more important for considerations of spatial skill, although Levy's (1974) theory assigns more importance to the existence of right-hemisphere speech in reducing the nonverbal specialization of the right hemisphere.

Could environment and experience affect lateralization by affecting preferred modes of processing? Bever and Chiarello (1974) reported that musicians showed right ear advantages on a dichotic listening test with musical stimuli, unlike nonmusicians, who showed the more usual left ear (right-hemisphere) advantage. Similar results have been obtained by Hirshkowitz, Earle, and Paley (1978) and Shanon (1980), although not by Zatorre (1979). Gaede, Parsons, and Bertera (1978) have argued against the interpretation that musical training changes the direction of ear advantages by obtaining evidence that musical *aptitude* rather than musical experience predicts lateralization patterns. However, Peretz and Morais (1980) have shown that nonmusicians reporting using analytic strategies are more likely to show right ear advantages than nonmusicians who do not report such strategies. The correlation of strategy use and laterality effects is of more interest than the effects of experience or aptitude *per se*, although the latter are important in so far as they affect strategy selection. The crucial question of whether strategy use produces laterality differences, or differences in cortical organization lead to use of certain strategies, can only be addressed through longitudinal study.

Very little evidence is currently available concerning the impact of experience with spatial tasks on hemispheric specialization for spatial stimuli. Mossip (1977) found positive correlations of the number of spatial toys (e.g., vehicles, construction sets) owned by 3- to 11-year-old boys and right-hemispheric advantages on a face perception task. The size of the correlations increased with age. Relationships for girls were puzzling: increasing negative correlations with age.

There is a clear need for further research on the effects of spatial experience on patterns of lateralization. This work should be experimental as well as correlational in nature, since people with certain cerebral and cognitive patterns may be attracted to certain activities (Gaede et al., 1978). There is also need for more direct assessment of the claim that females use verbal strategies for spatial tasks.

GROWTH CURVES AND DEVELOPMENTAL RELATIONSHIPS

A third issue for a lateralization–spatial ability theory concerns the developmental course of lateralization and of sex-related differences in lateralization. Specifically, do sex-related differences in lateralization appear as early as sex-related differences in spatial ability? Lenneberg's (1967) view that lateralization increased with age up to puberty has come under increasing attack. Studies of neonates have shown both electrophysiological (Davis & Wada, 1977; Molfese, Freeman, & Palermo, 1975) and behavioral (Entus, 1977) evidence of hemispheric asymmetry in processing of speech and nonspeech auditory stimuli and visual flash stimuli. Anatomical studies have supplemented this by showing structural asymmetries in the brains of infants (Wada, Clark, & Hamm, 1975; Witelson & Pallie, 1973). Hiscock and Kinsbourne (1977) provide convincing evidence that the right-ear advantage on dichotic listening tests does not change with age when care is taken to make tasks developmentally appropriate and to reduce variability in attentional and reporting strategies.

Lateral asymmetries have thus been demonstrated in infants and preschool children. Are sex-related differences also present at these early ages? Kimura's (1963, 1967) work with children is often cited as showing earlier right-ear advantages for girls than boys, on a dichotic listening task using words. But in an exhaustive review of subsequent developmental dichotic listening studies, Witelson (1977) has concluded that there is remarkably little evidence in the bulk of studies that girls show earlier or stronger asymmetries for speech than boys.

For infants, there is presently a more mixed picture, probably in large part because there are fewer studies. Entus (1977) found no sex differences using changes in sucking rate as a behavioral indication of response to dichotic stimuli. Using auditory evoked potentials, Molfese has reported sex differences or interactions involving sex (Molfese & Molfese, 1979a, 1979b; Molfese, Nunez, Seibert, & Ramanaiah, 1976), but not all of these interactions have been interpretable (Molfese & Molfese, 1979a) and apparently not all studies have found interactions (Molfese, 1977). Also using auditory evoked potentials, Shucard, Shucard, Cummins, and Campos (1981) have reported higher amplitude left than right responses by girls to *both* words and music, with boys showing greater amplitudes in the right hemisphere for both kinds of stimuli.

The main rationale for investigation of early sex-related differences in lateralization of speech is Levy's (1974) proposal that lower spatial ability results from bilateralization of speech functions to the right hemisphere. The available evidence shows either no early sex-related difference, or greater left-hemisphere responsiveness by girls. Neither is compatible with the idea of bilateralization, although the Molfese and the Shucard *et al.* data can be viewed as supportive of the Levy and Reid (1978) proposal that left-hemisphere functions mature earlier in girls, and subsequently bilateralize. Future research should explore, among other issues, why sex-related differences appear to be most common in studies of auditory evoked potential.

Developmental studies of right-hemisphere specialization relevant to spatial ability have largely focused on right hemisphere superiority in two tasks: visual recognition of faces or of spatial patterns and dichaphtic tasks involving recognition of nonsense shapes felt with the hands (e.g., Witelson, 1976). Unfortunately, the data emerging on age and sex differences in these studies are quite contradictory, as is the evidence that lateralization on them is significantly related to spatial ability (e.g., Newcombe & Bandura, in press).

Studies of visual field asymmetries in the accuracy of dot enumeration have found greater left-field advantages in 5-, 7-, and 11-year-old boys (Young & Bion, 1979) and adult males (Kimura, 1969; McGlone & Davidson, 1973) as compared with females of corresponding ages. However, reaction-time data gathered by Young and Bion showed that girls, but not boys, responded more quickly to left-field stimuli. This raises the issue of whether males and females differ primarily in the speed–accuracy tradeoffs made with left-field, as opposed to right-field, stimuli. These tradeoffs may have been present in the adult studies, since reaction time was not assessed.

Unfortunately, the above work in assessing hemispheric organization has generally not used spatial tasks that have shown sex-related differences in level of performance. One exception is work by Dawson, Warrenburg, and Fuller (1979) showing less left- and more right-hemispheric activation, as measured by EEG, in 7- to 13-year-old boys during performance of block design, mental rotation, and copy design tasks, as opposed to rote verbal memory, etc. But this study did not include girls as subjects!

The main support for the existence of early sex-related differences in right-hemispheric specialization comes from dichhaptic tasks, but these effects are confused by failures to replicate and by the facts that the task does not show sex-related differences in overall performance (except in the study of Dawson,

Farrow, & Dawson, 1980) and has not been shown to correlate with spatial ability tests.

Ratcliff's excellent discussion (Chapter 13, this volume) of problems in classifying spatial tasks and deficits, and in localizing lesions that can cause particular problems, should caution investigators against a global assumption that any apparently spatial task is right-hemispheric or that all right-hemisphere functions (e.g., musical perception, face perception) are necessarily relevant to spatial ability.

Socialization and Sex-Differentiated Experience

Despite considerable interest in the idea that sex differences in spatial abilities are related to environment and experiences, relatively little research has examined this hypothesis directly. Instead, work has focused on the personality and sex-typing correlates of spatial ability. Therefore, I will discuss this topic and sex differences in response to training and then review the few studies that have attempted to study spatial experiences themselves.

PERSONALITY AND SEX-TYPING CORRELATES

Maccoby (1966) argued that spatial ability was associated with cross-sex-typing, that is, with "masculine" traits in women and "feminine" traits in men. Kagan and Kogan (1970) suggested that this relationship may be stronger in females than in males. Maccoby and Jacklin (1974) suggested that this relationship might not, however, be specific to spatial ability, but occur for other intellectual abilities as well. In one of the few studies of correlations between traditional tests of spatial ability and personality, Ferguson and Maccoby (1966) identified fifth-grade children with uneven profiles of mental abilities, including groups high in spatial ability but low in either verbal or number ability. The high-space boys were low in aggression, maculinity, mastery, and sex-role acceptance, and high in withdrawal. High-space girls were low in aggression anxiety and tended to be high in aggression and masculinity, and low in sex-role acceptance. However, these groups were selected, not only on the basis of high spatial ability, but also for low verbal or numerical ability. Thus, the personality correlates and apparent cross sex-typing may be more associated with these children's deficiencies than with their high spatial abilities.

Nash (1975) found no relationship between global measures of sex-typing and DAT Space Relations Test scores in sixth and ninth graders. For ninth-grade boys rating one's actual-self in the masculine direction on 10 intellectually relevant traits was associated with high spatial ability ($r = .36$). For girls, ratings of ideal-self in the masculine direction were associated with high spatial ability ($r = .24$). Thus, viewing oneself (for boys) or desiring to be (for girls) active, independent, persistent, interested in math and science, and uninterested in art and literature are associated with high spatial ability.

Nash also asked the students whether they would prefer to be male or female. No ninth-grade boys said they would have preferred to be female, but the 7 (of 48) girls who would have preferred to be male had much higher spatial ability than the other girls, equivalent to that of the boys. In sixth grade, where there were no overall sex differences in spatial ability, the 23 girls who would have preferred to be boys had higher spatial scores than the 32 who preferred being girls, and the 4 boys who said they would rather be girls had very low spatial scores indeed. These results are, of course, compromised by the small Ns in many cells, but are, nevertheless, suggestive.

Signorella and Jamison (1978) gave eighth graders the Card Rotations Test as the primary measure of spatial ability, and also asked them to do a Piagetian horizontality task and complete an Embedded Figures Test. Sex-role orientation was measured using the Bem Sex-Role Inventory. Sex differences were observed on the first two spatial tests, but not on the EFT. Masculinity in girls was associated with better performance on the EFT and the Piagetian task, but not on the CRT. No associations of spatial ability and sex-role orientation were found for boys. Jamison and Signorella (1980) found that female undergraduates classified as masculine by the Bem Sex Role Inventory did better than other females on a Piagetian horizontality task; males did better who had either masculine or androgynous classifications.

In sum, the association of masculinity with high spatial ability in females seems reasonably well established, but the picture for males is more complex. This may be partly due to the fact that, especially in adolescence, some components of the male sex role may be inversely related. Although it is "masculine" to like math and science, be intellectually inquiring, and so on, such individuals may be perceived as "sissy" and may be less likely to be involved in sports or in heterosexual social activities.

The idea that masculine personality traits are correlated with spatial ability should be supplemented by specification of the mechanism involved. Since masculine activities and toys are more likely than feminine ones to be spatial in nautre, sex-typing measures may be just indirect assessments of the impact of experience. Another link might be that spatial ability is perceived as masculine, and that people strive to maintain congruence between their self-categorizations and their other attributes (Nash, 1979). They may well have different achievement standards for sex-typed areas (Lenney, 1977).

EFFECTS OF TRAINING

The hypothesis that differential experience accounts for sex differences in spatial ability predicts that "remedial" training or experience should differentially improve the performance of females, and reduce or eliminate sex differences. Connor, Serbin, and Schackman (1977) and Connor, Schackman, and Serbin (1978) have found evidence of this sort for preschool children's performance on the Preschool Embedded Figures Test, and Goldstein and Chance (1965) have observed such effects for adults' performance on the Embedded Figures Test. Vandenberg (1975) gave his Mental Rotations Test to sixth-graders before and after model building using blocks like those shown in the test. Both sexes showed better performance following model building, but only the increase for girls was significant. Other training efforts, however, have failed entirely, or failed to produce sex-differential results (see review by Harris, 1978). One striking instance of failure to eliminate sex differences by training or task adaptation is shown by studies concerned with understanding of Piagetian horizontality and verticality tasks. Liben (1978) and Liben and Golbeck (1980), and Thomas, Jamison, and Hummel (1973) had either very limited success in training people on these tasks, or success which was not differentially greater for females than males.

Of course, failures to train spatial ability or to eliminate sex differences by training could result from the treatment offered being insufficient or inappropriate. Although future efforts to identify how difficulties can be overcome certainly would have practical as well as theoretical value, such training studies seem quite unsuited for assessing the *causes* of sex-related differences, since whether training works or does not work could be unrelated to the original causes of the aptitude or deficiency.

EFFECTS OF EXPERIENCE

The hypothesis that differential experience with activities that develop spatial abilities cause sex differences in spatial abilities would be most directly tested by specifying these activities, showing that their frequency differs between the sexes, and showing that, within the sexes, engaging in such activities is correlated with higher spatial ability. Unfortunately, any results found might be due to self-selection: People with high spatial ability may choose (or be chosen) to engage in activities engaging these strengths rather than the strength being developed by the activity.

Nevertheless, the first step in investigating the effects of experience is to demonstrate the existence of correlations of activity and ability. For preschoolers, Connor and Serbin (1977) have shown that performance on WPPSI Block Design and the PEFT are positively correlated with masculine activity preferences and negatively correlated with feminine activity preferences for boys. No relationships were observed for girls. Masculine activities included playing with puzzles, building toys and vehicles, as well as large motor activities. Feminine activities included playing with dolls, sewing boards, and toy telephones, painting and crayoning, using the record player and musical instruments, and playing in a toy kitchen. The face validity of the masculine activities as being conducive to spatial abilities is farily obvious, as many involve three-dimensional manipulation and aiming. Serbin and Connor (1979) supplemented these findings by showing that, for both male and female preschoolers, those who clearly preferred the masculine to the exclusion of the feminine activities had higher spatial scores than those who preferred the feminine to the masculine. Coates, Lord, and Jakabovics (1975) found block play related to field independence for preschool girls.

Munroe and Munroe (1971) and Nerlove, Munroe, and Munroe (1971) report that spatial ability in an East African society is related to distance children go from home, and suggest that one interpretation is that experience in way-finding and maintaining a system of geographic landmarks stimulates spatial ability. Saegert and Hart (1978) review a variety of evidence showing that male children in the United States and other societies are allowed wider spatial ranges than female children.

Blade and Watson (1955) showed that coursework in engineering was related to gains in spatial ability for college men. They also found that the 10 highest scoring precollege men on the spatial relations test were more likely to have had mechanical or technical work or school experience than the 10 lowest scoring men. All of these experiences are more common for males than for females.

Tobin–Richards (1980) found no relationships between spatial ability and masculine or feminine activity preferences in male and female high-school seniors. It was not known, however, to what extent the activities were spatial in nature; despite a correlation between masculinity and spatiality of activities, there are surely many nonspatial masculine activities and many spatial feminine ones. Newcombe, Bandura, and Taylor (in press) found a significant correlation between spatial ability as measured by the Differential Aptitudes Test and participation in spatial activities. The 81 spatial activities were selected from a large pool of activities on the basis of agreement among 75% or more of judges that the activity was spatial in nature. Forty of the activities were sex-typed masculine, 21 feminine, with the remainder not sex typed.

Thus, several studies have identified experiences from the preschool through the college years that seem to be conducive to spatial ability and that boys participate in more than girls. The next step is to confront the issue of self-selection. Longitudinal data concerning the spatial activity and ability of children and adolescents are very much needed.

TIMING OF PUBERTY

Waber (1976, 1977a, 1977b) has combined hormonal (McGee, Chapter 9, this volume) and lateralization explanations in proposing that higher spatial ability is associated with later onset of puberty. Since boys reach puberty about

2 years later than girls on the average, such an effect might explain the sex-related difference in spatial ability. Using early- or late-maturing girls selected from fifth and eighth grades, and early- or late-maturing boys selected from eighth and tenth grades, she showed that the late maturers had higher spatial scores than the early maturers, for each age and sex. The groups did not differ in verbal skills.

Waber suggested that the difference between early and late maturers might be due to differences in patterns of hemispheric specialization. Late maturers (and males) might show greater functional hemispheric asymmetry than early maturers (and females); one reason might be that an ongoing process of lateral specialization is brought to a halt at puberty. She gave a dichotic listening test for speech stimuli to her subjects to test this hypothesis. No differences in ear advantage were found between early and late maturers for girls in fifth grade or boys in eighth grade, but significant differences emerged for girls in eighth grade and boys in tenth grade. For these older groups, late maturers showed greater right-ear advantages.

Controversy about Waber's hypothesis and results has focused on two issues: the replicability of the basic relationship between timing of puberty and spatial ability, and the proposed mechanism. The first question clearly has the logical priority. Although several studies with adolescents have replicated Waber's finding (Carey & Diamond, 1980; Herbst, 1980; Newcombe & Bandura, in press) others have not (Herbst & Petersen, 1979; Petersen, 1976; Rovet, 1979). Studies with adult women have also produced contradictory findings. Ray *et al.* (1981) found later maturers to do better on a test of spatial orientation, but not a test of spatial visualization. Strauss and Kinsbourne (1981) found no correlation of age at menarche with performance on Piaget's water-level task. Large-scale British (Douglas & Ross, 1964; Nisbet & Illesley, 1963) and American (Stone & Barker, 1939) studies have found seemingly opposite relationships between timing of puberty and cognitive ability among 7- and 8-year-olds: Early maturers do *better* than late maturers on standard intelligence tests and measures of school achievement. This may be due to the measures being more heavily loaded with verbal than with spatial items; perhaps early maturers (and girls) are better at verbal and worse at spatial skills.

The relationship may be real but relatively weak. Using 11-year-old girls, unselected for spatial ability or timing of puberty, Newcombe and Bandura found a small but significant relationship: Maturity status accounted for about 5% of the variance in spatial ability. Thus the association, although potentially of theoretical importance, may not be a large one, and may be hard to observe if samples are small or heterogeneous, or lack a range of early and late maturers. Furthermore, Waber's hypothesis may hold only for children from particular socioeconomic groups (Waber, Bauermeister, Cohen, Ferber, & Wolff, 1981).

If timing of puberty is related to spatial ability, is this assoication indeed mediated by differences in cerebral organization, as suggested by Waber? Waber's (1977b) proposal that lateralization increases with age until puberty has been increasingly questioned, and Waber (1979) has recently proposed a different model, in which cognitive functions undergo reorganization during the pubertal growth spurt. The nature of the reorganization is not precisely specified, however, and the proposal awaits empirical tests.

A second problem is that Waber as well as Rovet (1979) have found differences in ear advantage between early and late maturers only for the older subjects of each sex. The younger subjects did not show such a pattern, although they did already show maturity-related differences in spatial ability. Thus, Waber is in the position of arguing that something that appears later in development explains something that appears earlier.

Waber suggested that the dissociation might be explained by supposing that there is a temporary disruption of lateralization of function when children are undergoing rapid pubertal change. There is some support for this idea in the data of Leehey (1977), noted above, who found that left field advantages for perception of unfamiliar faces appear at age 10 for both sexes, but temporarily disappear in girls at 12, and in boys at 14. This modification of the hypothesis, however, opens up the question of whether the finding of a relationship between time of maturation and ear advantage among the older subjects could also be an artifact of pubertal variations in hemispheric specialization. Waber tested no subjects at an age where both early and later maturers would have completed puberty.

In the absence of convincing evidence for the lateralization explanation, an alternative explanation for the effect of timing of puberty deserves consideration. Timing of puberty is associated with certain personality characteristics (e.g., Jones & Bayley, 1950; Weatherley, 1964). The personality of the late maturer, and in particular his or her sex-role orientation, resembles that of the person with high spatial ability (Kagan & Kogan, 1970; Maccoby, 1966).

For males, the personality structure for late maturers involves traits generally characterized as less masculine and/or more feminine (Jones, 1957; Jones & Bayley, 1950; Mussen & Jones, 1957; Weatherley, 1964). For example, in their thirties, late-maturing males were less dominant and higher in need for succorance (Jones, 1957). Late-maturing females have traits that can be characterized as less feminine and/or more masculine than those of early maturers. For example, late maturers appear higher in need for recognition and achievement (Jones & Mussen, 1958). The conclusion that late maturers of both sexes may show cross-sex-typing dovetails nicely with findings discussed in the previous section that the same pattern may characterize males and females high in spatial ability, although the pattern there was clearer for females.

Newcombe and Bandura (in press) attempted to test both the lateralization and the sex-role explanations of the relationship between timing of puberty and spatial ability. A sample of 85 11-year-old girls was assessed in terms of their pubertal status, spatial ability, right-hemispheric lateralization for spatial perception (using the Flanery and Balling [1979] dichhaptic task) and personality traits related to masculinity, femininity, masculine intellectual interests, and wanting to be a boy. Subjects also reported on their spatial activity.

As noted above, timing of puberty was significantly correlated with spatial ability in this sample. Right-hemispheric lateralization as assessed by the dichhaptic task did not seem to be the key to the relationship, however, since laterality coefficients calculated from it were uncorrelated with either timing of puberty or spatial ability. Early maturers were more feminine than later maturers on a bipolar masculinity–femininity scale from the California Personality Inventory, as suggested by the literature review above. But they were not less masculine, or more feminine, on unipolar scales of these traits devised by Spence and Helmreich (1978). High spatial ability was correlated with masculinity and masculine intellectual interests, as in other studies, but not clearly correlated with feminity. Thus, sex role, as measured in the study, did not appear to provide an explanation for the relationship of timing of puberty and spatial ability either.

Future work to assess the mechanism of the effect of timing of puberty must address several issues. If hemispheric specialization is to be considered the causal mechanism, it will need to be shown, preferably in longitudinal samples, that sex- and maturity-related differences in specialization appear *before* sex- and maturity-related differences in spatial ability, even though pubertal dips may result in their temporary disappearance. Sex-role explanations of the effect may also be assessed by further work addressed to personality differences and to the daily spatial activities of adolescents; behavioral observation or time-budget studies may be preferable to paper-and-pencil personality measures.

INTERACTIONIST APPROACHES

The preceding sections have reviewed separately research on lateralization and experiential accounts of sex-related differences in spatial ability. Waber's (1977b) combination of the hormonal and lateralization hypotheses has a virtue lacking in much of the other research assessed: It provides a specific and testable hypothesis focused on the explanation of sex-related cognitive differences and it attempts to integrate a variety of previously isolated phenomena. There is currently increasing interest in the formulation of specific models explaining sex-related differences in spatial ability, especially those that combine consideration of biological and environmental forces in an interactionist manner (Parsons, 1980; Petersen, 1980). Two interactionist hypotheses are discussed briefly by Harris (1980). He suggests that small initial differences between the sexes, perhaps in visual versus auditory sensitivity (McGuinness & Pribram, 1979; cf. Hermelin, Chapter 2, this volume), could be magnified by environmental reinforcement and by self-selection of experiences to create sex-related differences in spatial ability. A second possibility is that biological variables could act in complex and socially-mediated ways: For instance, androgens, rather than acting directly on the brain to influence spatial ability, could affect energy expenditure (Ehrhardt & Baker, 1974), and this could in turn affect spatial ability by influencing the geographic range covered by the child in play.

A third interactionist hypothesis is that timing of maturation could cause variations in the type of activities people engage in and the personalities they develop, and thus in their spatial ability, independently or in addition to possible effects of timing of puberty on cortical organization. A large number of interactionist hypotheses can in fact be formulated, and it is important at this time to realize that we will have to go beyond pious subscription to the statement "biological and social forces interact" to the specification and testing of models.

Our understanding of sex-related differences in spatial ability can be enhanced in several ways. One is by greater specification of the term *spatial ability,* and careful cognitive-psychological study of the mental processes involved in spatial cognition. A second and key issue is the need for increased recognition of the correlational nature of most of the evidence that is presently available. Work designed especially to assess direction of causation and third-variable problems can help to cope with this problem. Experimental methodology should be used whenever feasible, for instance in studying lateralization-ability relationships in situations in which people are randomly assigned to programs designed to enhance ability levels. A third need is for model building and testing. When a chapter such as the present one is written in 10 or 20 years, perhaps it will be possible to organize it with headings other than a simple list of "genetic," "hormonal," "lateralization," and "experiential" factors.

ACKNOWLEDGMENTS

I would like to thank Lynn Liben and Michael Potegal for their very careful readings of drafts of this chapter.

REFERENCES

Allen, G. L., Kirasic, K. C., Siegel, A. W., & Herman, J. F. Developmental issues in cognitive mapping: The selection and utilization of environmental landmarks. *Child Development,* 1979, *50,* 1062–1070.

Berenbaum, S. A., & Harshman, R. A. On testing group differences in cognition resulting from differences in lateral specialization: Reply to Fennell et al. *Brain and Language*, 1980, *11*, 209–220.

Bergan, A., McManis, D. L., & Melchert, P. A. Effects of social and token reinforcement on WISC Block Design performance. *Perceptual and Motor Skills*, 1971, *32*, 871–880.

Bever, T. G., & Chiarello, R. V. Cerebral dominance in musicians and nonmusicians. *Science*, 1974, *185*, 587–589.

Birkett, P. Predicting spatial ability from hemispheric 'non-verbal' lateralisation: Sex, handedness and task differences implicate encoding strategy effects. *Acta Psychologica*, 1980, *46*, 1–14.

Blade, M., & Watson, W. S. Increase in spatial visualization test scores during engineering study. *Psychological Monographs*, 1955, *69* (Whole No. 397).

Bryden, M. P. Evidence for sex-related differences in cerebral organization. In M. A. Wittig & A. C. Petersen (Eds.), *Sex-related differences in cognitive functioning*. New York: Academic Press, 1979.

Carey, S., & Diamond, R. Maturational determination of the developmental course of face encoding. In D. Caplan (Ed.), *The biological bases of cognitive processes*. Cambridge: MIT Press, 1980.

Carroll, J. B., & Maxwell, S. E. Individual differences in cognitive abilities. In M. R. Rosenweiz & L. W. Porter (Eds.), *Annual Review of Psychology* (Vol. 30). Palo Alto: Annual Reviews, 1979.

Carter–Saltzman, L. Patterns of cognitive functioning in relation to handedness and sex-related differences. In M. A. Wittig & A. C. Petersen (Eds.), *Sex-related differences in cognitive functioning*. New York: Academic Press, 1979.

Coates, S. Sex differences in field independence among preschool children. In R. C. Friedman, R. M. Richart, & R. L. Vande Wiele (Eds.), *Sex differences in behavior*. New York: John Wiley, 1974.

Coates, S., Lord, M., & Jakabovics, E. Field dependence–independence, social–non-social play and sex differences in preschool children. *Perceptual and Motor Skills*, 1975, *40*, 195–202.

Cohen, R. *The influence of acquistion conditions and task demands on interlocation distance estimates*. Paper presented to the Society for Research in Child Development, Boston, April, 1981.

Cohen, R., & Weatherford, D. L. Effects of route traveled in the distance estimates of children and adults. *Journal of Experimental Child Psychology*, 1980, *29*, 403–412.

Cohen, R., Weatherford, D. L., & Byrd, D. Distance estimates of children as a function of acquisition and response activities. *Journal of Experimental Child Psychology*, 1980, *30*, 464–472.

Cohen, R., Weatherford, D. L., Lomenick, T., & Koeller, K. Development of spatial representations: Role of task demands and familiarity with the environment. *Child Development*, 1979, *50*, 1257–1260.

Connor, J. M., Schackman, M., & Serbin, L. A. Sex-related differences in response to practice on a visual–spatial test and generalization to a related test. *Child Development*, 1978, *49*, 24–29.

Connor, J. M., & Serbin, L. A. Behaviorally based masculine- and feminine-activity-preference scales for preschoolers: Correlates with other classroom behaviors and cognitive tests. *Child Development*, 1977, *48*, 1411–1416.

Connor, J. M., Serbin, L. A., & Schackman, M. Sex differences in children's response to training on a visual-spatial test. *Developmental Psychology*, 1977, *13*, 293–294.

Cooper, L. A. Individual differences in visual comparison processes. *Perception & Psychophysics*, 1976, *19*, 433–444.

Davis, A. E., & Wada, J. A. Hemispheric asymmetrics in human infants: Spectral analysis of flash and click evoked potentials. *Brain and Language*, 1977, *4*, 23–31.

Dawson, G., Warrenburg, S., & Fuller, P. *Task-related EEG asymmetry in children*. Paper presented to the Society for Research in Child Development, San Francisco, March, 1979.

Dawson, G. D., Farrow, B. J., & Dawson, W. E. *Sex differences and haptic cerebral asymmetry.* Paper presented to the Midwestern Psychological Association, St. Louis, May, 1980.

Douglas, J. W. B., & Ross, J. M. Age of puberty related to educational ability, attainment and school leaving age. *Journal of Child Psychology and Psychiatry,* 1964, *5,* 185–196.

Dreyer, A. S., Dreyer, C. A., & Nebelkopf, E. B. Portable rod-and-frame test as a measure of cognitive style in kindergarten children. *Perceptual and Motor Skills,* 1971, *33,* 775–781.

Droege, R. C. Sex differences in aptitude maturation during high school. *Journal of Counseling Psychology,* 1967, *14,* 407–411.

Ehrhardt, A. A., & Baker, S. W. Fetal androgens, human central nervous system differentiation, and behavior sex differences. In R. C. Friedman, R. M. Richart, & R. L. Vande Wiele (Eds.), *Sex differences in behavior.* New York: John Wiley, 1974.

Emmett, W. G. Evidence of a space factor at 11+ and earlier. *British Journal of Psychology: Statistical Section,* 1949, *2,* 3–16.

Entus, A. K. Hemispheric asymmetry in processing dichotically presented speech and non-speech stimuli by infants. In S. J. Segalowitz & F. A. Gruber (Eds.), *Language development and neurological theory.* New York: Academic Press, 1977.

Fairweather, H. Sex differences in cognition. *Cognition,* 1976, *4,* 231–280.

Fairweather, H. Sex differences: Still being dressed in the emperor's new clothes. *The Behavioral and Brain Sciences,* 1980, *3,* 234–235.

Feldman, A., & Acredolo, L. The effect of active versus passive exploration on memory for spatial location in children. *Child Development,* 1979, *50,* 698–704.

Fennell, E., Satz, P., Van Den Abell, T., Bowers, D., & Thomas, R. Visuospatial competency, handedness, and cerebral dominance. *Brain and Language,* 1978, *5,* 206–214.

Ferguson, L. R., & Maccoby, E. E. Interpersonal correlates of differential abilities. *Child Development,* 1966, *37,* 549–571.

Flanagan, J. C., Davis, F. B., Dailey, J. T., Shaycroft, M. F., Orr, D. B., Goldberg, I., & Newman, C. A., Jr. *Project talent: The identification, development, and utilization of human talents.* Pittsburgh: Project Talent Office, University of Pittsburgh, 1964.

Flanery, R. C., & Balling, J. D. Developmental changes in hemispheric specialization for tactile spatial ability. *Developmental Psychology,* 1979, *15,* 364–372.

Furst, C. J. EEG asymmetry and visuospatial performance. *Nature,* 1976, *260,* 254–255.

Gaede, S. E., Parsons, O. A., & Bertera, J. H. Hemispheric differences in music perception: Aptitude is experience. *Neuropsychologia,* 1978, *16,* 369–373.

Goldstein, A. G., & Chance, J. E. Effects of practice on sex-related differences in performance on embedded figures. *Psychonomic Science,* 1965, *3,* 361–362.

Guilford, J. P. (Ed.). *Printed classification tests* (Report No. 5, Army Air Force Aviation Psychology Program Research Reports) Washington D.C.: U.S. Printing Office, 1947.

Gur, R. C., & Gur, R. E. Handedness and individual differences in hemispheric activation. In J. Herron (Ed.), *Neuropsychology of left-handedness.* New York: Academic Press, 1980.

Gur, R. C., & Reivich, M. Cognitive task effects on hemispheric blood-flow in humans: Evidence for individual differences in hemispheric activation. *Brain and Language,* 1980, *9,* 78–92.

Harris, L. J. Sex differences in spatial ability: Possible environmental, genetic, and neurological factors. In M. Kinsbourne (Ed.), *Asymmetrical function of the brain.* Cambridge: Cambridge University Press, 1978.

Harris, L. J. Variances and anomalies. *Science,* 1979, *206,* 50–52.

Harris, L. J. Sex-related variations in spatial skill. In L. S. Liben, A. H. Patterson, and N. Newcombe (Eds.), *Spatial representation and behavior across the life span: Theory and application.* New York: Academic Press, 1981.

Hart, R. *Children's experience of place.* New York: Irvington, 1979.

Havighurst, R. J., & Janke, L. L. Relations between ability and social status in a Midwestern community. I: Ten-year-old children. *Journal of Educational Psychology,* 1944, *35,* 357–368.

Herbst, L. *Timing of maturation, brain lateralization and cognitive performance.* Paper presented to the American Psychological Association, Montreal, September, 1980.

Herbst, L., & Petersen, A. C. *Timing of maturation, brain lateralization and cognitive performance in adolescent females.* Paper presented to the Fifth Annual Conference on Research on Women and Education, Cleveland, November, 1979.

Herman, J. F. Children's cognitive maps of large-scale spaces: Effects of exploration, direction, and repeated experience. *Journal of Experimental Child Psychology,* 1980, *29,* 126–143.

Herman, J. F., & Siegel, A. W. The development of cognitive mapping of the large-scale environment. *Journal of Experimental Child Psychology,* 1978, *26,* 389–406.

Herron, J., Galin, D., Johnstone, J., & Ornstein, R. E. Cerebral specialization, writing posutre, and motor control of writing in left-handers. *Science,* 1979, *205,* 1285–1289.

Hirschkowitz, M., Earle, J., & Paley, B. EEG alpha asymmetry in musicians and nonmusicians: A study of hemispheric specialization. *Neuropsychologia,* 1978, *16,* 125–128.

Hiscock, M., & Kinsbourne, M. Selective listening asymmetry in preschool children. *Developmental Psychology,* 1977, *13,* 217–224.

Jamison, W., & Signorella, M. L. Sex-typing and spatial ability: The association between masculinity and success on Piaget's water-level task. *Sex Roles,* 1980, *6,* 345–353.

Jones, M. C. The later careers of boys who were early- or late-maturing. *Child Development,* 1957, *28,* 113–128.

Jones, M. C., & Bayley, N. Physical maturing among boys as related to behavior. *Journal of Educational Psychology,* 1950, *41,* 129–148.

Jones, M. C., & Mussen, P. H. Self-conceptions, motivations, and interpersonal attitudes of early- and late-maturing girls. *Child Development,* 1958, *29,* 491–501.

Kagan, J., & Kogan, N. Individuality and cognitive performance. In P. H. Mussen (Ed.), *Carmichael's Manual of Child Psychology* (3rd ed.). New York: John Wiley, 1970.

Kail, R., Pellegrino, J., & Carter, P. Developmental changes in mental rotation. *Journal of Experimental Child Psychology,* 1980, *29,* 102–116.

Kail, R. V., Jr., & Siegel, A. W. Sex differences in retention of verbal and spatial characteristics of stimuli. *Journal of Experimental Child Psychology,* 1977, *23,* 341–347.

Keogh, B. K. Pattern copying under three conditions of an expanded spatial field. *Developmental Psychology,* 1971, *4,* 25–31.

Keogh, B. K., & Ryan, S. R. Use of three measures of field organization with younger children. *Perceptual and Motor Skills,* 1971, *33,* 466.

Kimura, D. Speech lateralization in young children as determined by an auditory test. *Journal of Comparative and Physiological Psychology,* 1963, *56,* 899–902.

Kimura, D. Functional asymmetry of the brain in dichotic listening. *Cortex,* 1967, *3,* 163–178.

Kimura, D. Spatial localization in left and right visual fields. *Canadian Journal of Psychology,* 1969, *23,* 445–458.

Kinsbourne, M. If sex differences in brain lateralization exist, they have yet to be discovered. *The Behavioral and Brain Sciences,* 1980, *3,* 241–242.

Leehey, S. C. *A developmental change in right hemisphere specialization.* Paper presented to the Society for Research in Child Development, New Orleans, March, 1977.

Lenneberg, E. H. *Biological foundation of language.* New York: Wiley, 1967.

Lenney, E. Women's self-confidence in achievement settings. *Psychological Bulletin,* 1977, *84,* 1–13.

Levy, J. Psychobiological implications of bilateral asymmetry. In S. Dimond & J. B. Beaumont (Eds.), *Hemispheric function in the human brain.* New York: Halstead Press, 1974.

Levy, J., & Reid, M. Variations in cerebral organization as a function of handedness, hand posture in writing, and sex. *Journal of Experimental Psychology: General,* 1978, *107,* 119–144.

Liben, L. S. Performance on Piagetian spatial tasks as a function of sex, field dependence, and training. *Merrill-Palmer Quarterly,* 1978, *24,* 97–110.

Liben, L. S., & Golbeck, S. L. Sex differences in performance on Piagetian spatial tasks: Differences in competence or performance? *Child Development,* 1980, *51,* 594–597.

Lord, F. E. A study of spatial orientation of children. *Journal of Educational Research,* 1941, *34,* 481–505.

Maccoby, E. E. Sex differences in intellectual functiong. In E. E. Maccoby (Ed.), *The development of sex differences.* Stanford: Stanford University Press, 1966.

Maccoby, E. E., & Jacklin, C. N. *The psychology of sex differences.* Stanford: Stanford University Press, 1974.

McGee, M. Human spatial abilities: Psychometric studies and environmental, genetic, hormonal, and neurological influences. *Psychological Bulletin,* 1979, *86,* 889–918.

McGlone, J. Sex differences in human brain asymmetry: Crticial survey. *The Behavioral and Brain Sciences,* 1980, *3,* 215–227.

McGlone, J., & Davidson, W. The relation between cerebral speech laterality and spatial ability with special reference to sex and hand preference. *Neuropsychologia,* 1973, *11,* 105–113.

McGuinness, D., & Pribram, K. H. The origins of sensory bias in the development of gender differences in perception and cognition. In M. Bortner (Ed.), *Cognitive growth and development.* New York: Brunner/Mazel, 1979.

Mellone, M. A. A factorial study of picture tests for young children. *British Journal of Psychology,* 1944, *35,* 9–16.

Michael, W. B., Guilford, J. P., Fruchter, B., & Zimmerman, W. S. The description of spatial-visualization abilities. *Educational and Psychological Measurement,* 1957, *17,* 185–199.

Molfese, D. L. Infant cerebral asymmetry. In S. J. Segalowitz & F. A. Gruber (Eds.), *Language development and neurological theory.* New York: Academic Press, 1977.

Molfese, D., Freeman, R. B., & Palermo, D. S. The ontogeny of brain lateralization for speech and nonspeech stimuli. *Brain and Language,* 1975, *2,* 356–368.

Molfese, D. L., & Molfese, V. J. Hemisphere and stimulus differences as reflected in the cortical responses of newborn infants to speech stimuli. *Developmental Psychology,* 1979, *15,* 505–511. (a)

Molfese, D. L., & Molfese, V. J. VOT distinctions in infants: Learned or innate? In H. A. Whitaker & H. Whitaker (Eds.), *Studies in neurolinguistics* (Vol. 4). New York: Academic Press, 1979. (b)

Molfese, D. L., Nunez, V., Seibert, S. M., & Ramanaiah, N. V. Cerebral asymmetry: Changes in factors affecting its development. *Annals of the New York Academy of Sciences,* 1976, *280,* 821–833.

More, D. M. Developmental concordance and discordance during puberty and early adolescence. *Monographs of the Society for Research in Child Development,* 1953, *18,*(1, Ser. No. 56).

Moscovitch, M., & Smith, L. C. Differences in neural organization between individuals with inverted and noninverted handwriting posures. *Science,* 1979, *205,* 710–713.

Mossip, C. *Hemispheric specialization as seen in children's perception of faces.* Paper presented to the Eastern Psychological Association, Boston, April, 1977.

Munroe, R. L., & Munroe, R. H. Effect of environmental experience on spatial ability in an East African society. *Journal of Social Psychology,* 1971, *83,* 15–22.

Mussen, P. H., & Jones, M. C. Self-conceptions, motivations, and interpersonal attitudes of late- and early-maturing boys. *Child Development,* 1957, *28,* 243–256.

Nash, S. C. The relationship among sex-role stereotyping, sex-role preference, and the sex difference in spatial visualization. *Sex Roles,* 1975, *1,* 15–32.

Nash, S. C. Sex role as a mediator of intellectual functioning. In M. A. Wittig & A. C. Petersen (Eds.), *Sex-related differences in cognitive functioning.* New York: Academic Press, 1979.

Nerlove, S. B., Munroe, R. H., & Munroe, R. L. Effect of environmental experience on spatial ability: A replication. *Journal of Social Psychology,* 1971, *84,* 3–10.

Newcombe, N., & Bandura, M. M. The effect of age at puberty on spatial ability in girls: A question of mechanism. *Developmental Psychology,* in press.

Newcombe, N., Bandura, M. M., & Taylor, D. G. Sex differences in spatial ability and spatial activities. *Sex Roles*, in press.

Nisbet, J. D., & Illesley, R. The influence of early puberty on test performance at age 11. *British Journal of Educational Psychology*, 1963, 33, 169–176.

Oltman, P. K., Ehrlichman, H., & Cox, P. W. Field independence and laterality in the perception of faces. *Perceptual and Motor Skills*, 1977, 45, 255–260.

Packer, I. K. *The relationship between individual differences in hemispheric activation and cognitive performance*. Unpublished doctoral dissertation, University of Pennsylvania, 1979.

Parsons, J. E. Psychosexual neutrality: Is anatomy destiny? In J. E. Parsons (Ed.), *The psychobiology of sex differences and sex roles*. Washington, D.C.: Hemisphere, 1980.

Pellegrino, J. W., & Glaser, R. Cognitive correlates and components in the analysis of individual differences. In R. J. Sternberg & D. K. Detterman (Eds.), *Human intelligence: Perspectives on its theory and measurement*. Norwood, New Jersey: Ablex, 1979.

Peretz, I., & Morais, J. Modes of processing melodies and ear asymmetries in non-musicians. *Neuropsychologia*, 1980, 18, 477–489.

Petersen, A. C. Physical androgny and cognitive functioning in adolescence. *Developmental Psychology*, 1976, 12, 533–542.

Petersen, A. C. Biopsychosocial processes in the development of sex-related differences. In J. E. Parsons (Ed.), *The psychobiology of sex differences and sex roles*. Washington, D.C.: Hemisphere, 1980.

Pizzamiglio, L. Handedness, ear preference, and field dependence. *Perceptual and Motor Skills*, 1974, 38, 700–702.

Porteus, S. D. The measurement of intelligence: Six hundred and fifty-three children examined by the Binet and Porteus tests. *Journal of Educational Psychology*, 1918, 9, 13–31.

Porteus, S. D. *Porteus Maze Test: Fifty years application*. Palo Alto, California: Pacific Books, 1965.

Ray, W. J., & Newcombe, N. Interpreting sex differences in lateralization. *The Behavioral and Brain Sciences*, 1980, 3, 246.

Ray, W. J., Newcombe, N., Semon, J., & Cole, P. M. Spatial abilities, sex differences and EEG functioning. *Neuropsychologia*, 1981, 19, 719–722.

Richmond, P. G. A limited sex difference in spatial test scores with a preadolescent sample. *Child Development*, 1980, 51, 601–602.

Rovet, J. F. *Individual differences and rates of development: A study of pathologies*. Paper presented to the Society for Research in Child Development, San Francisco, March, 1979.

Saegert, S., & Hart, R. The development of environmental competence in girls and boys. In M. Salter (Ed.), *Play: Anthropological perspectives*. Cornwall, New York: Leisure Press, 1978.

Satz, P. Incidence of aphasia in left-handers: A test of some hypothetical models of cerebral speech organization. In J. Herron (Ed.), *Neuropsychology of left-handedness*. New York: Academic Press, 1980.

Serbin, L. A., & Connor, J. M. Sex-typing of children's play preferences and patterns of cognitive performance. *Journal of Genetic Psychology*, 1979, 134, 315–316.

Schiller, B. Verbal, numerical and spatial abilities of young children. *Archives of Psychology*, 1934, 24, No. 161.

Shanon, B. Lateralization effects in musical decision tasks. *Neuropsychologia*, 1980, 18, 21–31.

Shucard, J. L., Shucard, D. W., Cummins, K. R., & Campos, J. J. Auditory evoked potentials and sex-related differences in brain development. *Brain and Language*, 1981, 13, 91–102.

Siegel, A. W., Herman, J. F., Allen, G. L., & Kirasic, K. C. The development of cognitive maps of large- and small-scale spaces. *Child Development*, 1979, 50, 582–585. '

Siegel, A. W., & Schadler, M. The development of young children's spatial representations of their classrooms. *Child Development*, 1977, 48, 388–394.

Signorella, M. L., & Jamison, W. Sex differences in the correlations among field dependence, spatial ability, sex-role orientation, and performance on Piaget's water-level task. *Developmental Psychology*, 1978, *14*, 689–690.

Spence, J. T., & Helmreich, R. L. *Masculinity and feminity: Their psychological dimensions, correlates and antecedents*. Austin: University of Texas Press, 1978.

Stone, C. P., & Barker, R. G. The attitudes and interests of premenarcheal and postmenarcheal girls. *Journal of Genetic Psychology*, 1939, *54*, 27–71.

Strauss, E., & Kinsbourne, M. Does age at menarche affect the ultimate level of verbal and spatial skills? *Cortex*, 1981, *17*, 323–325.

Thomas, H., Jamison, W., & Hummel, D. D. Observation is insufficient for discovering that the surface of still water is invariantly horizontal. *Science*, 1973, *81*, 173–174.

Tobin–Richards, M. H. *The influence of experience and sex role identity on cognitive performance*. Paper presented to the American Psychological Association, Montreal, September, 1980.

Vandenberg, S. G. Sources of variation in performance on spatial tasks. In J. Eliot & N. J. Salkind (Eds.), *Children's spatial development*. Springfield: Charles C Thomas, 1975.

Vandenberg, S. G., & Kuse, A. R. Spatial ability: A critical review of the sex-linked major gene hypothesis. In M. A. Wittig & A. C. Petersen (Eds.), *Sex-related differences in cognitive functioning*. New York: Academic Press, 1979.

Vasta, R., Regan, K. G., & Kerley, J. Sex differences in pattern copying: Spatial cues or motor skills? *Child Development*, 1980, *51*, 932–934.

Waber, D. P. Sex differences in cognition: A function of maturation rate? *Science*, 1976, *192*, 572–574.

Waber, D. P. Biological substrates of field dependence: Implications of the sex difference. *Psychological Bulletin*, 1977, *84*, 1076–1087. (a)

Waber, D. P. Sex differences in mental abilities, hemispheric lateralization and rate of physical growth at adolescence. *Developmental Psychology*, 1977, *13*, 29–38. (b)

Waber, D. P. Cognitive abilities and sex-related variations in the maturation of cerebral cortical functions. In M. A. Wittig & A. C. Petersen (Eds.), *Sex-related differences in cognitive functioning*. New York: Academic Press, 1979.

Waber, D. P., Bauermeister, M., Cohen, C., Ferber, R., & Wolff, P. H. Behavioral correlates of physical and neuromotor maturity in adolescents from different environments. *Developmental Psychobiology*, 1981, *14*, 513–522.

Wada, J. A., Clark, R., & Hamm, A. Cerebral hemispheric asymmetry in humans. *Archives of Neurology*, 1975, *32*, 239–246.

Wapner, S. Age changes in perception of verticality and of the longitudinal body axis under body tilt. *Journal of Experimental Child Psychology*, 1968, *6*, 543–555.

Weatherley, D. Self-perceived rate of physical maturation and personality in late adolescence. *Child Development*, 1964, *35*, 1197–1210.

Widiger, T. A., Knudson, R. M., & Rorer, L. G. Convergent and discriminant validity of measures of cognitive styles and abilities. *Journal of Personality and Social Psychology*, 1980, *39*, 116–129.

Willis, S. G., Wheatley, G. H., & Mitchell, O. R. Cerebral processing of spatial and verbal–analytic tasks: An EEG study. *Neuropsychologia*, 1979, *17*, 473–484.

Wilson, J. R., DeFries, J. C., McClearn, C. E., Vandenberg, S. G., Johnson, R. C., & Rashad, M. N. Cognitive abilities: Use of family data to assess sex and age differences in two ethnic groups. *International Journal of Aging and Human Development*, 1975, *6*, 261–276.

Witelson, S. F. Sex and the single hemisphere: Specialization of the right hemisphere for spatial processing. *Science*, 1976, *193*, 425–427.

Witelson, S. F. Early hemisphere specialization and interhemispheric plasticity: An empirical and theoretical review. In S. J. Segalowitz & F. A. Gruber (Eds.), *Language development and neurological theory*. New York: Academic Press, 1977.

Witelson, S., & Pallie, W. Left hemisphere specialization for language in the newborn. *Brain*, 1973, *96*, 641–646.

Witkin, H. A., Goodenough, D. R., & Karp, S. A. Stability of cognitive style from childhood to young adulthood. *Journal of Personality and Social Psychology*, 1967, *7*, 291–300.

Witkin, H. A., Oltman, P. K., Raskin, E. R., & Karp, S. A. *Manual for the Embedded Figures Test.* Palo Alto, California: Consulting Psychologists Press, 1971.

Wittig, M. A., & Petersen, A. C. (Eds.). *Sex-related differences in cognitive-functioning.* New York: Academic Press, 1979.

Yen, W. Sex-linked major gene influence on selected types of spatial performance. *Behavior Genetics*, 1975, *5*, 281–298.

Young, A. W., & Bion, P. J. Hemispheric laterality effects in the enumeration of visually presented collections of dots by children. *Neuropsychologia*, 1979, *17*, 99–102.

Zatorre, R. J. Recognition of dichotic melodies by musicians and nonmusicians. *Neuropsychologia*, 1979, *17*, 607–617.

Zoccolotti, P., & Oltman, P. K. Field dependence and lateralization of verbal and configurational processing. *Cortex*, 1978, *14*, 103–155.

CORTICAL CONTRIBUTIONS TO
SPATIAL ABILITIES

ARTHUR BENTON 11

Spatial Thinking in Neurological Patients: Historical Aspects[1]

This chapter sketches the development of knowledge and ideas about diverse defects of spatial thinking in patients with brain disease. It covers the period beginning with the inception of these ideas somewhat over 100 years ago to the 1960s. The review is necessarily selective and incomplete since the literature on the topic is enormous. However, it should provide a historical background for discussion of the current status of the field.

HUGHLINGS JACKSON

The idea that a lesion in the brain could produce a specific defect in thinking, orientation, or action with respect to space arose within the context of the broader concept of agnosia. The origins of the latter concept can be traced in the evolving thought of Hughlings Jackson. As early as 1864, Jackson raised the question of whether perception, particularly visual perception, might not have its seat in the right cerebral hemisphere. Ten years later, in his famous paper on the nature of the duality of the brain, he discussed the possibility in greater detail, expressing the opinion that the posterior part of the right hemisphere

[1]The personal investigations cited in this chapter were supported by Research Grant NS-00616 from the National Institute of Neurological and Communicative Disorders and Stroke.

SPATIAL ABILITIES
Development and Physiological Foundations

served the function of visual recognition and memory (Jackson, 1874). As he phrased it, "the hinder part of the brain on the right side is the chief seat of the revival of images in the *recognition* of object, places, etc. [p. 101]." As an example of a defect in this capacity caused by disease, he cited the case of a patient who exhibited what is now called disorientation for place. Although she was in the London Hospital, she asserted that she was in a place in Holborn where she had worked for some years. But Jackson admitted that the evidence for locating "perception" in the right hemisphere was not strong, remarking that "as will be seen, my facts are very few [p. 103]."

Two years later, Jackson's ideas took more definite form when he described a patient who proved to have a tumor in the posterior right hemisphere and whose "first symptoms were those of what I call imperception. She often did not know objects, persons and places." The first indication that something was amiss with this 59-year-old woman was when she lost her way going from her home to a nearby park, a route she had been taking for 30 years. Over the course of the next few weeks, she showed odd lapses of behavior. For example, she made mistakes in dressing herself, a type of disability that later was given the name of "dressing dyspraxia." After admission to the hospital, she was not able to identify the different nurses attending her. Collectively these deficits in recognition, orientation, and action constituted what Jackson called *imperception,* "a defect as special as aphasia." However, his term was not adopted by other clinicans who later described the same constellation of deficits under tbe rubric of *mindblindness* or *visual agnosia.*

The next major contribution to this topic came from the physiological laboratory. Having shown that extensive destruction of the occipital lobes produced blindness in dogs and monkeys, Hermann Munk (1878) then described the effects of a more limited lesion in the same area. The animal (in this case, the dog) obviously could see since he ambulated freely and avoided obstacles. Yet he seemed to have lost the capacity to grasp the meaning of many visual stimuli that he appeared to perceive clearly. For example, he showed no signs of special recognition of his master, nor did he react appropriately to a threatening gesture or to the sight of a piece of meat. Munk called this disturbance in visual behavior *mindblindness* and he ascribed it to a loss of visual memory images that prevented the animal from relating current visual experience to past experience.

Munk's idea of a higher order disturbance in which visual associational or memory processes are impaired against a background of intact visuosensory capacity received a skeptical reception from his fellow physiologists, who were inclined to interpret mindblindness as an expression of defects in visual discrimination. Some clinicans also doubted the reality of mindblindness, feeling that the condition was essentially a reflection of defective visual acuity, and their position was supported by Munk's own contention that mindblindness resulted from a lesion in the same areas as served foveal vision. However, other clinical observers readily accepted the concept since they were convinced that they had seen patients with the same impairment Munk had described in dogs. Case reports, such as those by Wilbrand (1887) and Lissauer (1890), established that minblindness or visual object agnosia did exist in human subjects. The primary deficit shown by these patients was an inability to recognize objects or persons despite apparently adequate visual acuity. In addition, some showed spatial disorientation, as reflected in losing one's way along familiar routes both indoors and outdoors, not being able to describe the location of familiar streets and buildings or to describe the disposition of the furniture in one's living room or bedroom. Since not all patients with visual object agnosia showed these spatial disabilities, the designation of *visuospatial agnosia* was given to them to indicate their essential independence from disorders of object recognition.

BADAL AND SPATIAL THINKING

Another trend in clinical investigation emphasized impairment in spatial thinking independently of sensory modality. The pioneer study of this type was that of a Bordeaux ophthalmologist, Jules Badal (1888), who entitled his paper "Contribution to the Study of Psychic Blindness; Alexia; Agraphia; Inferior Hemianopsia; Disorder of the Sense of Space" (cf. Benton, 1969; Benton & Meyers, 1956). His patient showed a large number of diverse deficits but her chief complaint was spatial disorientation. She could not find her way in her apartment or in the immediate neighborhood and she would not leave home without a guide. She could not answer simple questions about the spatial relations of the main streets of Bordeaux although she had long been thoroughly familiar with them. Like Jackson's patient, she showed dyspraxia for dressing. "She could not dress herself alone without putting on her clothes in reverse, taking the left sleeve for the right. It took her an infinite time to determine the order in which the different articles of clothing should be put on." She showed faulty visual guidance of hand movements in reaching (cf. Damasio & Benton, 1979). "She also had great difficulty in finding objects which she needed, even when they were in front of her. She would reach for it and her hand would constantly pass over it as if influenced by a false projection due to paralysis of the eye muscles. However, there was no diplopia, no strabismus and the eye movements were executed in a normal manner [p. 101]." Her central visual acuity was excellent. If her attention could be directed to them—for she had great difficulty in maintaining and shifting ocular fixation—she could name the smallest letters on the eye chart and she showed no difficulty in recognizing or naming the smallest objects.

But, as suggested by the title of his paper, Badal's main point was that the patient suffered from an impairment in spatial perception that transcended the visual modality. Although her auditory acuity appeared to be intact, she made errors in identifying the direction of the source of sounds. Impairment of spatial perception in the somesthetic modalities was manifested in defective body schema performances. She showed gross finger agnosia, not being able to state which of her fingers had been touched, and she could no longer distinguish between the right and left sides of her body. These errors in localization were made whether her eyes were open or closed.

Similar cases with visual disorientation and difficulty in making spatial judgments but with intact visual acuity and visual object recognition were described by Foerster (1890), Dunn (1895), Peters (1896), Meyer (1900), and Lenz (1905). Discussing the fact that one of his patients showed a serious disability in reading, Meyer pointed out that this type of reading failure, which reflected a disturbance in spatial orientation, was not to be confused with true alexia which was of a linguistic character.

In 1909 Balint described a patient with a rather distinctive assemblage of deficits associated with inaccurate visual judgments. The most prominent features of the syndrome, which Balint described as "psychic paralysis of gaze, optic ataxia, and spatial disturbance of attention [p. 51]" and which is now called *Balint's syndrome,* were maintenance of ocular fixation on a point in the visual field with seeming inability to shift fixation, inattention to objects and events in other parts of the visual field, and misreaching for objects. These impairments in oculomotor and visuomotor function were accompanied by a number of perceptual deficits, such as difficulty in estimating distances, in reading long polysyllabic words (although simpler words could be read), and in identifying figures such as a hexagon or an octagon although simpler figures such as a triangle or square were readily identified. Attention was generally fixated on a point in the right visual field with consequent neglect of stimuli in the left field. There were many expressions of impairment in the execution of visually guided movements. Besides misreaching for objects, the patient was unable to draw or copy or to use a knife accurately in cutting. But he could recognize objects and read single letters.

An interesting development of this early period is seen in Rieger's (1909) hypothesis that the brain contains two distinct but interacting apparatuses, one serving verbal–conceptual functions and the other spatial–practical functions. His ideas were developed further by Reichardt (1918) who described visuospatial and visuoconstructional defects in association with lesions of the right hemisphere and contrasted them with the linguistic impairments associated with left-hemisphere disease.

Thus, by 1910, a fairly large number of specific deficits in performance, indicative of imapirment of spatial thinking, had been described. As Table 11.1

TABLE 11.1
Performance Deficits Described in the Literature as of 1910

1. Inability to follow routes
2. Defective topographic or geographic memory
3. Difficulty in judging distance
4. Difficulty in reading
5. Inattention in left visual field
6. Misreaching for objects
7. Dyspraxia for dressing
8. Difficulty in locating objects
9. Disturbances in ocular fixation

shows, these deficits could be perceptual, amnesic, or praxic in nature. Thinking about their neurological basis followed two lines. The dominant point of view was that these disabilities were the product of bilateral disease involving the occipital and posterior parietal and temporal areas, and this position received substantial support from autopsy study. Thus, the brain of Foerster's (1890) patient showed bilateral softening confined to the occipital and parietal lobes (Sachs, 1895) and the brain of one of Meyer's (1900) patients showed bilateral occipital lobe disease. Autopsy study of Balint's (1909) patient showed bilateral atrophy of the occipital, parietal, and temporal lobes with the most marked changes evident in the posterior parietal and temporal areas. The prevailing view was that this extensive destruction of visual association cortex resulted in a loss of visual memory images.

A second line of thought emphasized the role of the right hemisphere in the mediation of spatial thinking. As has been seen, Jackson considered the posterior part of the right hemisphere to be a center for visual recognition and memory, including spatial memory. In the 1890s, a Philadelphia physician, T. D. Dunn (1895), specifically localized spatial orientation in the right hemisphere. Having described a patient who could not remember the spatial relations of familiar streets or even of the rooms in his house but who could still recognize objects, faces, and simple words, he postulated the existence of a "geographic center" in the right hemisphere for what he called "the sense of location." Since his patient did not have visual object agnosia or facial agnosia, Dunn denied that it was necessary to invoke a generalized loss of visual memory images as the underlying disability, the patient having only loss of "optical images of locality."

Further support for a specific association between spatial disorientation and disease of the right hemisphere came from the observations of Peters (1896) and Lenz (1905), each of whom called attention to the relatively high frequency of spatial disorientation in patients with left homonymous hemianopia. However,

in contrast to Dunn, both authors were extremely cautious in making inferences from their observations. Peters declined to draw a conclusion and
Lenz only raised the question of whether "the right occipital lobe is perhaps
more strongly related to orientation than the left."

THE CONTRIBUTION OF WALTHER POPPELREUTER

Study of patients with traumatic head wounds during and directly after
World War I added further knowledge about the visuospatial defects associated
with brain injury. The most comprehensive and detailed investigation was that
of Poppelreuter (1917), who was the first to apply experimental–psychological
methods such as tachistoscopy and the instrumental measurement of depth
perception in the evaluation of these patients. As a consequence, he was able to
show that defective depth perception was extraordinarily frequent in patients
with occipital wounds. About one-third of them (with or without visual-field
defects) proved to have some degree of impairment when acuity of depth perception was quantitatively assessed. Poppelreuter further pointed out that most
of these patients did not complain of having difficulties in depth perception in
everyday life.

He also studied accuracy in reaching for objects and found that misreaching
could occur with different degrees of severity, that it might be present in the
whole visual field or only in the periphery, and that patients with intact fields,
as well as hemianopics, showed the defect. He emphasized the motor or behavioral aspects of misreaching and believed that the traditional explanation of
it in terms of a disturbed "sense of locality" was incorrect. Rather it was a
disturbance of coordination between sensory and motor processes, that is, an
apraxia and not a disturbance in perception. Visuosensory capacity and distance
judgment can be completely intact in a patient who misreaches. Thus, misreaching or defective visual guidance of hand movements reflects "not an alteration in the content of perception but a disturbance of the mechanism of
localization."

Poppelreuter studied the exploration of visual space in his patients by means
of his "field of search" test in which 57 diverse visual stimuli (numbers, letters,
geometric figures) were presented in an irregular array and the subject was
required to point to stimuli named or shown by the examiner. Relying on
quantitative measures of performance, Poppelreuter was able to demonstrate
that visual exploration was significantly prolonged in many patients, particularly those with occipital or bilateral wounds. Hemianopic defect per se did not
appear to have an important influence on performance. Decades later Teuber,
Battersby, and Bender (1949; Teuber, 1964) utilized this procedure to investigate the exploration of visual space in both adults and children with brain

damage and found prolongation in searching time not only in those with parieto–occipital lesions but also in those with frontal damage. In some patients with unilateral injury the deficit was particularly severe in the contralateral visual field.

Misreaching for objects or inaccurate pointing to them was only one form of what Poppelreuter called *visual apraxia*. Other forms of the disability were awkwardness in manipulating objects (e.g., cutting with scissors), defective visually guided locomotion (e.g., rail-walking), and visuoconstructional defects such as poor copying of designs, failure in block designs tasks, and inaccurate construction of block models. In discussing these performances he noted once again that visual-field defects were not closely related to them and he emphasized that these visuomotor deficits were independent of both disturbances in form perception, as disclosed by tachistoscopic or "mixed figures" tests, and impairment in distance judgment.

KARL KLEIST AND CONSTRUCTIONAL APRAXIA

One of Poppelreuter's more important contributions was his emphasis on visuoconstructive disabilities as a distinctive type of disturbance that could be shown by patients with brain disease. This theme was then taken up by Karl Kleist (1922/1934), who earlier had made similar observations (Kleist, 1912) and who now coined the term *constructional apraxia* for the disability he defined as a disturbance "in formative activities such as assembling, building, and drawing, in which the spatial form of the product proves to be unsuccessful, without there being an apraxia of single movements."

As a neurologist, Kleist was particularly interested in the site of the lesions that led to constructional apraxia. He agreed with Poppelreuter that, at least in its pure form, constructional apraxia represented a failure in integration between perceptual and motor processes rather than a perceptual defect per se. Thus it was, in modern terminology, a "disconnection symptom" (Geschwind, 1965). At the same time, Kleist was aware that impaired visuoperceptual capacity could also be expressed in defective visuoconstructive performances and he conceded that it was often difficult to decide whether an observed visuoconstructive disability was a "true" constructional apraxia or the direct reflection of visuoperceptive impairment. In any case, what *he* meant by *constructional apraxia* was the disconnection symptom, not the motoric expression of a perceptual deficit. Indeed, that is why he called it an *apraxia,* that is, a disturbance in purposeful motor activity within a setting on intact sensory and motor capacity. With respect to the neurological basis for the disability, Kleist placed the crucial lesion in the parieto-occipital territory of the left or dominant hemisphere, the functional effect of such a lesion being to disconnect the visual association areas (Areas 18, 19, 39) from anterior motor and preomotor regions and hence to prevent visual information from reaching executive motor centers.

Kleist's description and particularly that of his student Hans Strauss (1924) established constructional apraxia as a form of motoric visuospatial disorder associated with brain disease. However, his insistence that it was purely a praxic or "executive" disability was for the most part ignored and sometimes actively challenged. For example, Kroll (1929), Kroll and Stollbun (1933), Lange (1936), and Schlesinger (1928) pointed out that disturbances in spatial orientation and visual perception were almost always part of the clinical picture of "constructional apraxia." Consequently, the term was generally applied to any visuoconstructional failure, whether or not it occurred within a setting of perceptual and orientational disability.

On the neurological side, Kleist's localization of the crucial lesion underlying constructional apraxia was also challenged by subsequent clinical study which indicated that in fact the deficit was more frequently shown by patients with right-hemisphere disease than by those with left-hemisphere damage (Benton, 1962; Benton & Fogel, 1962; Dide, 1938; Hécaen, Ajuriaguerra, & Massonet, 1951; Lange, 1936; McFie, Piercy, & Zangwill, 1950; Paterson & Zangwill, 1944; Piercy, Hécaen, & Ajuriaguerra, 1960; Piercy & Smyth, 1962). As has been mentioned, this point had already been made by Reichardt (1918), who constrasted visuospatial and visuoconstructive defects associated with right-hemisphere disease with the linguistic defects produced by lesions of the left hemisphere. Kleist took account of Reichardt's observations but questioned their validity on the grounds that his cases may have had left-hemisphere lesions that were not evident on macroscopic examination and also that some of his patients may have been left-handed.

Despite the accumulating evidence in its favor, the idea that there is a close association between constructional apraxia and disease of the right hemisphere was not readily accepted by neurologists. Admittedly, the early contributions to the topic, consisting of single case studies or global clinical impressions, could easily be discounted. Only after the more systematic studies of Zangwill, Hécaen, and their co-workers (1944–1951) were published was it necessary to consider the reported correlation seriously. However, even when the correlation was accepted, its neuropsychological meaning was not clear. It was not possible to consider the right hemisphere to be "dominant" for visuoconstructive or visuospatial activity in the same sense that the left hemisphere was considered to be dominant for language functions, for the two situations were not at all comparable. Aphasic disorder resulting from disease of the nondominant right hemisphere in a right-handed patient (i.e., so-called crossed aphasia), was recognized as a highly unusual phenomenon, occurring in not more than 1 or 2% of cases (Russell & Espir, 1961; Zangwill, 1960). But, of course, constructional apraxia resulting from disease of the left hemisphere in right-handed patients was not at all unusual. It was only a matter of a bias toward a somewhat higher frequency in patients with lesions of the right hemisphere (Benton, 1962; Benton & Fogel, 1962).

This empirical finding that, although it is more frequent and more severe in patients with right-hemisphere disease, constructional apraxia is also shown by a substantial proportion of patients with disease of the left hemisphere posed a problem of interpretation. One proffered explanation more or less reflected the observed facts. It hypothesized a partial "dominance" of the right hemisphere for visuoconstructive functions in the form of their "bilateral but unequal representation" in the two cerebral hemispheres (Piercy & Smyth, 1962). Another explanation, first advanced by Duensing (1953), held that there are two distinctive types of visuoconstructive disability, of fundamentally different character. One type, equivalent to the "true" constructional apraxia of Kleist, is of an executive or motor nature and is related to the disruption of mechanisms in the left hemisphere. The other type results from a basic impairment in visuospatial thinking and is related to disruption of mechanisms in the right hemisphere.

Piercy and Smyth (1962) undertook to evaluate the cogency of the two competing explanations by investigating the relationship between constructional performance and the status of visuoperceptive ability in patients with right- and left-hemisphere disease. A number of tests requiring the copying of designs and the construction of block and matchstick models were employed to assess constructional ability and Raven's Progressive Matrices provided the measure of visuoperceptive capacity. It was reasoned that if the hypothesis of two distinctive types of visuoconstructive disability in relation to side of lesion is correct, then a close association between defective constructional performance and visuoperceptual impairment should be seen in patients with right-hemisphere lesions but not in those with disease of the left hemisphere, at least some of whom should show constructional apraxia within a context of intact visuoperceptive capacity. The results were unequivocal. Constructional performances and score on the Raven test were very closely associated in both hemispheric groups. The two groups differed only in respect to level of performance, the patients with right-hemisphere damage showing more severe impairment on both the constructional and perceptual tasks. Piercy and Smyth concluded that the theory of two distinctive types of visuoconstructive disability could not be supported and that the concept of "bilateral but unequal representation" in the two hemispheres was the more likely possibility.

However, a study the same year by Costa and Vaughn (1962) generated different results. Giving the WAIS block designs and the Raven matrices to patients with unilateral lesions, they found a close correlation ($r = .83$) between the two performances in those with right-hemisphere disease but only a minimal positive association ($r = .33$) in those with left-hemisphere disease.

Although the authors did not draw a specific conclusion, their findings were consonant with the existence of discrete visuoperceptual and executive types of constructional impairment.

Qualitative analysis of the constructional performances provided another approach to the question of whether there existed distinctive types of disabilities that were differentially related to side of lesion. It was argued that impaired visuospatial functions should be reflected in productions characterized by distortion of the spatial relations among the elements of the construction. Construction from a model should be as defective as construction to verbal command because of the pervasive visuoperceptual disability. In contrast, executive or psychomotor impairment with attendant awkwardness in manipulation of the pencil or blocks should be reflected in constructions that are coherent and spatially correct but simplified and lacking in elaboration. Constructions from a model should be superior to those from verbal command since the model could facilitate slavish copying and the making of corrections when indicated. Fragmentary evidence in favor of this distinction was adduced by a number of investigators, among them Hécaen et al. (1951), Duensing (1953), McFie and Zangwill (1960), Piercy et al. (1960), and Arrigoni and De Renzi (1964). However, the detailed comparative study by Warrington, James, and Kinsbourne (1966) of drawing performance in relation to side of lesion failed to confirm most of the hypothesized qualitative differences in the productions.

Thus, by the 1960s there were a number of unresolved questions about constructional apraxia, the primary ones being whether it represented one or more than one disability and what was the exact nature of its relationship to side of lesion in patients with unilateral brain disease. In addition, the question of whether or not constructional apraxia was merely a reflection of general mental impairment was raised (Benton, 1967, 1969).

TYPES OF IMPAIRMENT

The casualties of World War I provided an abundance of case material for detailed study and this led to the development of classifications of types of visual disorientation. Gordon Holmes (1918; Holmes & Horrax, 1919) divided into two categories the disorders of visual orientation and attention that he observed: (1) disturbances in localization, orientation and distance perception, as reflected in misreaching for objects, inability to learn routes, and difficulty in reading connected text; (2) oculomotor defects, such as the inability to shift fixation on verbal command and absence of the blink reflex.

On his part, Kleist (1922/1934) distinguished between disturbances of visuospatial *perception* (as reflected in defective bisection of lines, pointing to objects, or distance judgments) and visuospatial *agnosia* or *amnesia* (as reflected in the forgetting of geographic and topographic relationships). The neurological significance of the distinction was that the perceptual disorders were related to lesions in the calcarine region and its immediate vicinity (Areas 17 and 18). These lesions might be unilateral (producing, for example, mislocalization in one visual half-field) or bilateral (producing, for example, defective depth perception). In contrast, the agnosic or amnesic disorders were related to lesions of the outlying Area 19 primarily involving the left hemisphere.

Russell Brain (1941) made still another distinction between defective localization of stimuli within grasping distance (as in misreaching for objects) and defective localization of stimuli beyond arm's reach (as in poor distance judgments). Some of his patients showed impairment in one form of localization and not in the other, dissociations that he attributed to differences in the site of the causative lesions. He suggested that lesions involving the connections between visual cortex and the hand and arm area of the parietal somatosensory cortex produced mislocalizations within arm's reach whereas those affecting the connections between the visual cortex and the leg area resulted in impaired localization of more distant stimuli within "walking distance," as he phrased it. Later Birkmayer (1951) made a similar distinction.

Brain also offered a classification of forms of visual disorientation. The first form was defective localization of objects in space and this could be subdivided into three types, depending on whether it resulted from impaired visual acuity, hemianopia, or impairment of cerebral integrative processes. This last type was true agnosia for spatial relationships and could be exhibited in either one half-field, as Riddoch (1935) had shown, or in the entire field. Patients with visuospatial agnosia, particularly if it is present throughout the visual field, "cannot find their own way round objects when they run into them, set out in the wrong direction to get to others which they clearly see, and have difficulty in finding their way about and learning the topography of a room [p. 268]."

A second type of disorientation was loss of stereoscopic vision, an uncommon disorder previously described by Riddoch (1917) and Holmes and Horrax (1919). Another rare disorder was visual allesthesia, that is, the referral of visual stimuli to the opposite half-field, comparable to the more extensively described tactile allesthesia in which stimuli applied to one side of the body are referred to the oppsoite side (cf. Benton, 1959). Brain had not seen such a case and mentioned it only for the sake of completeness in listing.

A more important type of disorientation was neglect or unawareness of the left half of visual space shown by patients with extensive lesions of the parieto-occipital territory of the right hemisphere. Comparable to the neglect of the left side of the body seen in patients with right parietal lesions, this unawareness of the left half of space produces an inability to follow routes because of a tendency to avoid making turns to the left when indicated.

Finally, Brain listed loss of topographical memory as an independent form of visual disorientation, pointing out that a patient may have serious visuospatial problems on the perceptual level and yet be able to describe routes and spatial interrelationships accurately. Although most of the cases of this type described in the literature had proven to have bilateral disease, Brain expressed the opinion that the crucial lesion was in the left occipital area.

DISTURBANCES OF THE BODY SCHEMA

As has been noted, Badal included his patient's inability to discriminate the left and right sides of her body and to identify her fingers as expressions of a "disorder of the sense of space." Many clinical investigators followed his example in viewing these and other disturbances of the body schema as reflecting a basic impairment in spatial thinking. After Gerstmann (1924, 1927) described finger agnosia as "a circumscribed disturbance of orientation towards one's body," Lange (1930, 1933) and Stengel (1944) suggested that this rather peculiar deficit as well as the other elements of the Gerstmann syndrome (right–left disorientation, agraphia, acalculia) should be considered as part of a more comprehensive syndrome of spatial disorientation involving external space and constructional praxis as well as the body schema.

There were, however, competing explanations. Head (1926) interpreted impairment in right–left discrimination to be a specific expression of the same disturbance in "symbolic formulation and expression" that defined aphasia. Consequently, he included tests of right–left orientation in his battery for the assessment of aphasic patients. Moreover, these "hand, eye and ear" tests, as he called them, included nonverbal imitation tasks as well as tasks requiring naming or understanding the names of body parts. Supporting Head's interpretation, Benton (1959) cited the extraordinarily high frequency with which aphasic patients showed disturbed left–right awareness and finger recognition as evidence that the occurrence of these deficits in patients with unilateral brain disease reflects a disturbance of language or symbolic thinking.

Ajuriaguerra and Hécaen (1949) and Benton (1959) pointed out that disorientation with respect to external space did not necessarily imply disorientation toward one's own body and, conversely, that disturbances in the body schema did not necessarily imply the occurrence of conventional visual disorientation as described by Poppelreuter and Kleist. However, the frequent concurrence of the two types of orientational disability, particularly in the form of neglect of the left half of space *and* of one's body, was noted by Pinéas (1931), Brain (1941), Critchley (1953), and other observers. Brain believed that visuospatial orientation and body awareness were intimately related and that the posterior parietal region connecting the visual cortex with the somatosensory area in the postcentral gyrus provided an anatomical basis for this association.

AUTITORY-SPATIAL DISCRIMINATION

Apart from single case reports, such as those of Badal (1888), Penfield and Evans (1934), Ross and Fountain (1948), and Wortis and Pfeffer (1948), the spatial aspects of audition in patients with brain disease were not investigated until the 1950s when Sanchez–Longo and Forster (1958; Sanchez–Longo, Forster, & Auth, 1957) reported their studies of auditory localization. In the first study the performance of five patients with unilateral temporal lobe lesions were compared to a group of control patients with the finding that four of the five patients were grossly inaccurate, their mean error of localization being larger than the poorest of the 20 control subjects. Moreover, in analogy with visual localization, every patient showed a larger error of localization in the auditory field contralateral to the side of lesion. The results were confirmed in a second study of a larger number of patients. In addition, a group of patients with lesions not involving the temporal lobe were assessed with the finding that their overall performance level was within the normal range. The observation that patients with unilateral temporal lobe injury often show particularly severe impairment in localization in the contralateral field was consonant with the earlier case reports of Penfield and Evans (1934) and Wortis and Pfeffer (1948).

However, Walsh (1957), presenting click stimuli through earphones and manipulating time and intensity differences between the clicks as cues to localization of the apparent direction of the source of sound, reported that patients with temporal lobe lesions performed quite adequately. A second study, in which sounds were transmitted through air to assess "real" rather than "apparent" localization, produced the same results. The three patients with temporal lobe lesions were able to localize the source of the sound. Walsh's treatment of his data was rather impressionistic (in contrast to the quantitative analyses of Sanchez–Longo and Forster) and it is not clear that patients performed on a normal level of accuracy. In any case, his results were in accord with those earlier reported by Ross and Fountain (1948).

The study of Shankweiler (1961) introduced some new elements into an already rather confused picture. Two of his basic findings were negative. There was no evidence for particularly severe impairment of localization in the contralateral auditory field in patients with unilateral lesions nor did those with temporal lobe disease show more severe defects than those with parieto-occipital or frontal lesions. However, he did find that the patients with right-hemisphere lesions were poorer in performance than those with left-hemisphere lesions, significantly on one auditory localization task (pointing to the source of a sound) and nonsignificantly on a second task (discrimination of angular differences). Pointing out that auditory localization takes place in a visually organized space, Shankweiler considered that the observed between-hemispheres difference in accuracy of localization was not unexpected in view of the established association between visuospatial deficits and disease of the right hemisphere. Another finding, which perhaps explains some of the disocrdant results of previous studies, was that for both control subjects and patients the correlation between performances on the two auditory localization tasks (pointing versus verbal judgment) was not significant.

Teuber (1962), employing dichotic clicks with manipulation of time and intensity differences to assess auditory localization, reported results that were in accord with those of Shankweiler. Patients with right-hemisphere lesions performed more poorly than those with left-hemisphere disease. In addition, performance level tended to be specific to the task. A patient might perform adequately on the intensity difference task but not on the time difference task or vice versa. Teuber concluded that "brain injury can impair binaural localization based on time and intensity differences, but the effects are dissociable, suggesting at least partial separation of neural mechanisms underlying these two forms of localication [p. 154]."

TACTILE–SPATIAL PERFORMANCES

The concept that the tactile and kinesthetic senses have a spatial component as well as methods for measuring that component arose from the inspired work of Ernst Heinrich Weber (see Ross & Murray, 1978), two of whose studies have recently appeared in English translation (Weber, 1978). He devised the tactile compass to measure the accuracy of discrimination of two points on the skin and later developed the procedure of having a subject indicate a point on the skin that had been touched in order to measure the accuracy of tactile localization. The two-point threshold is primarily a measure of tactile acuity, analogous to visual acuity. Nevertheless, that there is a spatial component in the task of discriminating two points on the skin surface is indicated by the significant correlation between this performance and the error of tactile localization (Boring, 1930). However, it was generally agreed that the task of localizing a stimulated point on the skin was the more direct measure of the "sense of space."

Beginning in the 1870s, two-point discrimination and single point localization were intensively studied both in patients with lesions of the central nervous system and in those with peripheral nerve injuries. It was readily established that impaired performance could occur as a consequence of disease of the somatosensory system at any level from the peripheral nerve to the cerebral cortex. Thereafter attention was focused on a number of questions. One of these was whether these tactile "spatial" deficits could occur independently of more basic defects in tactile sensitivity or were merely expressions of these defects. A second question concerned the neurological basis of impaired tactile–spatial performances.

With regard to the question of the relationships of tactile–spatial defects to impaired sensitivity, there was a gradual accumulation of evidence that defective two-point discrimination and point localization could be shown in the absence of defects in tactile sensitivity and, conversely, that patients with severely impaired tactile sensitivity (i.e., raised thresholds to pressure, pain, and temperature) could show intact localizing capacity. In 1901 Otfrid Foerster reviewed the pertinent literature and described the findings in a large sample of his own patients to establish these facts. However, at the same time he pointed to the close association between defects in tactile localization and disturbances in kinesthetic sensitivity, as reflected in raised thresholds for the detection of passive movements of the fingers, hands and arms.

Recalling the longstanding controversy between proponents of nativist and empiricist theories of the genesis of space perception, Foerster adduced clinical evidence to support the proposition that the apprehension of spatial characteristics and relations is not an inherent attribute of sensory experience but instead gradually develops as associations between different sensory experiences are formed. Thus, the tactile sense of space is the resultant of the establishment of connections between tactile, visual, and kinesthetic experience. If kinesthetic experience is impoverished, as in cases of congenital paralysis, or is impaired as a consequence of acquired disease of the nervous system, then tactile experience is deprived of its spatial component and this is reflected in defective performance on spatial tasks such as point localization.

At the same time, other somatosensory tests of a spatial character were developed and applied for clinical and investigative purposes. Position sense, that is, a patient's awareness of the position in which his arm, finger, or toe has been placed, was assessed by his verbal report or his imitation of the position. Recognition of the direction of lines drawn on the skin surface was another procedure used to assess the spatial aspects of tactile perception. A more complicated maneuver was to draw numerals or letters on the forearm, palm, or finger tips and determine whether or not the patient could identify them. His failure to do so was labeled *agraphesthesia*.

Over the decades there was much discussion about the meaning of failing performance on these spatial tasks. Broadly speaking, there were two schools of thought on the subject. One group saw these deficits as expressions of a primary disorder of spatial thinking in which the patient is unable to organize sensory information into a coherent spatial framework. Head (1920), for example, ascribed inaccurate point localization and position sense to the impairment of spatially organized "schemata" relating to the body surface and to posture. The other group insisted that failing performance was merely the result of more basic disorders in sensitivity such as raised or unstable sensory thresholds, disturbances in sensory processes over time, and fatigability, so that in fact the patient did not receive the sensory information required to localize accurately or discriminate between two stimulated points. The controversy is still not resolved.

However, all neurologists agreed that failure on these somatosensory spatial tasks was a sensitive indicator of the presence of brain disease and, specifically, of disease of the parietal lobes. The detailed investigation of sensory disturbances from cerebral lesions carried out by Head and Holmes (1911; Head, 1920; Holmes, 1927) showed this clearly and later work only confirmed their findings. The study of Shy and Haase (1957), in which 25 patients with verified focal lesions in the parietal lobe were given a battery of 12 sensory tests, found that four of the five most frequently encountered defects were of a spatial character, namely, impaired position sense, inaccurate point localization, raised thresholds for two-point discrimination and inability to identify numerals drawn on the skin. The one nonspatial deficit in this set of five was a raised threshold for the perception of passive movement which, as Foerster (1901) had shown decades earlier, was closely correlated with tactile–spatial defects. In contrast, sensory thresholds for pressure, pain, and temperature were altered in relatively few patients and all of these showed one or more spatial deficits. Shy and Haase suggested that somatosensory deficits could be arranged in a hierarchical order in relation to overall severity of impairment, as shown in Table 11.2. They conceded, however, that there were cases whose pattern of deficits did not conform to the schema.

With respect to the neurological basis of these somatosensory deficits, clinical investigators such as Foerster (1901) and Head (1920) demonstrated that impairment could be produced by lesions at any level of the nervous system. Head and Sherren (1905), for example, found that defective two-point discrimination was closely associated with impaired sensitivity to tactile pressure in patients with peripheral nerve injuries and indeed was a remarkably persisting deficit that might be present after all other sensory capacities had returned to normal. Head and Thompson (1906) found that defective two-point discrimination was a frequent consequence of spinal cord disease but in this instance it could appear independently of any important disturbances of tactile sensitivity. Nevertheless, Head believed that lesions below the level of the cortex were more likely to result in basic sensory impairment (i.e., raised or unstable thresholds to light, pressure, pain, and temperature) than were cortical lesions, which primarily impaired the discriminative aspects of sensory experience.

TABLE 11.2
Levels of Severity of Defect in Parietal Lobe Disease[a]

Level	Deficits
I	Impaired position sense
II	I plus slowed perception of passive movement (normal threshold)
III	II plus raised threshold for passive movement and defective point localization
IV	III plus raised two-point threshold and agraphesthesia
V	IV plus astereognosis, defective roughness discrimination and temperature sense
VI	V plus raised threshold for pressure and pain

[a]Adapted from Shy and Haase, 1957.

Clinical study during the last 2 decades of the nineteenth century established that the cortical mechanisms medating somatosensory responses were primarily located in the parietal lobe. However, the precentral gyrus, which today is usually designated as a motor area, was often included in the somatosensory region, the combined precentral and postcentral gyri constituting a sensorimotor region. Moreover, it was found that lesions in posterior part of the parietal lobe, as well as in the postcentral gyrus, could produce various forms of impairment. Thus, Head (1920) concluded that "loss of sensation of the cortical type may be produced by a lesion of the pre- and post-central convolutions, the anterior part of the superior parietal lobule, and the angular gyri. These portions of the hemisphere contain the sensory centres [p. 759]." However, it was not possible to relate different forms of impairment to lesions in specific loci within this extended somatosensory cortical region.

It was understood that, in consonance with the facts of contralateral innervation, unilateral parietal disease would produce tactile impairment on the opposite side of the body. However, beginning in 1906 and extending through the 1930s, a number of clinicians reported instances of bilateral impairment in patients with apparently unilateral lesions. The earliest papers described bilateral defects in tactile object identification in association with unilateral lesions but later studies, such as those of Bychowsky and Eidinow (1934) and Koerner (1938), reported bilateral disturbances of sensitivity to pressure, pain, and vibration. Little attention was paid to these scattered papers until the publication of the large-scale study of Semmes, Weinstein, Ghent, and Teuber (1960) who demonstrated that bilateral impairment in pressure sensitivity, two-point discrimination, point localization, and the detection of passive movement was not a rare finding in patients with unilateral lesions. Subsequent studies generally confirmed their finding. A second observation that these bilateral defects were encountered with particularly high frequency in patients with left-hemisphere disease was not consistently supported. Subsequently, Carmon and Benton (1969) found bilateral impairment in the perception of the direction of punctate stimuli applied to the skin in patients with right-hemisphere disease but not in those with left-hemisphere lesions. They interpreted their findings as indicating that the right hemisphere plays a particularly important role in the mediation of behavior requiring the appreciation of spatial relations in the tactile as well as the visual modality. The significance of a supramodal spatial factor in the mediation of tactile form discrimination was emphasized in studies by Semmes (1965), De Renzi and Scotti (1969), and Dee and Benton (1970) which found a close relationship between defect on the tactile task and corresponding defect in visuospatial performances.

CONCLUDING COMMENTS

Disorders in spatial thinking associated with disease of the nervous system were recognized by clinicans as early as the 1880s. These disorders were typically studied within a single sensory modality. Vision was investigated in great detail and types of visuospatial disorder were formulated by Kleist, Brain, and other neurologists. The occurrence of specific disorders, which were often defined in terms of level (e.g., sensory, integrative, mnesic), was then related to the locus and extent of the causative lesion. Spatial deficits in the somatosensory modality, such as inaccurate tactile localization and impaired position sense, were also investigated, mainly in relation to their value in diagnosis but also as a vehicle for studying integrative processes in the central nervous system. In contrast, the spatial aspects of audition received relatively little attention. The concept that brain disease can produce a generalized impairment in spatial thinking that affects performance in all sensory modalities was advanced. Over the decades there was a slow accumulation of evidence pointing to the paramount importance of the right hemisphere in mediating spatial performances.

The exact nature of the interrelations among the diverse performance deficits that are regarded as indicative of "spatial disorientation" remains to be determined. Clinical observation attests that patients may show dissociated deficits, that is, failure on one spatial task but not the other. For example, although perceptumotor spatial disability, as reflected in failing visuoconstructional performance, is usually associated with visuoperceptual spatial disability, as reflected in failing performance on discrimination tasks, some patients show one type of deficit but not the other (Costa & Vaughn, 1962; Dee, 1970). There is also evidence that the neuropathological bases for distinctive types of disabilities may differ. Defective three-dimensional block construction, a perceptuomotor spatial disability, is shown by a substantial proportion of aphasic patients with left-hemisphere disease but impaired appreciation of the direction of lines, a visuoperceptual–spatial disability, is not characteristic of these patients (Benton, 1973; Benton, Hannay, & Varney, 1975; Benton, Varney, & Hamsher, 1978). Moreover, despite theorizing about the dependence of orientation in external space on the integrity of the body schema, clinical observation discloses numerous instances of dissociated impairment, as documented in the section on disturbances of the body schema in this chapter.

The intriguing concept of a supramodal spatial disability deserves further exploration. An association of visuospatial (and visuoconstructive) performance deficits with tactile–spatial deficits has been described by a number of investigators (Corkin, 1965; Dee & Benton, 1970; DeRenzi & Scotti, 1969; Milner, 1965). It seems likely that the basis for the relationship is that in sighted persons all spatial performances—visual, auditory, somesthetic—take place within a visually organized spatial matrix. However, the observed occurrence of dissociated visual and tactile spatial deficits raises the question of whether the determining factors may not be neuropathological rather than behavioral in nature.

REFERENCES

Ajuriaguerra, J. de., & Hécaen, H. *Le cortex cérébral*. Paris: Masson et Cie, 1949.

Arrigoni, G., & De Renzi, E. Constructional apraxia and hemispheric locus of lesion. *Cortex*, 1964, *1*, 170–197.

Badal, J. Contribution a l'étude des cécités psychiques: Alexie, agraphie, hémianopsie inférieure, trouble du sens de l'espace. *Archives d'Ophtalmologie*, 1888, *8*, 97–117.

Balint, R. Seelenlähmung des Schauens, optische Ataxie, räumliche Störung der Aufmerksamkeit. *Monatsschrift fuer Psychiatrie und Neurologie*, 1909, *25*, 51–81.

Benton, A. L. *Right–left discrimination and finger localization: Development and pathology*. New York: Hoeber–Harper, 1959.

Benton, A. L. The visual retention test as a constructional praxis task. *Confinia Neurologica*, 1962, *22*, 141–155.

Benton, A. L. Constructional apraxia and the minor hemisphere. *Confinia Neurologica*, 1967, *29*, 1–16.

Benton, A. L. Disorders of spatial orientation. In P. L. Vinken & G. W. Bruyn (Eds.), *Handbook of clinical neurology* (Vol. 3, Ch. 13). Amsterdam: North Holland, 1969.

Benton, A. L. Visuoconstructive disability in patients with cerebral disease: Its relationship to side of lesion and aphasic disorder. *Documenta Opthalmologica*, 1973, *34*, 67–76.

Benton, A. L., & Fogel, M. L. Three-dimensional constructional praxis. *Archives of Neurology*, 1962, *7*, 347–354.

Benton, A. L., Hannay, H. J., & Varney, N. R. Visual perception of line direction in patients with unilateral brain disease. *Neurology*, 1975, *25*, 907–910.

Benton, A. L., & Meyers, R. An early description of the Gerstmann syndrome. *Neurology*, 1956, *6*, 838–842.

Benton, A. L., Varney, N. R., & Hamsher, K. Visuospatial judgment: A clinical test. *Archives of Neurology*, 1978, *35*, 364–367.

Birkmayer, W. *Hirnverletzungen: Mechanismus, Spaetkomplikationen, Funktionswandel*. Wien: Springer–Verlag, 1951.

Boring, E. G. The two-point limen and the error of localization. *American Journal of Psychology*, 1930, *42*, 446–449.

Brain, W. R. Visual disorientation with special reference to lesions of the right cerebral hemisphere. *Brain*, 1941, *64*, 244–272.

Bychowsky, G., & Eidinow, M. Doppelseitige Sensibilitaetstoerungen bei einseitigen Gehirnherden. *Nervenarzt*, 1934, *7*, 498–506.

Carmon, A., & Benton, A. L. Tactile perception of direction and number in patients with unilateral cerebral disease. *Neurology*, 1969, *19*, 525–532.

Corkin, S. Tactually-guided maze learning in man: Effects of unilateral cortical excisions and bilateral hippocampal lesions. *Neuropsychologia*, 1965, *3*, 339–351.

Costa, L. D., & Vaughan, H. G. Performances of patients with lateralized cerebral lesions: Verbal and perceptual tests. *Journal of Nervous and Mental Disease*, 1962, *134*, 162–168.

Critchley, M. *The parietal lobes*. London: Edward Arnold, 1953.

Damasio, A. R., & Benton, A. L. Impairment of hand movements under visual guidance. *Neurology*, 1979, *29*, 170–178.

Dee, H. L. Visuoconstructive and visuoperceptive deficit in patients with unilateral cerebral lesions. *Neuropsychologia*, 1970, *8*, 305–314.

Dee, H. L., & Benton, A. L. A cross-modal investigation of spatial performances in patients with unilateral cerebral disease. *Cortex*, 1970, *6*, 261–272.

De Renzi, E., & Scotti, G. The influence of spatial disorders in impairing tactual discrimination of shapes. *Cortex*, 1969, *5*, 53–62.

Dide, M. Les désorientations temporo-spatiales et la préponderance de l'hémisphère droit dans les agnoso-akinésies proprioceptives. *Encéphale*, 1938, *33*, 276–294.

Duensing, F. Raumagnostische und ideatorisch-apraktische Stoerung des gestaltenden Handelns. *Deutsche Zeitschrift fuer Nervenheilkunde*, 1953, *170*, 72–94.

Dunn, T. D. Double hemiplegia with double hemianopsia and loss of geographic centre. *Transactions, College of Physicians of Philadelphia*, 1895, *17*, 45–56.

Foerster, O. Untersuchungen ueber das Localisationsvermoegen bei Sensibilitaetsstoerungen: Ein Beitrag zur Psychophysiologie der Raumvorstellung. *Monatsschrift fuer Psychiatrie und Neurologie*, 1901, *9*, 31–42.

Foerster, R. Ueber Rindenblindheit. *Graefes Archiv fuer Ophthalmologie*, 1890, *36*, 94–108.

Gerstmann, J. Fingeragnosie: Eine umschriebene Stoerung der Orientierung am eigenen Koerper. *Wiener Klinische Wochenschrift*, 1924, *37*, 1010–1012.

Gerstmann, J. Fingeragnosie und isolierte Agraphie-ein neues Syndrom. *Zeitschrift fuer die gesamte Neurologie und Psychiatrie*, 1927, *108*, 152–177.

Geschwind, N. Disconnexion syndromes in animals and man. *Brain*, 1965, *88*, 237–294; 586–644.

Head, H. *Studies in neurology*. London: Oxford University Press, 1920.

Head, H. *Aphasia and kindred disorders of speech*. Cambridge: The University Press, 1926.

Head, H., & Holmes, G. Sensory disturbances from cerebral lesions. *Brain*, 1911, *34*, 102–254.

Head, H., & Sherren, J. The consequences of injury to the peripheral nerves in man. *Brain*, 1905, *28*, 116–338.

Head, H., & Thompson, T. The grouping of afferent impulses within the spinal cord. *Brain*, 1906, *29*, 537–741.

Hécaen, H., Ajuriaguerra, J. de, & Massonet, J. Les troubles visuoconstructifs par lésion pariéto-occipitale droite. *Encéphale*, 1951, *40*, 122–179.

Holmes, G. Disturbances of visual orientation. *British Journal of Ophthalmology,* 1918, *2,* 449–468; 506–516.

Holmes, G. Disorders of sensation produced by cortical lesions. *Brain,* 1927, *50,* 413–427.

Holmes, G., & Horrax, G. Disturbances of spatial orientation and visual inattention, with loss of stereoscopic vision. *Archives of Neurology and Psychiatry,* 1919, *1,* 385–407.

Jackson, J. H. Clinical remarks on defects of expression (by words, writing, signs, etc.) in diseases of the nervous system. *Lancet,* 1864, *1,* 604–605.

Jackson, J. H. On the nature of the duality of the brain. *Medical Press and Circular,* 1874, *17,* 19. (Reprinted in *Brain,* 1915, *38,* 80–103.)

Jackson, J. H. Case of large cerebral tumour without optic neuritis and with left hemiplegia and imperception. *Royal Ophthalmic Hospital Reports,* 1876, *8,* 434–444.

Kleist, K. Der Gang und der gegenwaertige Stand der Apraxieforschung. *Ergebnisse der Neurologie und Psychiatrie,* 1912, *1,* 342–452.

Kleist, K. Kriegsverletzungen des Gehirns in ihrer Bedeutung fuer die Hirnlokalisation und Hirnpathologie. In O. von Schjerning (Ed.), *Handbuch der Aerztlichen Erfahrung im Weltkriege,* Bd. 4. Leipzig: Barth, 1922/1934.

Koerner, S. C. Die Beeinflusskeit der Sensibilitaet an symmetrischen Hautgebieten bei einseitiger Hirnschaedigung und bei Gesunden. *Deutsche Zeitschrift fuer Nervenheilkunde,* 1938, *145,* 116–130.

Kroll, M. *Die neurologischen Symptomenkomplexe.* Berlin: Springer, 1929.

Kroll, M., & Stollbun, D. Was ist konstruktive Apraxie? *Zeitschrift fuer die gesamte Neurologie und Psychiatrie,* 1933, *148,* 142–158.

Lange, J. Fingernosie und Agraphie. *Monatsschrift fuer Psychiatrie und Neurologie,* 1930, *76,* 129–188.

Lange, J. Probleme der Fingeragnosie. *Zeitschrift fuer die gesamte Neurologie und Psychiatrie,* 1933, *147,* 594–620.

Lange, J. Agnosien und Apraxien. In O. Bumke & O. Foerster (Eds.), *Handbuch der Neurologie,* Bd. 6. Berlin: Springer–Verlag, 1936.

Lenz, G. *Beitraege zur Hemianopsie.* Stuttgart, 1905.

Lissauer, H. Ein Fall von Seelenblindheit nebst einem Beitrag zur Theorie derselben. *Archiv fuer Psychiatrie und Nervenkrankheiten,* 1890, *21,* 222–270.

McFie, J., Piercy, M. F., & Zangwill, O. L. Visual–spatial agnosia associated with lesions of the right cerebral hemisphere. *Brain,* 1950, *67,* 167–190.

McFie, J., & Zangwill, O. Visual–constructive disabilities associated with lesions of the left hemisphere. *Brain,* 1960, *83,* 243–260.

Meyer, O. Ein- und doppelseitige homonyme Hemianopsie mit Orientierungstoerungen. *Monatsschrift fuer Psychiatrie und Neurologie,* 1900, *8,* 440–456.

Milner, B. Visually-guided maze learning in man: Effects of bilateral hippocampal, bilateral frontal, and unilateral cerebral lesions. *Neuropsychologia,* 1965, *3,* 317–338.

Munk, H. Weitere Mittheilungen zur Physiologie der Grosshirnrinde. *Archiv fuer Anatomie und Physiologie,* 1878, *2,* 162–178.

Paterson, A., & Zangwill, O. L. Disorders of visual space perception associated with lesions of the right cerebral hemisphere. *Brain,* 1944, *67,* 331–358.

Penfield, W., & Evans, J. P. Functional defects produced by cerebral lobectomies. *Research Publications, Association for Research in Nervous and Mental Disease,* 1934, *13,* 352–377.

Peters, A. Ueber die Beziehungen zwischen Orientierungsstoerungen und ein- und doppelseitige Hemianopsie. *Archiv fuer Augenheilkunde,* 1896, *32,* 175–187.

Piercy, M. F., Hécaen, H., & Ajuriaguerra, J. de. Constructional apraxia associated with unilateral cerebral lesions—left and right sided cases compared. *Brain,* 1960, *83,* 225–242.

Piercy, M. F., & Smyth, V. Right hemisphere dominance for certain nonverbal intellectual skills. *Brain*, 1962, *85*, 775–790.

Pinéas, H. Ein Fall von raeumliche Orientierungsstoerung mit Dyschirie. *Zeitschrift fuer die gesamte Neurologie und Psychiatrie*, 1931, *133*, 180–195.

Poppelreuter, W. *Die psychischen Schaedigungen durch Kopfschuss im Kriege 1914–1916: die Stoerungen der niederen und hoeheren Sehleistungen durch Verletzungen des Okzipitalhirns.* Lepizig: Voss, 1917.

Reichardt, M. *Allgemeine und spezielle Psychiatrie: ein Lehrbuch fuer Studierende und Aerzte*, II Aufl., Jena, Germany: G. Fischer, 1918.

Riddoch, G. Dissociation of visual perceptions due to occipital injuries, with special reference to appreciation of movement. *Brain*, 1917, *40*, 15–57.

Riddoch, G. Visual disorientation in homonymous half-fields. *Brain*, 1935, *58*, 376–382.

Rieger, C. Ueber Apparate in dem Hirn. *Arbeiten aus der Psychiatrischen Klinik Wuerzburg*, 1909, *5*, 176–197.

Ross, H. E., & Murray, D. J. Introduction. In E. H. Weber, *The sense of touch*. London: Academic Press, 1978.

Ross, S. J., & Fountain, G. Phenomenon of cutaneous sensory extinction. *Archives of Neurology and Psychiatry*, 1948, *59*, 107–113.

Russell, W. R., & Espir, M. L. E. *Traumatic aphasia*, London: Oxford University Press, 1961.

Sachs, H. Das Gehirn des Foerster'schen Rindenblinden. *Arbeiten aus der Psychiatrischen und Nervenklinik Breslau*, 1895, *2*, 55–122.

Sanchez–Longo, L. P., & Forster, F. M. Clinical significance of impairment of sound localization. *Neurology*, 1958, *8*, 119–125.

Sanchez–Longo, L. P., Forster, F., & Auth, T. L. A clinical test for sound localization and its applications. *Neurology*, 1957, *7*, 655–663.

Schlesinger, B. Zur Auffassung der optischen und konstruktiven Apraxie. *Zeitschrift fuer die gesamte Neurologie und Psychiatrie*, 1928, *117*, 649–697.

Semmes, J. A non-tactual factor in stereognosis. *Neuropsychologia*, 1965, *3*, 295–315.

Semmes, J., Weinstein, S., Ghent, L., & Teuber, H. L. *Somatosensory changes after penetrating brain wounds in man.* Cambridge: Harvard University Press, 1960.

Shankweiler, D. Performance of brain-damaged patients on two tests of sound localization. *Journal of Comparative and Physiological Psychology*, 1961, *54*, 375–381.

Shy, G. M., & Haase, G. R. Sensorische Stoerungen bei Scheitellappenlaesionen. *Deutsche Zeitschrift fuer Nervenheilkunde*, 1957, *176*, 519–542.

Stengel, E. Loss of spatial orientation, constructional apraxia and Gerstmann's syndrome. *Journal of Mental Science*, 1944, *90*, 753–760.

Strauss, H. Ueber Konstruktive Apraxie. *Monatsschrift fuer Psychiatrie und Neurologie*, 1924, *63*, 739–748.

Teuber, H. L. Effects of brain wounds implicating right or left hemisphere in man: Hemisphere differences and hemisphere interaction in vision, audition and somesthesis. In V. B. Mountcastle (Ed.), *Interhemispheric relations and cerebral dominance*. Baltimore: Johns Hopkins Press, 1962.

Teuber, H. L. The riddle of frontal lobe function in man. In J. M. Warren & K. Akert (Eds.), *The frontal granular cortex and behavior*. New York: McGraw–Hill, 1964.

Teuber, H. L., Battersby, W., & Bender, M. B. Changes in visual searching performance follow- ing cerebral lesions. *American Journal of Physiology*, 1949, *159*, 592.

Walsh, E. G. An investigation of sound localization in patients with neurological abnormalities. *Brain*, 1957, *80*, 222–250.

Warrington, E. K., James, M., & Kinsbourne, M. Drawing disability in relation to laterality of cerebral lesion. *Brain*, 1966, *89*, 53–82.

Weber, E. H. *The sense of touch: De tactu* translated by H. E. Ross: *Der Tastsinn* translated by D. J. Murray. London: Academic Press, 1978.

Wilbrand, H. *Dei Seelenblindheit als Herderscheinung und ihre Beziehungen zur homonymen Hemianopsie.* Wiesbaden: J. F. Bergmann, 1887.

Wortis, S. B., & Pfeffer, A. Z. Unilateral auditory–spatial agnosia. *Journal of Nervous and Mental Disease,* 1948, *108,* 181–186.

Zangwill, O. L. *Cerebral dominance and its relation to psychological function.* Edinburgh: Oliver & Boyd, 1960.

MICHAEL E. GOLDBERG

12

Moving and Attending in Visual Space: Single-Cell Mechanisms in the Monkey

Much of the activity of higher organisms involves physically interacting with objects in the space around them. Animals see stimuli, they make eye movements toward them, they reach out and touch them, they run toward or away from them, or they merely notice them. Such behavior requires several kinds of analysis before the requisite movements can be organized. The first analysis is a purely visual one: What is the stimulus and where is it on the retina? The second analysis answers a visuobehavioral question: Does this stimulus attract the attention of the organism? The third question requires a visuoproprioceptive analysis: What is the relationship of the retinal stimulus to the body? The final analysis deals with the visuomotor problem: How do we translate the visual stimulus into data useful in programming relevant movement?

Much research over the past decade has gone into the investigation of the brain's methods of answering these questions, and from this certain anatomic and physiological correlations have begun to be made. It is growing clear that detailed visual analysis is the province of the geniculostriate system (Hubel & Wiesel, 1968), the prestriate cortex (Baizer, Robinson, & Dow, 1976; Zeki, 1975), and the inferotemporal cortex (Gross, Rocha–Miranda, & Bender, 1972). Single neurons in these areas demonstrate extraordinary mechanisms for extracting feature information from the visual world. Cells show orientation specificity, selectivity for stimulus length, direction of movement, color, and

277

SPATIAL ABILITIES
Development and Physiological Foundations

ISBN   0-12-563080-8

binocular retinal disparity. These mechanisms have been discussed elsewhere and will be assumed here.

The loci of the other kinds of analysis are more obscure, but it is equivalently clear that these analyses are performed to some extent in the cortical and subcortical projection areas of the tecto-pulvinar projection system. This system takes a retinal projection from the more primitive visual tectum, the superior colliculus, and projects it to cortical association areas via the more recently evolved pulvinar and lateral posterior nuclei of the thalamus, and thence to cerebral cortical association areas (Diamond, 1973; Goldberg & Robinson, 1978). Two cortical areas related to this system, the frontal eye fields (Bushnell & Goldberg, 1979; Goldberg & Bushnell, 1981; Mohler, Goldberg, & Wurtz, 1973; Mohler & Wurtz, 1976b) and the posterior parietal cortex (Goldberg & Robinson, 1977; Hyvärinen & Poranen, 1974; Lynch, Mountcastle, Talbot, & Yin, 1977; Mountcastle, Lynch, Georgopoulos, Sakata, & Acuna, 1975; Robinson, Goldberg, & Stanton, 1978) have been investigated in great detail in the last 10 years. From study of these extra geniculostriate visual areas it is becoming possible to begin to understand some of the brain mechanisms involved in behaving in visual space.

BASIC METHODS

Traditional neurophysiology has centered around the study of brains of anesthetized or paralyzed preparations, and this approach has yielded a vast amount of basic knowledge about the sensory and motor systems. However, to study brain mechanisms relevant to higher behavioral processes, it is necessary to use a preparation in which behavior is relatively undisturbed. To accomplish this, techniques for recording the activity of single neurons in awake behaving animals have evolved, starting with the work of Evarts (1966) who developed methods of recording from tbe brain in monkeys trained to perform motor tasks. Wurtz (1969) combined the physiological techniques of Evarts (1966) with behavioral techniques for training monkeys to control their eye position. This made it possible to apply the strategy of neurophysiological recording in awake monkeys to the visual system.

Since much of the work in these areas has involved the use of awake monkeys, it is helpful to review the basic behavioral and physiological methods in some detail, to clarify the subsequent discussion. A rhesus monkey is first trained to fixate a spot of light. He learns to press a lever and bring a spot of light on to a tangent screen. Several seconds later the light dims, and the monkey learns to release his lever in response to the dimming of the light to earn a reward. The spot of light is so small (usually a quarter of a degree of arc) and the dimming so subtle that the monkey must fixate it with his fovea. He is

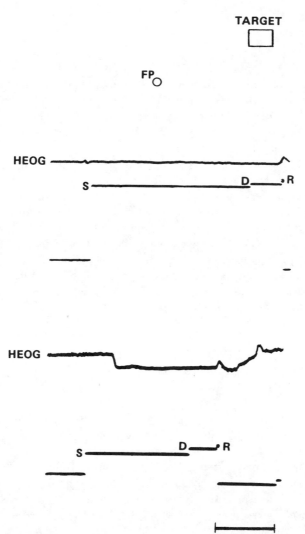

Figure 12.1. Monkey eye movements recorded during behavioral tasks. At the beginning of each trial the animal brings a fixation point on the screen and looks at it. The fixation point is marked FP. Two pairs of horizontal electrooculograms (labeled HEOG) and behavioral artifact traces (labeled S D R) are shown. In the top pair the monkey continues to look at the fixation point even though at the time marked by the artifact labeled S the target appears on the screen. The monkey continues to look at the fixation until it dims (artifact D) and the monkey releases the bar and gets a reward (artifact labeled R). The fixation point goes off at the reward. After the reward the monkey breaks fixation because the trial is over and his eye movements are irrelevant to the reward or task. The second pair of traces shows a saccadic eye movement. The monkey begins by looking at the fixation point. When the target comes on (artifact S) the fixation point goes out, and several hundred milliseconds later the animal makes a saccadic eye movement to fixate the target. The target then dims (artifact D) and the monkey releases the bar and gets a reward (artifact R). Here too the monkey makes spontaneous eye movements after the trial is over.

279

so interested in succeeding in his task that he does not break fixation to look at a second spot of light, should it be flashed on the screen. If the fixation point goes out when the second spot appears, the monkey makes a rapid saccadic eye movement to fixate the second spot. Thus the second spot can be made either irrelevant to the monkey's behavior, or quite important, the target for an eye movement. Figure 12.1 shows eye movements of a monkey during a fixation task and a saccade task.

After behavioral training is completed, the monkeys are prepared for neurophysiological recording by surgical implantation of recording cylinders, electrooculogram electrodes, and head holding apparatus. During each experimental session a microelectrode is placed through the recording cylinder into the brain without anesthesia (a painless procedure), and the activity of single neurons are recorded while the monkey performs its various conditioned tasks. A digital computer is used for behavioral control and on-line data for analysis. The technique can be used to work out sensory properties of neurons, such as their receptive fields, or it can be used to work out the relationship of cell discharge to various behaviors.

BEHAVIORAL MODULATION OF VISUAL
RESPONSES: SPATIAL ANALYSIS FOR
EYE-MOVEMENT CONTROL

The first extrageniculostriate brain region to which these techniques were applied was the superior colliculus (Goldberg & Wurtz, 1972a, 1972b). This nucleus is a multilayered structure with six laminae situated on the dorsal surface of the midbrain (Figure 12.2). Both retinas project the contralateral visual field to the superficial layers of the colliculus through a direct pathway. Striate cortex projects congruently to these layers so that a part of striate cortex occupied with a given area of the visual field will project to the area of the colliculus that receives a retinal projection from the same area of visual field. (For a more detailed discussion of the superior colliculus see Goldberg & Robison, 1978; Wurtz & Albano, 1980.) The vast majority of neurons in the superficial layers of the superior colliculus can be driven by small spots of light in particular small areas (receptive fields) in the contraleteral field. Most of these cells do not have requirements for stimulus orientation, direction of movement, or color, although many of them have suppressive surrounds (Goldberg & Wurtz, 1972a; Wurtz, Richmond, & Judge, 1980) and complicated within-field structure that makes them more responsive to small spots than to large ones. This lack of visual sophistication is markedly different from the organization found in the geniculostriate system, where most cells require stimuli of particu-

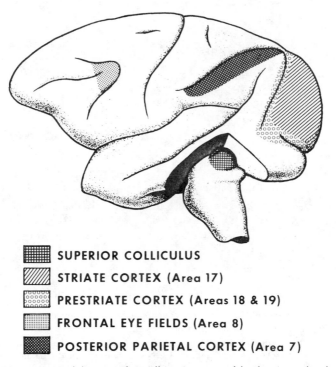

SUPERIOR COLLICULUS

STRIATE CORTEX (Area 17)

PRESTRIATE CORTEX (Areas 18 & 19)

FRONTAL EYE FIELDS (Area 8)

POSTERIOR PARIETAL CORTEX (Area 7)

Figure 12.2.   Anatomical diagram of visually active areas of the rhesus monkey brain. Various areas discussed are shaded in different patterns. The temporal lobe is cut away to reveal the underlying superior colliculus and brain stem. (Reproduced with permission from Wurtz *et al.,* 1980.)

lar orientation, direction of movement, or color (Baizer, *et al.,* 1976; Hubel & Wiesel, 1968; Zeki, 1975).

However, half of the cells in the superficial layers of the colliculus have a behavioral responsiveness that renders this area quite different from the geniculostriate system. The firing rate of the response of these neurons depends on whether or not the animal is going to use the stimulus in the receptive field as the target for an eye movement (Goldberg & Wurtz, 1972b). Figure 12.3 illustrates a demonstration of this behavioral modulation of a visual response in a neuron in the monkey superior colliculus. Figure 12.3(a) shows that the cell has a visual receptive field and responds to a stimulus in the field while the animal performs a fixation task. Figure 12.3(b) shows that the response of the cell to the stimulus is markedly enhanced when the animal makes a saccadic eye movement to the stimulus in the receptive field. Roughly half of the neurons in the superficial gray and optic layers of the monkey superior col-

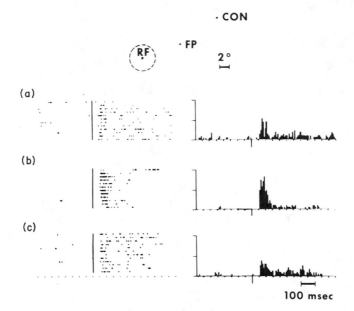

Figure 12.3.   Spatially specific enhancement of visual response in monkey superior colliculus. The dot pattern in (a) shows the response of the neuron to the onset of a visual stimulus consisting of two spots, one in the receptive field of the neuron (labeled RF) and the other a control spot outside the receptive field (labeled CON). Each line of the dot pattern corresponds to a single trial and each dot corresponds to a cell action potential. Successive trials are aligned on the onset of the visual stimulus, signified by a vertical line near the center of each diagram. The histograms at the right sum the activity in the dot patterns. The downward line at the base of each histogram signals stimulus onset. Part (b) shows the enhanced response of the same neuron to the same stimuli when the animal makes a saccade to the stimulus in the receptive field. Part (c) shows the baseline response of the same neuron to the same stimuli when the animal makes a saccadic eye movement to the control stimulus (CON). The neuron does not respond at all if the control stimulus appears alone, without the receptive field stimulus. (Reproduced with permission from Wurtz & Mohler, 1976b.)

liculus give a similar enhanced response (Goldberg & Wurtz, 1972b; Wurtz & Mohler, 1976a). There are many nonspecific factors that might be responsible this enhancement; for example, the animal's level of arousal might be greater during the saccade task or the pupil might be dilated. However, such nonspecific effects can be ruled out by the fact that this enhancement is spatially selective: When the animal makes a saccade to a stimulus outside the receptive field, the response to the stimulus in the receptive field is no greater than when the animal makes no eye movement at all. Figure 12.3(c) shows the response to the stimulus in the receptive field when the animal makes an eye movement to a stimulus outside the receptive field. Thus the superior colliculus has a mechanism whereby a response to a target on the retina is selectively enhanced when the animal makes an eye movement to that target, but not when the animal makes a different eye movement. It is important to emphasize that the enhancement is not a simple response to an eye movement. The effect is a modulation of a response to a stimulus: If the stimulus is absent the neuron will not discharge, even though the eye movement takes place. Thus these neurons do not discharge when the animal makes an equivalent eye movement in total darkness. This stimulus requirement differentiates neurons in the superficial layers of the superior colliculus from neurons found in more oculomotor areas of the brain such as the brain stem reticular formation (Luschei & Fuchs, 1972) or the intermediate layers of the superior colliculus itself (Wurtz & Goldberg, 1971, 1972a; Schiller and Koerner, 1971). In these areas the cells discharge whenever the eye movement is made, in light or darkness or even during vestibular nystagmus. In the superficial layers the cells require the stimulus and have their discharge modulated according to the relevance of the stimulus for an eye movement.

The collicular neurons do not show this enhancement when the animal used the target in the receptive field for a behavior other than an eye movement. Wurtz and Mohler (1976b) trained monkeys to respond to a peripheral target without making an eye movement to it by making either the dimming of the central fixation point or the peripheral stimulus the cue for the monkey's lever release. Since the fixation point was small and the peripheral stimulus large, the monkeys could perceive the dimming of the peripheral target with their peripheral retinas, and they tended not to make an eye movement to the stimulus. The neurons in the superior colliculus do not give an enhanced response to the stimulus when it is the cue for this task, although they clearly do give the enhanced response when the stimulus is the target for an eye movement. Thus the superficial layers of the superior colliculus have a mechanism that links visual information to an impending eye movement relative to that information. As such the area seems dedicated to transmitting visual information to the oculomotor system, the fourth question mentioned earlier.

A similar mechanism is present in the frontal eye fields (Figure 12.2). This is a region that yields eye movements on electrical stimulation (Robinson & Fuchs, 1969) and receives visual information via a strong projection from the medial pulvinar (Trojanowski & Jacobson, 1974) and the prestriate and inferior temporal cortices (Chavis & Pandya, 1976). The first neurons recorded here were a small population that discharged during and after eye movements (Bizzi, 1968). These neurons could not drive the eye movements, so the relationship of frontal eye fields to eye movement so clearly indicated by electrical stimulation data was not apparent. However, subsequently a population of neurons which had visual receptive fields was described. These receptive fields resembled those in the superior colliculus in their large size and freedom from stimulus requirements (Mohler, Goldberg, & Wurtz, 1973). They also resembled the colliculus neurons in that roughly half of them yielded enhanced responses before eye movements to stimuli in their receptive fields (Wurtz & Mohler, 1976b). These neurons only show the enhancement when the animal makes an eye movement to the receptive field, not when it makes a response to a stimulus that does not include an eye movement. When the frontal eye fields are stimulated electrically through the recording microelectrode, the resulting eye movement is made to the area of the visual field that is within the receptive field of the neuron at the stimulation site (Goldberg & Bushnell, 1981). These neurons, like those in the superior colliculus, only discharge when the animal makes an eye movement to a visual stimulus, and are silent when the animal makes the same eye movement in a dark environment.

These areas, the frontal eye fields and the superior colliculus, perform an analysis of visual space to find the target for an eye movement. Enhanced activity in these areas occurs only if there is going to be an eye movement. The analysis performed here therefore is the fourth one listed earlier: the transformation of a visual spatial location into data for a specific behavior. Destroying either of the two areas leaves the oculomotor system slightly less accurate and responsive than before; eye movements after superior colliculus lesions are slower to begin and tend to slightly undershoot the target (Wurtz & Goldberg, 1972b). However, destruction of both of these areas results in a devastating inability of the monkey to make saccades to visual targets in the periphery (Schiller, True, & Conway, 1979). Presumably the destruction of both loci of presaccadic visual enhancement destroys the oculomotor system's ability to find visual targets, although the mechanism to move the eyes remains intact, as is shown by the preservation of the animal's ability to generate the saccadelike quick phases of vestibular nystagmus.

ENHANCEMENT OF VISUAL RESPONSES IN POSTERIOR PARIETAL CORTEX: STIMULUS SELECTION FOR ATTENTION

Eye movements are an important component of behavior in visual space, but spatial behavior must be dissociable from eye movements, for otherwise we could never attend to and act on objects in our peripheral retina without making an eye movement to them. The mechanism of behavioral enhancement of sensory input, so important to the oculomotor process, could easily serve as a general mechanism for the organization of spatial behavior, as long as the behaviors under which the enhancement takes place are not limited to a specific kind of response. Such response modality independent enhancement is found in the posterior parietal cortex, area 7.

The posterior parietal cortex has long been associated with the process underlying visual attention (Critchley, 1953; Lynch *et al.*, 1977; Lynch, 1980; Wurtz *et al.*, 1980). Patients with lesions in the posterior parietal cortex show visual and somatosensory neglect and do not perform the movements or form the perceptions in complicated environments that they can easily do in simple environments (Heilman, 1979). Similarly, monkeys with unilateral parietal lesions show deficits in responding to the contralateral field when presented with double simultaneous stimulation (Heilman, Pandya, Karol, & Geschwind, 1971). Monkeys with bilateral parietal lesions cannot perform tasks that require them to discriminate the spatial relationships among visual simuli (Pohl, 1973; Ungerleider & Brody, 1977). These deficits are typical of an inability to pick a stimulus out of the field. This process of stimulus selection is consistent with the process of visual attention (Wurtz *et al.*, 1980) and recent work has begun to unravel the physiological mechanisms underlying visual attention.

The first evidence that such a mechanism may lie in area 7 came from the work of Hyvärinen and Poranen (1974), who showed that neurons in area 7 of untrained monkeys discharged when the animals looked at object of interest. Mountcastle and his group studied area 7 in great detail in animals trained to perform a series of visuomotor tasks similar to those described above (Lynch *et al.* 1977; Mountcastle, 1976; Mountcastle *et al,* 1975). They described a series of neurons discharging in association with visually guided saccades, visual fixation, and smooth pursuit, and suggested that these neurons might be playing an important role in directing visual attention by commanding eye movements to stimuli of interest to the animal. They claimed that these neurons were not responsive to sensory stimuli alone, although they subsequently described a class of light sensitive neurons in area 7 (Yin & Mountcastle, 1977; Yin & Mountcastle, 1978).

We studied these neurons from a somewhat different point of view (Goldberg & Robinson, 1977; Robinson, Bushnell, & Goldberg, 1980; Robinson *et al.*, 1978). As in the superficial layers of the superior colliculus, parietal neurons seem to require that the animal respond to a stimulus. The neurons do not discharge when the animals make the equivalent movement in total darkness (Lynch *et al.*, 1977). It therefore seemed reasonable to suggest not that the parietal neurons functioned as an input independent command mechanism, but rather that the discharge seen in association with visually guided eye movements was a behavioral enhancement of a visual response. In the population of neurons studied (Robinson *et al.*, 1978), every neuron that could be associated with an eye movement could be driven by some stimulus in the absence of that eye movement.

Thus neurons associated with saccadic eye movements have visual receptive fields in the periphery (Figure 12.4). Figure 12.4(a) shows a parietal neuron discharging before a visually guided eye movement. Figure 12.4(b) shows the same responses of the neuron synchronized on the stimulus onset rather than

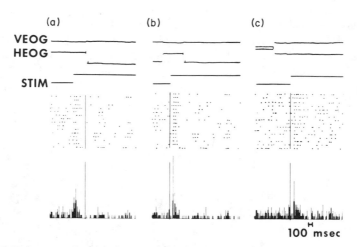

Figure 12.4. Presaccade enhancement of visual response in monkey area 7. Part (a) shows the response of a single neuron in area 7 synchronized on the onset of eye movement to a visual stimulus 10 degrees contralateral and 2 degrees up. Sample traces show vertical (VEOG) and horizontal (HEOG) electrooculograms, and a stimulus artifact. The target light appears at the break in the stimulus line. A shows a series of trials synchronized on the beginning of the eye movement (marked by the vertical line through the histogram and dot pattern). The computer recognition ticks for beginning and end of the eye movements are visible on the horizontal trace. Part (b) shows the same trials as in (a), synchronized on the stimulus onset rather than on the eye movement. The vertical line depicts stimulus onset. The neuronal response appears more sharply synchronized to the response onset than to the stimulus onset. Part (c) shows the response of the same neuron to the same stimulus in a series of fixation trials in which the animal does not break fixation to look at the stimulus. There is a smaller but noticeable response to the stimulus. (Reproduced with permission from Robinson et al., 1978.)

on the movement, to show that the neuronal response is better linked to the stimulus onset than to the movement. In one trial (the ninth) the neuron responded to the stimulus but the animal did not make the movement. Figure 12.4(c) shows the weaker response that the neuron makes to the stimulus in the receptive field in the absence of the eye movement. The response, although not as strong as when the eye movement occurred, is present. For half of these neurons the visual response to the stimulus is quite weak, although a visual response can always be found. The neuron's response to the stimulus is enhanced when the animal makes the saccadic eye movement.

Figure 12.5. Graded response of area 7 neuron to stimuli of increasing size. The diagram at the top of the figure shows the size and position of different size stimuli relative to the fixation point (FP). Each label dot pattern and histogram shows the response of a parietal neuron to the correspondingly labeled stimulus. The briskest response is to the largest stimulus, and the response decreases with decreasing stimulus size. Each stimulus was two log units brighter than the background of 1 cd/m². Vertical lines depict the onset of the stimulus. Time dots are 200 msec apart. (Reproduced with permission from Robinson et al., 1978.)

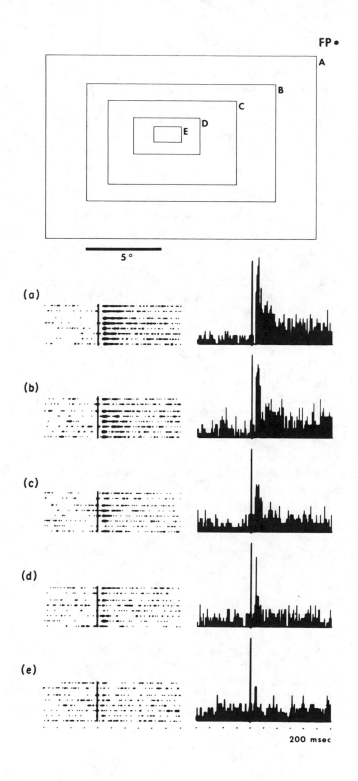

FP •

A

B

C

D

E

5°

(a)

(b)

(c)

(d)

(e)

200 msec

287

The visual properties of these neurons resemble those of neurons in the superior colliculus. The receptive fields are large and the cells do not have striking requirements for stimulus properties. In general cells respond best to large, bright stimuli. Figure 12.5 shows the discharge of a parietal neuron in response to visual stimuli of increasing size. Like that seen in the superior colliculus, the enhancement is spatially specific, occurring only when the animal makes an eye movement to fixate the stimulus in the receptive field, not when the animal makes an eye movement to an object outside of the receptive field. Figure 12.6 illustrates spatial selectivity in area 7. Figure 12.6(a) shows the response of the neuron to a large spot within its receptive field. Figure 12.6(b) shows the response of the same neuron to the onset of two spots, one in and one outside the receptive field. Figure 12.6(c) shows the response of the neuron when the animal makes the eye movement to the stimulus in the receptive field, and figure 12.6(d) shows the response of the same neuron when the animal makes the eye movement outside the receptive field.

Similarly, parietal neurons that discharge when an animal gazes at a target anywhere can be shown to respond to light stimuli in the fovea (Robinson *et al.*, 1978), and neurons that discharge in association with tracking eye movements of a given direction can be shown to discharge to stimuli moving in tbe appropriate direction when the animal holds his eyes still (Robinson *et al.*, 1978; Sakata, Shibutani, & Kawano, 1978). Thus it is highly unlikely that the discharge of these neurons can signify anything but the presence of the stimulus—either a moving stimulus in the case of neurons that discharge in association with smooth pursuit, or a foveal stimulus in the case of neurons that discharge in association with visual fixation. Thus for all classes of eye movements, neurons in area 7 that discharge in association with the movement actually dishcarge in response to the stimulus evoking the movement. The response may be enhanced when the animal makes the eye movement to the stimulus, relative to the case when the animal ignores the stimulus.

When examined in terms of only eye movements and visual stimuli, the neurons in area 7 closely resemble the cells in the superior colliculus and the frontal eye fields. However, in the frontal eye fields and the superior colliculus and enhancement occurs only when the animal actually makes the eye movement. In the posterior parietal cortex the enhancement occurs whenever the animal attends to the stimulus, regardless of the way in which he will respond to it. Since almost by definition a monkey attends to a stimulus when he makes an eye movement to it, the enhancement in the parietal cortex occurs not because of the eye movement, but because of the attention to the stimulus. To show that parietal enhancement is not specifically locked to eye movements, we trained monkeys to do a peripheral attention task similar to the one used by Wurtz & Mohler (1976b). The monkeys performed series of trials in which they had to respond only to a central fixation point, not to a peripheral stimulus

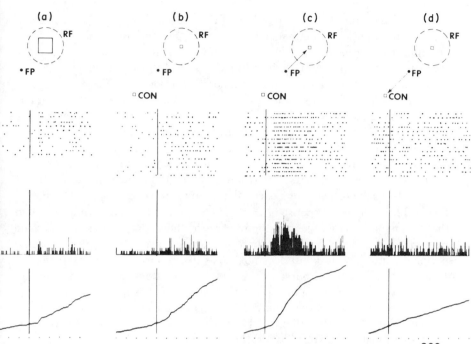

Figure 12.6.   Spatial selectivity of response enhancement in area 7. Part (a) shows the response of a neuron in area 7 to a large stimulus flashed in its receptive field (RF) while the monkey looked at the fixation point (FP). The display beneath the post stimulus histogram is a cumulative histogram computed by displaying time on the abscissa and the sum of all neuronal discharges from the beginning of the display on the ordinate. Part (b) shows the faint response of the same neuron to the onset of two small stimuli, one inside the receptive field and one control stimulus (CON) outside the receptive field while the monkey fixated. Part (c) shows the enhanced response to the stimulus when the animal makes a saccade to the small stimulus in the receptive field. Part (d) shows the response when the animal makes a saccadic eye movement to the control stimulus. The enhancement only occurs when the animal makes the eye movement to the spot in the receptive field. Each vertical line signifies stimulus onset. Time dots are 200 msec apart. (Reproduced with permission from Robinson et al., 1978.)

(the fixation case), and then they performed series of trials in which they had to monitor the same stimulus, because they would have to respond to it if it should dim (the peripheral attention case). The implicit assumption was that since the monkeys had to respond to the stimulus in the periphery, they had to attend to it. Since they did not respond to the stimulus when it was presented as an irrelevant stimulus in the fixation task, they did not have to attend to it, although of course one can never tell when a monkey does not attend to a stimulus to which he is not asked to respond. We then compared the response of the neuron to the same stimulus in the fixation case and the peripheral attention case. Slightly fewer than half of the neurons in area 7 that respond to the stimuli at all give an enhanced response to the stimulus in the peripheral attention case, even though the monkeys do not make eye movements to the stimulus. The response of neurons in the saccade task correlates with the response of the same neuron in the peripheral attention task. Neurons that give an enhanced response to stimuli used as targets for eye movements inevitably give enhanced responses when the same targets are used as cues in the peripheral attention task. Conversely, neurons that fail to show an enhanced response in the one task do not give an enhanced response in the other. Figure 12.7 shows the response of a parietal neuron to the same stimulus in the fixation task, the peripheral attention task, and the saccade task. Figure 12.7(a) shows the response of a parietal neuron to an irrelevant visual stimulus. Figure 12.7(c) shows response of the same neuron to the same stimulus when the animal makes a saccade, and Figure 12.7(b) shows the response of the same neuron to the same stimulus when the animal uses the stimulus in the peripheral attention task but does not make an eye movement to the stimulus. Note that the enhanced burst in the eye movement and peripheral attention cases are quite similar. Thus the enhancement of visual responses in the posterior parietal cortex does not depend upon the exact motor nature of the animal's response to the stimulus, but rather depends upon the fact that the animal will make some response to the stimulus.

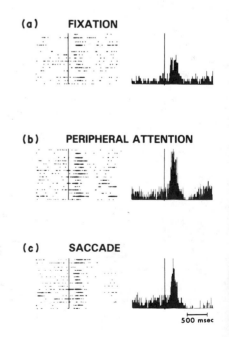

Figure 12.7. Enhancement of posterior parietal visual response in association with saccade and peripheral attention tasks. Part (a) shows the response of the neuron to the peripheral stimulus when the animal fixates and does not respond to the peripheral stimulus. Part (b) shows the response of the same neuron to the same stimulus when the animal must attend to the peripheral stimulus but may not make an eye movement to it. Part (c) shows the response of the neuron to the same stimulus when the animal makes the eye movement. The responses are enhanced in (b) and (c) relative to (a). The response is foreshortened in (c) because the eye movement takes the stimulus out of the receptive field. In each case the vertical line signifies the onset of the visual stimulus. (Reproduced with permission of the publisher from The role of posterior parietal visual attention, by D. L. Robinson, M. C. Bushnell, & M. E. Goldberg, *Progress in oculomotor research.* Copyright 1980 by Elsevier Science Publishing Co., Inc.)

Like the oculomotor enhancement, enhancement in the peripheral attention task is spatially specific. It occurs whenever the monkey attends to the stimulus in the receptive field, but not when he performs a task that requires him to ignore the stimulus in the receptive field and respond to a stimulus outside the receptive field.

In a confirmatory experiment, we measure responses when the animal reached out and touched a lit panel without making an eye movement to the stimulus. The enhancement occurred in this case also (Robinson et al., 1980). Again, neurons that show hand-movement-related enhancement also show eye-movement-related enhancement.

Thus this class of neurons in the parietal cortex signals the retinal position of a stimulus, and for neurons which give an enhanced discharge, the intensity of the discharge specifies whether or not the stimulus is important to the organism. It is not difficult to propose that absence of this enhancement mechanism will result in neglect for contralateral stimuli: If there is no brain mechanism to say that an object is important, the remainder of the nervous system may not respond to that stimulus. However, the absence of response is not absolute. Thus parietal lobe patients show deficits in complicated situations, such as double simultaneous stimulation, but do not show deficits if presented with similar stimuli under less confusing circumstances. This mechanism could be related to spatial attention in that significant stimuli evoke an enhanced response relative to behaviorally insignificant stimuli. This is a more general application of the enhancement principle than found in the frontal eye fields and superior colliculus, and it suggests that modulation of sensory input is a phenomenon that may be applied by the brain for many purposes of which attention and targeting movement are the two best studied.

This physiological demonstration of the dissociability between attention and any specific orienting movement agrees with the psychological observations of Posner (1980) and Klein (1979). These workers used reaction time as an indication of attention and showed that reaction time to the response of flashed stimuli could be independent of the direction of cued eye movements. However, under ordinary circumstances attention and eye movements do move in parallel, as is shown by Posner's demonstration that in a simple eye movement paradigm attention tends to move before the eyes, but in the direction of the eye movement. Another example of the usual linkage between eye movements and spatial perception is the demonstration of Jones and Kabanoff (1975) that auditory localization in the dark is quicker if one allows the subjects to move their eyes to the target. Similarly, children can localize auditory objects more quickly in the light than in the dark (Jones, 1975), although monkeys seem to be able to make eye movements to auditory stimuli more rapidly than they can to visual targets (Whittington & Hepp–Raymond, 1977). The neuronal activity described here includes two separate systems: an attentional one in the parietal lobe that functions independently of movement, and an eye movement one in the frontal eye fields linked to eye movement. One would assume that the parietal system was participating in the eye movement independent phenomena described by Posner and Klein, and that both participate in eye movements. It is important to emphasize that ordinarily the two systems function together, and indeed there are reciprocal anatomical projections between the two systems (Chavis & Pandya, 1976). However, attention and eye movements can be separated psychologically, just as they can be separated physiologically, and it would be an oversimplification to construct a theory of spatial attention that embedded the process in an oculomotor framework (cf. Rudel, Chapter 6, this volume).

Area 7 may generate this enhancement of visual responses by combining visual inputs with inputs from the limbic system. Area 7 receives presumably visual connections by projections of the prestriate cortex to the intraparietal sulcus (Kuypers, Schwarcbart, Mishkin, & Rosvold, 1965). In addition it receives a projection from the medial pulvinar and the lateral posterior nucleus of the thalamus, both of which have visual activity. It also receives a rich projection from limbic structures: the substantia innominata, and the cingulate gyrus (Mesulam, Van Hoesen, Pandya, & Geechwind, 1977; Stanton, Cruce, Goldberg, & Robinson, 1977). Activity in these neurons may represent an association between the presumably affective and appetitive limbic system and the visual system. The result of this association could be to specify the location in the visual field of stimuli that serve the needs of the organism. Because the enhancement occurs in a variety of behavioral situations requiring a variety of motor responses, it is unlikely that these neurons specify any one of the responses. It is much more likely that other brain structures transform this spatially selective, movement independent information into the requisite movement.

PROPRIOCEPTIVE–VISUAL ASSOCIATION AREA 7: CORRELATING LIMB AND EYE POSITION WITH RETINAL INFORMATION

Patients with lesions in the posterior parietal cortex show visual and somatosensory neglect, spatial disorientation, constructional apraxia, and many other deficits that could be construed as problems with behavior in visual and body space (Benton, Chapter 11, this volume, Critchley, 1953). From anatomic data, it was assumed (Jones & Powell, 1970) that area 7 in the monkey was used as a higher order somatosensory area, bearing the same relation to the somatosensory system that the prestriate cortex bears to the visual system. However, one of the striking deficits in monkeys with posterior parietal cortex lesions is an inability to perform accurate visually guided reaching tasks

(Faugier–Grimaud, Frenois, & Stein, 1978; Hartje & Ettlinger, 1973; LaMotte & Acuna, 1978). This deficit is seen most dramatically in the contralateral hand. Petrides and Iversen (1979) showed that monkeys with parietal lesions could not perform a task that required them to guide their hands along a bent wire to remove a reward. The hallmark of these deficits is an inability to correlate visual information with limb position or with the location of objects in visual space.

The previous discussion has dealt with mechanisms that select significant target areas in the retina for visual attention or visually guided movement. However, the retina is a notoriously mobile device, and to correlate between the retina and the world one needs to know where the eyes are in the orbit, where the orbit is in the head, and where the head is in the world. Alternatively, to move a limb to a visual image, one must know the correspondence between limb position and the position of a retinal image. There is good evidence that neurons in area 7 can contribute to this process. Hyvärinen and Poranen (1974) described neurons that discharged in association with touching and looking and postulated that area 7 was associating visual and somatosensory information to generate some brain scheme of the position of the eyes and limbs in visual space. Subsequent studies have shown that this is accomplished by several mechanisms. The first mechanism is one of association between visual and limb somatosensory information. When an animal reaches for a visual target, many neurons in area 7 discharge in association with this behavior, but the bulk of these neurons do not discharge when the same movement is made in total darkness (Mountcastle, Lynch, Georgopoulos, Sakata, & Acuna, 1975; Robinson et al., 1978). These neurons can all be driven by the proper visual stimuli (Robinson et al., 1978). However, some neurons that respond to visual stimuli also respond when the animal makes the same movement in the dark without any visual cues. The response of such a neuron in association with a visually cued hand movement is shown in Figure 12.8.

Monkeys were trained to begin trials as ordinary fixation trials. In half of the trials the fixation light went out and a touch panel was illuminated by a projector. The monkeys had to reach out and touch the panel to obtain a reward. Figure 12.8(a) shows the response of the neuron when the monkey reaches out and touches the illuminated panel. There is a biphasic response in association with each movement, the first part of which is synchronized with the onset of the light illuminating the panel. Figure 12.8(b) shows the response of the same cell when the animal makes the same movement in total darkness. In this case the signal to make the movement was the extinction of the fixation point, but the stimulus did not come on and the animal had to make the movement by remembering where the panel was. The response of the neuron resembles the second part of the biphasic response in 12.8(a). Figure 12.8(c) shows the response of the neuron to the illumination of the touch panel during

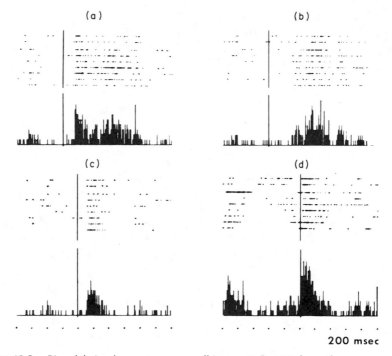

Figure 12.8. Bimodal visual–somatosensory cell in area 7. Part (a) shows the response of the neuron when the animal reaches out to touch a panel illuminated by a projector. The vertical line indicates the time of the signal to touch the stimulus, which was the simultaneous offset of the fixation point and onset of the panel illumination. Part (b) shows the response of the same neuron to the same hand movement made in total darkness. The vertical line shows the signal to make the hand movement, the extinction of the fixation point. Part (c) shows the response of the neuron to the illumination of the panel while the animal fixated and did not make the hand movement. Vertical line shows the onset of the stimulus. Part (d) shows the response of the neuron to tactile stimulation of the monkey's hand. The displays are synchronized on an artifact from the stimulating probe. Time dots are 200 msec apart. (Reproduced with permission from Robinson *et al.*, 1978.)

fixation trials in which the fixation point did not go out and the animals did not make the movement. This response resembles the first part of the biphasic response. Figure 12.8(d) shows the response of the neuron to tactile stimulation of the same hand used to touch the panel.

Therefore the neuron, which gives its best discharge during a visually guided reaching movement, actually is responding to the visual stimulus that will be touched and the tactile stimulation of the hand that touches the stimulus. This class of cell provides an ambiguous message: It describes the existence of a visual or tactile stimulus and its retinal or somatic location, but can give no other detail. This kind of associative mechanism could be instrumental in creating a supramodal representation of spatial location in which the original sensory signal would be subordinate to a more general schema of the environment (cf. Hermelin & O'Connor, Chapter 2, this volume).

In a similar way, area 7 has a mechanism for associating information about eye position with visual information. Hyvarinen and Poranen (1974) showed that neurons in area 7 discharge when the animals look at certain objects, and Mountcastle and his group, described neurons that discharged when animals look in some areas of the field and not others. Robinson *et al.* (1978) established that the great majority of these neurons that discharged in association with specific directions of gaze could be influenced by specific visual events such as altering background illumination or stimulating a visual receptive field. Figure 12.9 shows such a neuron. In 12.9(a) the neuron gives a definite discharge when the monkey makes a leftward eye movement in light, and maintains that discharge. In 12.9(b) the same eye movement in the dark does not result in a similar discharge. Sakata, Shibutani, and Kawano (1980), studying a generally more caudal part of area 7, described a population of neurons that reliably give a discharge with increasing eccentricity of eye position. The majority of these neurons had visual responses, some yielding different discharges at different

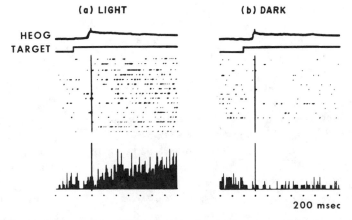

Figure 12.9. Variation of eye position response of parietal neuron with background illumination. Part (a) shows the response of a parietal neuron as the animal makes a saccade between two points in a dimly lit room. The neuron discharges in a tonic manner when the animal is looking at the saccade target but not when the animal looks at the fixation point. Part (b) shows the changed response to the same eye positions when the same movements are made in total darkness except for the dim fixation point or saccade target. The neuron fails to discharge at the same eye positions. Each dot pattern is synchronized on the end of the eye movement. Representative stimulus artifacts and horizontal electrooculogram traces are shown above the dot patterns. Time dots are 200 msec apart. (Reproduced by permission of Cambridge University Press from M. Goldberg & D. L. Robinson, The significance of enhanced visual responses in posterior parietal cortex, *Behavioral and Brain Sciences,* 1980, *3,* 505-507.)

background illuminations, and others requiring both certain visual stimuli and certain eye position. This association of eye position and visual activity could clearly help to locate the position of an object in space, since the location of the stimulus on the retina, crucial to all the neurons discussed previously, is less important here than is the position of the stimulus in space.

It is not yet clear if the eye position information comes from proprioceptors in the orbit or ocular muscles, or from a corollary discharge or efferent copy that tells the parietal lobe where the oculomotor system is going to move the eye. Since the neurons that discharge in association with visual fixation initiate their activity after the visual target has been acquired and the fixation has begun, either source is possible. Because the information about limb position clearly comes from proprioceptive afferents, one could make a more parsimonious system by assuming that the eye position information is proprioceptive in nature. However, no evidence exists for either alternative. In any event, the abundance of visual neurons with either limb proprioceptive or eye position information make it reasonable to expect that a very important function for area 7 is this association of visual and somotosensory information for the organization of spatial behavior.

The single-cell studies outlined here provide certain useful principles for the analysis of brain mechanisms for spatial behavior. The first is that to some extent the mechanisms of all sorts of spatial behavior are handled independently from, and in parallel with, the mechanisms of more sophisticated visual analysis. Thus, the receptive fields of behaviorally sensitive neurons are crude, and not particularly useful for fine visual analysis, whereas neurons such as those in the striate and prestriate cortex which are visually sophisticated are not particularly behaviorally sensitive (Robinson, Baizer, & Dow, 1980; Wurtz & Mohler, 1976). In a similar way, spatial behavior and fine visual cognition can be segregated. Thus Dee (1970) showed that a few patients who had parietal lesions and constructional apraxia had intact visual perception. Newcombe and Russell (1969) showed that patients with right posterior parietal lesions had difficulty with visual maze learning but not with a visual perceptual task. Conversely, their patients with right posterior temporal lesions could solve the spatial task, but not the perceptual task. Similarly, patients with striate lesions can perform eye movements into their hemianopic field, yet they are severely visually impaired (Weiskrantz, Warrington, Sanders, & Marshall, 1974). These clinical findings can be explained if one appreciates the differences between neuronal properties of geniculostriate and extrageniculostriate systems. Patients with striate or posterior temporal lesions would not have neurons capable of discriminating color or fine form, and their deficits are those of fine discrimination. Patients with parietal lesions may have their visual discriminative processes intact, but because they do not have the neurons associating visuospatial or visual attentional factors, they cannot place their visual capacities in a spatial framework. A second principle is that although the analysis of space involves only simple visual information, this crude visual information must be combined with other sensory modalities, of which the best studied is somatosensory. Thus for right-hemisphere parietal patients, visual–spatial deficits correlate with tactile–spatial deficits (Dee & Benton, 1970), yet both can be separated with the fine discrimination deficits seen in either modality (Semmes, 1968). At least some aspects of spatial integration are done by a supramodal or haptic processing which although grounded in primary sensory modalities is ultimately not linked to any single modality. One would assume that auditory, vestibular, and even olfactory influences would enter into such a supramodal construct, but these have not been as intensely studied as vision and proprioception.

CONCLUSION

In summary, the mechanisms for behaving in visual space that have described by single-unit studies in three nongeniculostriate visual areas have shown a general pattern of sophisticated association of information. In all the areas that have been studied visual processing occurs, but this visual processing is not dependent upon aspects of the stimulus other than its spatial location and occasionally on the direction of movement. Each of these areas adds to this primitive visual information some other dimension of neural information. Thus visual responses in the superior colliculus and the frontal eye fields are enhanced when the stimulus is going to be the target for an eye movement, thus transforming the visual information into a form that is particularly useful for motor processing. In the posterior parietal cortex this process of enhancement occurs whenever the animal attends to the stimulus, not merely when it makes an eye movement to it, and is therefore useful for the processing underlying selective visuospatial attention. Finally, the visual information in posterior parietal cortex is combined with information about somatosensory events and eye position. The combination of these mechanisms enables the organism to see a stimulus, locate it in visual and real space, and, possibly, orient to it. In these areas there are now convincing single neuron mechanisms to fulfill the functions that have been ascribed on clinical and psychological grounds.

ACKNOWLEDGMENTS

The author is grateful to Sandy Catron for typing this manuscript, Guy Bateman and Mark Boehme for preparing the illustrations, John Watts and John Hamilton for electronic help, David Lee Robinson and M. Catherine Bushnell for years of heated discussion in a dark noisy room, and Darrel N. McIndoe, Paul Tyler, and Charles Bonney of the Armed Forces Radiobiology Research Institute for their generous support.

REFERENCES

Baizer, J. S., Robinson, D. L., & Dow, B. S. Receptive fields in areas 18 and 19 of the awake, behaving monkey. *Journal of Neurophysiology*, 1976, *40*, 1024–1037.

Bizzi, E. Discharge of frontal eye field neurons during saccadic and following eye movements in unanesthetized monkeys. *Experimental Brain Research*, 1968, *6*, 69–80.

Bushnell, M. C., & Goldberg, M. E. The frontal eye fields have a signal that precedes visually guided saccades. *Abstracts of Society for Neuroscience*, 1979, *5*, 779.

Bushnell, M. C., Robinson, D. L., & Goldberg, M. E. Dissociation of movement and attention: Neuronal correlates in posterior parietal cortex. *Abstracts of Society for Neuroscience*, 1978, *4*, 621.

Chavis, D., & Pandya, D. N. Further observations on corticofrontal connections in the rhesus monkey. *Brain Research*, 1976, *117*, 369–386.

Critchley, M. *The parietal lobe*. London: Arnold, 1953.

Dee, H. L. Visuoconstructive and visuoperceptive deficit in patients with unilateral cerebral lesions. *Neuropsychologia*, 1970, *8*, 305–314.

Dee, H. L., & Benton, A. D. Spatial performances in unilateral cerebral disease. *Cortex*, 1970, *6*, 261–272.

Diamond, I. T. The evolution of the tectal–pulvinar system in mammals: Structural and behavioral studies of the visual system. *Symposium of the Zoological Society of London*, 1973, *33*, 205–233.

Evarts, E. V. Methods for recording activity of individual neurons in moving animals. In R. F. Rushmer (Ed.), *Methods for medical research* (Vol. 2). Chicago: Year Book Medical Publishers, 1966. Pp. 241–250.

Faugier–Grimaud S., Frenois, C., & Stein, D. G. Effects of posterior parietal lesions on visually guided behavior in monkeys. *Neuropsychologia*, 1978, *16*, 151–168.

Goldberg, M. E., & Bushnell, M. C. (1981). The role of the frontal eye fields in visually guided saccades. In A. Fuchs & W. Becker (Eds.), *Progress in oculomotor research*. New York: Elsevier–North Holland, pp. 185–193.

Goldberg, M. E., & Robinson, D. L. Visual mechanisms underlying gaze: The cerebral cortex. In R. Baker & A. Berthoz, (Eds.), *Control of gaze by brainstem interneurons*. Amsterdam: Elsevier North Holland, 1977.

Goldberg, M. E., & Robinson, D. L. The superior colliculus. In R. B. Marleston (Ed.), *Handbook of behavior neurobiology* (Vol. 1: Sensory integration). New York: Plenum, 1978. Pp. 119–164.

Goldberg, M. E., & Robinson, D. L. The significance of enhanced responses in monkey parietal cortex. *Behavioral and Brain Science*, 1980, *3*, 503–505.

Goldberg, M. E., & Wurtz, R. H. Activity of superior colliculus in behaving monkey: I. Visual receptive fields of single neurons. *Journal of Neurophysiology*, 1972, *35*, 560–574. (a)

Goldberg, M. E., & Wurtz, R. H. Activity of superior colliculus in behaving monkey: II. Effect of attention on neuronal responses. *Journal of Neurophysiology*, 1972, *35*, 560–574. (b)

Gross, C. G., Rocha–Miranda, C. E., & Bender, D. B. Visual properties of neurons in inferotemporal cortex of the macaque. *Journal of Neurophysiology*, 1972, *35*, 96–111.

Hartje, W., & Ettlinger, G. Reaching in light and dark after unlatteral posterior parietal ablations in the monkey. *Cortex*, 1973, *9*, 346–354.

Heilman, K. M. Neglect and related disorders. In K. M. Heilman & E. Valenstein (Eds.), *Clinical neuropsychology*. New York: Oxford University Press, 1979. Pp. 268–307.

Heilman, K. M., Pandya, D. N., Karol, E. A., & Geschwind, N. Auditory inattention. *Archives of Neurology*, 1971, *24*, 323–325.

Hubel, D. H., & Wiesel, T. N. Receptive fields and functional architecture of monkey striate cortex. *Journal of Physiology*, 1968, *195*: 215–243. (London).

Hyvärinen, J., & Poranen, A. Function of the parietal associative area 7 as revealed from cellular discharges in alert monkeys. *Brain*, 1974, *67*, 673–692.

Jones, B. Visual facilitation of auditory localization in school children: A signal detection analysis. *Perception and Psychophysics*, 1975, *17*, 217–220.

Jones, B., & Kabanoff, B. Eye movements in auditory space perception. *Perception & Psychophysics*, 1975, *17*, 241–245.

Jones, E. G., & Powell, T. P. S. An anatomical study of converging sensory pathways within the cerebral cortex of the monkey. *Brain*, 1970, *93*, 793–820.

Klein, R. *Does oculomotor readiness mediate cognitive control of visual attention? Attention and performance VIII*. Hillsdale, New Jersey: Lawrence Erlbaum and Associates, 1979.

Kuypers, H. G. J. M., Scwarcbart, M. K., Mishkin, M., & Rosvold, H. E. Occipitotemporal corticocortical connections in the rhesus monkey. *Experimental Neurology*, 1965, *11*, 245–262.

LaMotte, R. H., & Acuna, C. Defects in accuracy of reaching after removal of posterior parietal cortex in monkeys. *Brain Research*, 1978, *139*, 309–326.

Luschei, E. S., & Fuchs, A. F. Activity of brain stem neurons during eye movements of alert monkeys. *Journal of Neurophysiology*, 1972, *35*, 445–461.

Lynch, J. C. The functional organization of posterior parietal association cortex. *Behavioral and Brain Sciences*, 1980, *3*, 485–499.

Lynch, J. C., Mountcastle, V. B., Talbot, W. H., & Yin, T. C. T. Parietal lobe mechanisms for directed visual attention. *Journal of Neurophysiology*, 1977, *40*, 362–389.

Mesulam, M.-M., Van Hoesen, G. W., Pandya, D. N., & Geschwind, N. Limbic and sensory connections of the inferior parietal lobule (area PG) in the rhesus monkey: A study with a new method for horseradish peroxidase histochemistry. *Brain Research*, 1977, *136*, 393–414.

Mohler, C. W., Goldberg, M. E., & Wurtz, R. H. Visual receptive fields of frontal eye field neurons. *Brain Research*, 1973, *61*, 385–389.

Mountcastle, V. B. The world around us: Neural command functions for selective attention (The F. O. Schmitt Lecture in Neuroscience, 1975). *Neurosciences Research Program Bulletin 14*, 1976, supplement.

Mountcastle, V. B., Lynch, J. C., Georgopoulos, A., Sakata, H., & Acuna, C. Posterior parietal association cortex of the monkey: Command functions for operations within extrapersonal space. *Journal of Neurophysiology*, 1975, *38*, 871–908.

Newcombe, F., & Russell, W. R. R. Dissociated visual perceptual and spatial deficits in focal lesions of the right hemisphere. *Journal of Neurology, Neurosurgery, and Psychiatry*, 1969, *32*, 73–81.

Petrides, M., & Iversen, S. E. Restricted posterior parietal lesions in the rhesus monkey and performance on visuospatial tasks. *Brain Research*, 1979, *161*, 63–77.

Pohl, W. Dissociation of spatial discrimination deficits following frontal and parietal lesions in monkeys. *Journal of Comparitive Physiology*, 1973, *82*, 227–648.

Posner, M. I. Orienting of attention. *Quarterly Journal of Experimental Psychology*, 1980, *32*, 3–25.

Robinson, D. L., Baizer, J., & Dow, B. M. Behavioral enhancement of visual responses of pre-striate neurons of the rhesus monkey. *Investigative Ophthalmology and Visual Science*, 1980, *19*, 1120–1123.

Robinson, D. L., Bushnell, M. C., & Goldberg, M. E. The role of posterior parietal cortex in selective visual attention. In W. Becker & A. Fuchs, *Progress in oculomotor research*. The Hague: Elsevier, 1980, in press.

Robinson, D. L., & Fuchs, A. F. Eye movements evoked by stimulation of frontal eye fields. *Journal of Neurophysiology*, 1969, *32*, 637–648.

Robinson, D. L., Goldberg, M. E., & Stanton, G. B. Parietal association cortex in the primate: Sensory mechanisms and behavioral modulations. *Journal of Neurophysiology*, 1978, *41*, 910–932.

Sakata, H., Shibutani, H., & Kawano, K. Parietal neurons with dual sensitivity to real and induced movements of visual target. *Neuroscience Letters*, 1978, *9*, 165–169.

Sakata, H., Shibutani, H., & Kawano, K. Spatial selectivities of visual fixation neurons in the posterior parietal association cortex of the monkey. *Journal of Neuophysiology*, 1980, in press.

Schiller, P. H., & Koerner, F. Discharge characteristics of single units in the superior colliculus of the alert rhesus monkey. *Journal of Neurophysiology*, 1971, *34*, 920–936.

Schiller, P. H., True, S. D., & Conway, J. L. Effects of frontal eye field and superior colliculus ablations on eye movements. *Science*, 1979, *206*, 590–592.

Semmes, J. Hemispheric specialization: A possible clue to mechanisms. *Neuropsychologia*, 1968, *6*, 11–26.

Stanton, G. B., Cruce, W. L. R., Goldberg, M. E., & Robinson, D. L. Some ipsilateral projections to areas PF and PG of the inferior parietal lobule in the monkey. *Neuroscience Letter*, 1977, *6*, 243–250.

Trojanowski, J. Q., & Jacobson, S. Medial pulvinar afferents to frontal eye fields in rhesus monkey demonstrated by horseradish peroxidase. *Brain Research*, 1974, *80*, 395–411.

Ungerleider, L. G., & Brody, B. A. Extrapersonal spatial orientation: The role of posterior parietal, anterior frontal, and inferotemporal cortex. *Experimental Neurology*, 1977, *56*, 265–280.

Weiskrantz, L., Warrington, E. K., Sanders, M. D., & Marshall, J. Visual capacity in the hemianopic field following a restricted occipital ablation. *Brain*, 1974, *97*, 709–728.

Whittington, D. A., & Hepp-Raymond, M. C. Eye and head movements to auditory targets. *Abstracts of the Society for Neuroscience*, 1977, *3*, 158.

Wurtz, R. H. Visual receptive fields of striate cortex neurons in awake monkeys. *Journal Neurophysiology*, 1969, 32, 727–742.

Wurtz, R. H., & Albano, J. E. Visuomotor function of the primate Superior Colliculus. *Annual Review of Neuroscience*, 1980, 3, 189–226.

Wurtz, R. H., & Goldberg, M. E. Superior colliculus cell responses related to eye movements in awake monkeys. *Science*, 1971, *171*, 82–84.

Wurtz, R. H., & Goldberg, M. E. Activity of superior colliculus in behaving monkey. III. Cells discharging before eye movements. *Journal of Neurophysiology*, 1972, 35, 575–586. (a)

Wurtz, R. H., & Goldberg, M. E. Activity of superior colliculus in behaving monkey. IV. Effects of lesions on eye movements. *Journal of Neurophysiology*, 1972, 35, 587–596. (b)

Wurtz, R. H., & Mohler, C. W. Organization of monkey superior colliculus: Enhanced visual response of superficial layer cells. *Journal of Neurophysiology*, 1976, 39, 745–762. (a)

Wurtz, R. H., & Mohler, C. W. Enhancement of visual responses in monkey striate cortex and frontal eye fields. *Journal of Neurophysiology*, 1976, 39, 745–762. (b)

Wurtz, R. H., Richmond, B. J., & Judse, S. J. Vision during saccadic eye movements. III. Visual interactions in monkey superior colliculus. *Journal of Neurophysiology*, 1980, *43*, 1168–1181.

Yin, T. C. T., & Mountcastle, V. B. Visual input to the visuomotor mechanisms of the monkey's parietal lobe. *Science*, 1977, *197*, 1381–1383.

Yin, T. C. T., & Mountcastle, V. B. Mechanisms of neural integration in the parietal lobe for visual attention. *Fed. Proc.*, 1978, 37, 766–772.

Zeki, S. M. The function organization of projections from striate to prestriate visual cortex in the rhesus monkey. *Cold Spring Harbor Symposium on Quantitative Biology*, 1975, *40*, 591–600.

GRAHAM RATCLIFF 13

Disturbances of Spatial Orientation Associated with Cerebral Lesions

Since Hughlings Jackson described his case of left hemiplegia with "imperception" in 1876 there have been many more reports that confirm that disorders of visual perception and spatial orientation can follow damage to the brain, particularly when the lesion involves the posterior half of the right cerebral hemisphere. Virtually all the data collected before the middle of this century (see Benton, Chapter 11, this volume) consisted of detailed studies of individual cases. More recently, single case reports have been supplemented by studies in which groups of patients are defined in terms of some independent variable, usually locus of lesion, and their performance is compared on some perceptual or spatial task. Both methods have yielded valuable information but the data are not easy to evaluate.

PROBLEMS OF INTERPRETATION AND TERMINOLOGY

Individual case studies bear witness to the clinical acumen and acute observation of their authors but the boundary between observation and interpretation is not always obvious and the different methods and terminology used by different authors make their reports difficult to compare. Consequently, a variety of symptoms and syndromes have come to be recognized but the criteria

301

SPATIAL ABILITIES
Development and Physiological Foundations

for their diagnosis are frequently neither explicit nor generally agreed. It is not always clear to what extent the traditional categories of disorder represent unitary phenomena nor is it clear whether different terms denote different disorders or different manifestations of the same underlying deficit, or whether different authors have simply used different names for identical phenomena.

Consider, for example, the disorder traditionally called *constructional apraxia* which is one of the commoner forms of spatial deficit. This term refers to a difficulty in free drawing or in copying stimuli that are usually visually presented and consist of line drawings of geometric forms or patterns made from blocks or sticks. Kleist (1934), who is usually credited with the first description of the disorder, conceived of it as a specific perceptuomotor phenomenon distinct from other visual–perceptual deficits on the one hand and apraxia on the other.

Later authors have given the term a wider reference and suggested that their broader category of constructional apraxia can be divided into two subtypes, one reflecting a perceptual or gnostic deficit, the other resulting from an executive or praxic disability (McFie & Zangwill, 1960; Piercy, Hécaen, & Ajuriaguerra, 1960; Warrington, 1969; Warrington, James, & Kinsbourne, 1966). Lange (1936) recognized the importance of both components in his use of the term *apractognosia for spatial articulation* whereas Benton (1973, 1979) recommends the term *visuo-constructive disability* to avoid the implication that either component is crucial.

Other authors have described constructional disability along with topographical disorientation, unilateral neglect, and innaccurate visually guided reaching under the headings *visual–spatial agnosia* and *disturbances of visual space perception,* suggesting that they regard them simply as different forms of a more general spatial disorder (Ettlinger, Warrington, & Zangwill, 1957; McFie, Piercy, & Zangwill, 1950; Paterson & Zangwill, 1944, 1945). However, each of these other manifestations of disturbed space perception has also been regarded as an independent phenomenon that can exist in its own right, though not necessarily under the same name. De Renzi, Faglioni, and Villa (1977) make this claim for "topographical amnesia," Heilman (1979) for unilateral neglect, and Ratcliff and Davies–Jones (1972) and Rondot, Recondo, and Ribadeau Dumas (1977) for inaccurate visually-guided reaching, the former authors calling it *defective visual localisation* and the latter using the term *visuomotor ataxia.*

This confusing state of affairs makes it difficult to answer even such basic questions as whether a given disorder is more frequently associated with damage to one or other of the cerebral hemispheres. Constructional apraxia, for example, was originally ascribed to left-hemisphere damage (Kleist, 1934) but in its wider and more conventional sense is generally (though not universally) thought to be more frequent and more severe after right-hemisphere lesions

(Critchley, 1953; Hécaen, 1969). However, Benton (1967) has shown that although this is true for some constructional tasks it is not true for others. Furthermore the difference in incidence between right- and left-hemisphere cases in Benton's series decreased as he employed less stringent criteria for failure.

The interpretation of data from laboratory studies of groups of patients is subject to similar difficulties. Until very recently most authors were content to describe rather than analyze the deficits they studied and relate their conclusions to the gross anatomy of the brain, but the different tasks used by different authors and differences in their criteria for grouping patients make their studies difficult to compare. The result is that poor performance on a number of spatial tasks is now known to be reliably associated with damage to the right hemisphere but much less is known about the nature of the deficits revealed by the poor performance.

A further problem arises when we try to relate the results of group studies to the conclusions suggested by individual case reports. Group studies of face recognition, for example, suggest that the neural mechanisms subserving this ability are located in the right hemisphere (Benton & Van Allen, 1968; De Renzi, Faglioni, & Spinnler, 1968; De Renzi & Spinnler, 1966a; Warrington & James, 1967b; but see also Hamsher, Levin, & Benton, 1979). However, Meadows's (1974) review of the literature on prosopagnosia (a severe disorder of face recognition) shows that bilateral lesions are present in the great majority of reported cases, suggesting that both hemispheres contribute to the task of recognizing faces.

This apparent anomaly can be resolved if we remember that patients with bilateral lesions are usually excluded from group studies and that prosopagnosia is extremely rare. It is quite likely that no prosopagnosic patient has been included in the majority of group studies of face recognition and the deficit detected in these group studies may be qualitatively different from the deficit (or deficits) responsible for prosopagnosia. Some support for this hypothesis comes from a prosopagnosic patient who performed normally on Tzavaras, Hécaen, and Le Bras's (1970) face recognition tasks.

Similar anomalies appear in connection with other spatial deficits. I have examined a patient with severe topographical disorientation who was unable to find her own bed in the ward or to learn the way from the ward to the neuropsychology unit although she walked over this route many times during her stay in the hospital. Yet she performed normally on most of the spatial tasks used in our laboratory including a visually guided stylus maze and a locomotor maze task that requires the patient to navigate a route with the aid of a simple map. Both these tasks are thought to be sensitive to disorders of spatial orientation but presumably the deficits they detect are not necessarily those that cause topographical disorientation. One occasionally sees patients whose ability to carry out spatial tasks under visual guidance is severely disrupted, although their performance is near normal on similar haptic tasks when vision is excluded. Yet group studies suggest that spatial deficits are supramodal, performance on both visual and haptic tasks being adversely affected by right posterior lesions (De Renzi, Faglioni, & Scotti, 1968, 1970; De Renzi & Scotti, 1969; Ratcliff 1970, but see also Chedru, 1976).

It will be apparent that considerable caution should be exercised in drawing any but the most general conclusions about the effects of cerebral lesions on spatial orientation. But if the clinical literature continues to be restricted to variations on the theme suggested by Hughlings Jackson, namely that spatial and perceptual disorders are commonly associated with damage to the posterior part of the right hemisphere, its contribution to our understanding of the neural basis of spatial orientation will be limited. Accordingly, in the interests of furthering our understanding, and despite the uncertainties involved, it seems reasonable to adopt a more analytic approach and look for the underlying causes of the spatial disorders encountered in the clinic, paying particular attention to the question of whether distinct and independent types of disorder can be distinguished. Only after this has been done will it be possible to make educated guesses about the nature of the neural mechanisms involved.

LOW-LEVEL SENSORY ANALYSIS AND
HIGH-LEVEL PERCEPTUAL INTEGRATION

One important distinction has been made in two recent reviews. Moscovitch (1979) contrasted the extraction of low-level stimulus features, which may be presumed to occur early in the information processing sequence, with later higher order processes that integrate sensory features into more abstract categorical properties. He pointed out that hemispheric asymmetries are rarely reported for the extraction of low level features such as color, brightness, and contour, suggesting that this is accomplished in a similar manner by both hemispheres, although the functions of the hemispheres clearly diverge at later stages of processing.

A similar conclusion is implicit in the distinction made by Ratcliff and Cowey (1979) between disorders attributable to defective sensory analysis of a single stimulus feature, which can occur after damage to either hemisphere, and the more complex perceptual disorders that are predominantly associated with right-hemisphere pathology. I will argue here that disorders that reflect defective sensory analysis of the spatial properties of a stimulus are modality specific and can plausibly be interpreted as the result of damage to specific neural structures beyond the striate cortex. They are different in these respects from disorders that are more frequently associated with damage to a specific hemisphere and that often manifest themselves in more than one sensory modality. These presumably result from damage to the specialized mechanisms of each hemisphere which operate on spatial data at later, more abstract stages of processing and whose existence is still a matter of inference.

DISORDERS OF SENSORY ANALYSIS

The spatial disorders that fall into this category are those in which the patient's ability to locate a single visual stimulus in space is impaired although other stimulus features, such as color, may be perceived normally.

Visual Localization

The most dramatic examples of defective visual localizaiton occur in the syndrome that has come to be known as *visual disorientation* or *Bálints syndrome* (Bálint, 1909). In the classic accounts of this condition Holmes and his contemporaries (Holmes, 1918, 1919; Holmes & Horrax, 1919; Riddoch, 1917; Smith & Holmes, 1916; Yealland, 1916) described patients with bilateral penetrating missile wounds of the posterior part of the brain whose "most obvious symptom was the inability to determine the position in space, in relation to themselves, of objects which they saw distinctly [Holmes, 1919, p. 231]." The clearest evidence for this disability was the gross inaccuracy of visually guided reaching. Holmes reported, "When I held up a knife in front of one man he said at once 'That's a pocket-knife' but though his eyes were directed on it he stretched out his arm in a totally wrong direction when he was told to take hold of it. Another man struck my face with his hand as he attempted to point to a pencil which I held two or three feet to my right side [p. 231]." The patients had difficulty in redirecting their gaze to bring peripherally exposed stimuli to fixation although "none of the ocular muscles were paralysed." They also made errors in the verbal estimation of distance, "a tall chimney about 500 yards away seemed 'perhaps twenty yards' from one man [p. 231]," and experienced difficulty in locomotion, "they frequently collided with large obstacles as they failed to realize their direction or distance from themselves [p. 231]."

The disorder was specifically visual: Sound and touch were localized normally by most patients and when they were permitted to explore their environment tactually orientation was promptly restored. They all exhibited some topographical memory loss, which Holmes regarded as an integral part of the syndrome but which can certainly occur in the absence of other manifestations of visual disorientation, and other symptoms that he discounted as explanations for the defective visual localization on the grounds that they were also commonly found in patients who could localize visual stimuli normally.

The disorder of visual localization in Holmes's patients affected stimuli in all parts of the visual field, including fixation, but a form of the disorder has since been described in which the innacurate reaching is limited to the half field contralateral to a unilateral parietal lesion (see Figures 13.1 and 13.2). This can occur in the absence of visual field defect, gross oculomotor disturbance, hemiplegia, or other spatial disorder (Benton, 1969; Cole, Schutta, & Warrington, 1962; Ratcliff & Davies–Jones, 1972; Riddoch, 1935). The unilateral disorder is much more common than the generalized visual disorientation described by Holmes and occurs equally frequently after right- and left-hemisphere lesions, the errors predominantly reflecting underestimation of the distance of the target from the fixation point (Ratcliff & Davies–Jones, 1972).

It is clear that the inaccurate reaching is a result of failure to specify the position of the target to which the arm is to be directed and not a general inability to put the arm ʌn the desired position: Patients with unilateral lesions are generally able to reach accurately with either arm for visual targets exposed at fixation or in the ipsilateral half field and for targets defined by tactile stimulation on the body, but they are inaccurate with both arms when reaching for contralateral visual targets (Cole, Schutta, & Warrington, 1962; Ratcliff & Davies–Jones, 1972) although cases have been described in which the disorder affects only one arm or one arm-field combination (Bálint, 1909; Rondot *et al.*, 1977). It also appears that patients with unilateral lesions, unlike those with bilateral lesions as described by Holmes, are able to direct their eyes accurately to targets to which they cannot accurately direct their arms (Ratcliff & Davies–Jones, 1972; Riddoch, 1935). This conclusion must be tentative as eye

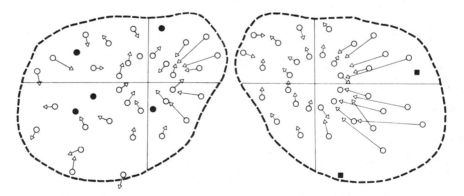

Figure 13.1. Grossly defective visual localization in the right hononymous half field of a patient with a left parietal missile wound. The results for each eye are shown separately. Open circles show target positions; open triangles show position indicated by the patient; solid circles indicate direct hits; squares show positions in which target was not detected. (Adapted from Ratcliff & Davies–Jones, 1972.)

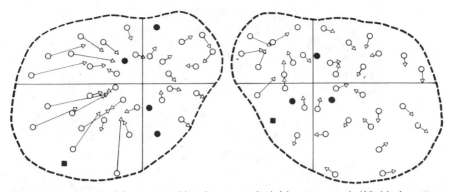

Figure 13.2. Grossly defective visual localization in the left hononymous half field of a patient with a right parietal missile wound. Symbols as in Figure 13.1. (Adapted from Ratcliff & Davies–Jones, 1972.)

movements were not formally recorded in these studies but it is unlikely that gross inaccuracies in redirection of gaze would have remained unnoticed.

Some authors have sought to explain defective visual localization on the basis of a disconnection between the visual cortex and the sensory–motor cortex (Boller, Cole, Kim, Mack, & Patawaran, 1975; Rondot *et al.,* 1977; Russell, 1951). Alternatively it may be the result of damage to a cortical structure that participates in the extraction of information about position or the mapping of visual onto manual space (Ratcliff & Cowey, 1979; Riddoch, 1935). Damasio and Benton (1979) thought that damage to Brodmann's areas 7, 19, and 39 (or their motor outflow) was responsible for the impairment in their case and this fits well with the location of the lesions in most of the other reported cases of bilateral visual disorientation. The unilateral lesions in Ratcliff and Davies–Jones's (1972) patients, however, were higher in the parietal lobe (Figure 13.3) and this led Ratcliff and Cowey (1979) to place more emphasis on area 7, where neurones with properties that suggest they may be involved in visually guided reaching have recently been described.

Although there has been dispute about the physiological properties and behavioral function of area 7 neurones, there is agreement on some of their general characteristics (see Goldberg, Chapter 12, this volume for a review). Two types of cell seem particularly relevant to visual orientation: "saccade neurones" and "hand projection neurones" which fire prior to visually guided eye movements and reaching movements respectively. Robinson, Goldberg, and Stanton (1978) have denied the suggestion that these neurones have a motor command function (Mountcastle, 1975; Mountcastle, Lynch, Georgopoulos, Sakata, & Acuna, 1975) emphasizing instead their involvement in the direction of attention to the stimulus that can reasonably be expected to precede the execution of a hand or eye movement toward it.

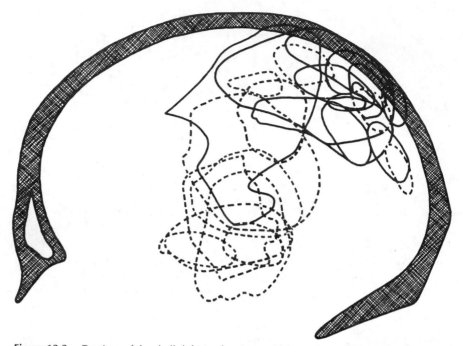

Figure 13.3. Tracings of the skull defects of patients with posterior missle wounds in Ratcliff and Davies–Jones (1972) series. Solid lines show the defects in patients with grossly defective localization in the contralateral half field.

   This argument is convincing but it seems possible that these neurons may also be important in providing the motor system with information about the position of the visual stimulus to which a movement is required, without necessarily commanding it to execute that movement. The results of lesion studies support this view: Bilateral ablation of posterior parietal cortex in the monkey causes bilateral misreaching and general disorientation very similar to that after bilateral lesions in human beings, as described by Holmes; unilateral lesions cause a predominantly contralateral disorder as they do in human beings but reaching with the contralateral arm is affected in the monkey rather than reaching toward the contralateral half-field as is the case in man (Hartje & Ettlinger, 1973; Mountcastle, 1975). In the monkey, as in human beings, it is quite clear that the misreaching is not a trivial consequence of sensory loss or limb apraxia (Hartje & Ettlinger, 1973).
   To summarize, it seems that there is an area in the monkey's parietal lobe in which neurons have properties consistent with their being involved in visual orientation. Destruction of this area causes a disorder of visual orientation that is very similar to the consequence of damage in a similar place in the human

brain. It seems reasonable to speculate, therefore, that a similar structure plays a similar role in the maintenance of visual orientation in human beings. A fuller discussion of the cortical control of visually guided reaching is available in Humphrey (1979).

The Role of the Superior Colliculus

Before leaving the topic of visually guided reaching a further phenomenon must be considered. It is now thought that a visual pathway that does not involve striate cortex can guide behavior in humans. The most impressive demonstration of this is provided by the preservation of various visual functions in the perimetrically blind part of the field of a patient (D.B.) who had most of his right striate cortex surgically removed (Weiskrantz, Warrington,. Sanders, & Marshall, 1974). Visually guided reaching was assessed in a manner very similar to that used by Ratcliff and Davies–Jones (1972) and D.B. was able to localize with remarkable precision targets that he could not consciously "see."

Area 7 was presumably intact in D.B. but it may have been partially deafferented as it seems likely that some of its visual input comes indirectly via striate cortex although there is also a considerable subcortical input (see Humphrey, 1979). Could his accurate reaching have been mediated by some other structure?

One possibility is that the remaining visual capacity was subserved by the superior colliculus, which raises the question of how damage to the colliculi affects visual orientation in human beings. Since isolated damage to the colliculus is extremely rare this question cannot be answered with certainty, although visually guided reaching has been tested in one patient (P.L.) following surgical treatment of a small angioma centered on the right superior colliculus (Heywood & Ratcliff, 1975). There has been no postmortem verification of the locus of lesion in this case, but it has been assumed that there was severe unilateral collicular damage. Three to four weeks after the operation visually guided reaching was within normal limits throughout the visual field (see Figure 13.4) and although P.L. was slightly less accurate in the contralateral half field he was far more accurate than Ratcliff and Davies–Jones's (1972) patients with parietal lesions.

The observation that an intact colliculus is not necessary for accurate reaching in the contralateral half field reinforces the suggestion that area 7 is crucial. Area 7 was presumably intact in both P.L. and D.B. although the source of visual input was presumably different in the two cases, the striate route being available to P.L. and an alternative route, possibly via colliculus and pulvinar (Bayledier, 1977; Trojanowski & Jacobson, 1976; Yin & Mountcastle, 1977) in D.B. Perenin and Jeannerod (1978) have demonstrated accurate localization of stimuli within about 45 degrees of fixation in hemidecorticate subjects but

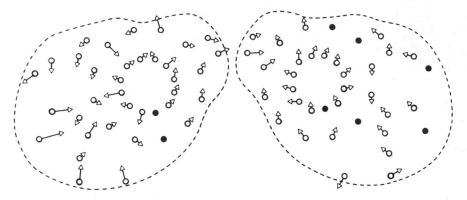

Figure 13.4. Visual localization by P.L. Symbols as in Figure 13.1.

localization broke down for more peripheral stimuli. Lesions requiring such drastic treatment have usually been sustained early in life and some functional reorganization may have taken place.

The human superior colliculus may be more important for the oculomotor exploration of space. Initially P.L. had a hemianopia and gross left-sided neglect (possibly as a result of retraction of the right hemisphere during the operation) but three weeks later, when these symptoms had cleared, eye movements to targets he could readily detect were still markedly slowed and often led by a head movement. Seventeen weeks after the operation P.L.'s visual and oculomotor status was normal on clinical examination apart from limitation of upward gaze, but formal recording of his eye movements revealed other impairments. There was a marked reduction in the frequency of spontaneous saccadic eye movements to the side contralateral to the lesion when he sat in the dark and a greater number of correction saccades were required before he could fixate a target exposed in the left (contralateral) field although leftward eye movements now had a normal latency (Heywood & Ratcliff, 1975). These findings would be consistent with impairment of the ability to specify the position of a left-sided target sufficiently accurately to bring it to fixation in a single movement, or loss of the motor skill involved in making the required movement, or both.

Finally, P.L. was impaired in a saccade matching task in which he sat in the dark, with no fixation mark, and was asked to look to the left or right, pause, and return to the starting position. The accuracy with which the subject can do this has been said to reflect his or her ability to use information about the extent of his or her outward eye movement to control the return movement (Heywood, 1973a, 1973b). Like those of most normal subjects, P.L.'s returns from the left were less accurate than his returns from the right but the difference was greater than for normal subjects (Heywood & Ratcliff, 1975).

It would not be legitimate to draw definite conclusions about the function of the human superior colliculus on the basis of a less than exhaustive study of a single patient whose lesion may not have totally destroyed that structure and was almost certainly not restricted to it. We can only say that the long-term deficits in P.L. appear to have been limited to oculomotor function whereas the manual indication of position seems to have been more affected in Ratcliff and Davies–Jones's (1972) patients with parietal lesions. Selective impairment on tasks involving visual searching and scanning ability has also been reported in progressive supranuclear palsy, a condition associated with neuropathological changes in several structures including the superior colliculi but excluding the cerebral cortex (Kimura, Barnett, & Burkhart, 1979).

Stereoacuity

Although formal tests of stereoscopic vision were not included in any of the studies of patients with defective visual localization discussed earlier, the fact that these patients make gross errors of direction as well as distance in reaching for visual stimuli effectively rules out impaired stereopsis as the sole explanation for the disorder. But cerebral lesions can affect stereopsis in at least two ways; stereoacuity can be disturbed by damage to either hemisphere (Danta, Hilton, & O'Boyle, 1978; Lehman & Wälchli, 1975; Rothstein & Sacks, 1972) whereas global stereopsis appears to be selectively affected by right hemisphere lesions (Benton & Hécaen, 1970; Carmon & Bechtoldt, 1969; Hamsher, 1978).

Stereoacuity, or local stereopsis, is the ability to detect small differences in depth. It is usually tested either by exposing clearly visible stimuli that actually differ in their distance from the observer or by the use of conventional stereograms in which well-defined forms, visible monocularly in each of the stereo pair, are so arranged that their images fall on noncorresponding points in the two retinae and yield an impression of depth when they are fused in a stereoscope. In this situation there is no problem about detecting the form of the stimulus but the ability to tell whether it is at a different distance from its surroundings depends on the detection of retinal disparity.

Global stereopsis, however, is tested with stereograms of the kind devised by Julesz (1971). These are pairs of matrices, each with a large number of cells, in which each cell is randomly filled or left blank. The two matrices are identical except that in one of them a section is laterally displaced with respect to the corresponding section of the other. Typically the displacement is great enough to yield a disparity that is well above threshold. Viewed monocularly the stimuli look randomly speckled and the displaced section cannot be discerned until they are viewed stereoscopically when it is perceived as a surface standing out (or back) in depth. Local stereopsis, the simple detection of disparity, is not sufficient to achieve this percept because for each cell in one matrix there are many potentially corresponding cells in the other, each with a different disparity value. Instead, the cerebral lesion evidence is consistent with the view that global stereopsis requires the additional cooperation of higher-order perceptual processes which, in humans, are lateralized to the right hemisphere.

Stereopsis has also been investigated in the rhesus monkey and again local and global stereopsis have been dissociated. The striate cortex of the rhesus monkey projects separately to several secondary, prestriate visual areas in which the retina is topologically represented (see Cowey,1979 for a review). In the first of these, known as V2, neurons respond to stimuli falling on noncorresponding points in the two retinae (Zeki, 1978). Bilateral destruction of the part of V2 corresponding to the central retina impairs stereoacuity in the monkey (Cowey & Wilkinson, personal communication) although global stereopsis, which can be disturbed by inferotemporal lesions, is unaffected (Cowey & Porter, 1979). These results, demonstrating that the integrity of V2 is necessary for normal stereoacuity in the monkey, led Ratcliff and Cowey (1979) to suggest that damage to a similar structure might account for impaired stereoacuity in humans.

Unfortunately, what little we know of the location of the lesions responsible for impaired stereoacuity in human beings does not unequivocally support this hypothesis. Although the damage usually involves the posterior half of the brain, the lesions are not all concentrated in one specific area (Danta *et al.,* 1978) and many of them are more anterior than one would have expected if V2 were involved. However, another study in which the location of the lesions was not reported found a much lower incidence of impaired stereoacuity (Hamsher, 1978) so the rejection of Ratcliff and Cowey's hypothesis on anatomical grounds may be postponed.

Defective visual localization and impaired stereoacuity have the following characteristics: They are modality specific; they are not significantly more frequently associated with damage to one hemisphere than to the other; they selectively affect processing of information about one feature of the stimulus; and a case has been made for attributing each of them to damage to the human equivalent of cortical areas in which it is known from studies of infrahuman primates that neurons can be driven by appropriate visual stimuli. A similar case was made for achromatopsia (the central disturbance of color vision) by Ratcliff and Cowey (1979).

In Moscovitch's (1979) terms these disorders could be said to occur at early, sensory stages in the information-processing sequence before the point at which functional hemispheric asymmetry emerges. The anatomical analysis offered here implies that the structural locus at which the functions of the hemispheres diverge in visual processing is anterior to the prestriate visual "association" areas, probably in the posterior temporal and parietal cortex. It is with lesions in this region (and parts of the frontal lobe) that the most obvious differences between the effects of damage to the right and left hemispheres are seen. As Moscovitch points out, the posterior Sylvian region is also the part of the human brain in which morphological differences between the hemispheres are most frequently reported (see Geschwind, 1974; LeMay, 1976; Rubens, 1977; and Witelson, 1977 for reviews).

DISORDERS OF PERCEPTUAL INTEGRATION

The disorders that are typically associated with damage to a particular hemisphere are more complex. In attempting to analyze them a distinction will first be made between spatial disorders and other disorders of perception. Then the spatial disorders associated with damage to the left hemisphere will be distinguished from those that more frequently follow right-hemisphere lesions. Finally the symptoms of spatial disorientation associated with right-hemisphere lesions will be examined with a view to specifying the nature of the factors that limit patients' performance on spatial tasks.

Perceptual versus Spatial Disorder

All visual stimuli have position and extension in space. They are all, therefore, spatial in a strict sense but it does seem reasonable to distinguish those deficits in which difficulty with the analysis of the spatial aspects of stimuli seems to be crucial from other disorders in which this does not seem to be the limiting factor.

These two types of deficit can be dissociated and appear to result from damage to different parts of the right hemisphere. Newcombe and Russell (1969) showed that performance on a spatial task, visually guided stylus maze learning, and another perceptual task, Mooney's visual closure test, were both significantly impaired following right-hemisphere missile wounds. However, performance on the two tasks was not significantly correlated, different subsets of the right-hemisphere group being impaired on each.

Several other authors have also found that patients with right-hemisphere lesions have difficulty in identifying stimuli that, like the faces in Mooney's visual closure test, have been degraded in some way—for example, the Ghent–Poppelreuter overlapping figures and the Street completion figures (De Renzi & Spinnler, 1966b), the Gollin figures (Warrington & James, 1967a), and letters written in unconventional script and crossed through (Faglioni, Scotti, & Spinnler, 1969). Although the evidence is by no means conclusive, or even entirely consistent, there are some grounds for suggesting that the visual recognition deficit detected in these studies follows damage to the posterior temporal lobe near the parieto–temporo–occipital junction whereas the parietal lobe is more involved in spatial tasks (Meier & French, 1965; Newcombe & Russell, 1969).

Hemispheric Asymmetry in Spatial Tasks

As pointed out earlier, the lack of agreed criteria and terminology makes it more difficult than is sometimes supposed to decide which disorders typically follow damage to which hemisphere. However some general conclusions can be drawn if we assume that consistent criteria have been adopted by a given author. The consensus of opinion in three sets of studies that have considered the question (Gloning, Gloning, & Hoff, 1968; Hécaen, 1969; McFie & Zangwill, 1960) is that disorders of awareness of the position, orientation and name of parts of one's body (Gerstmann's syndrome, autotopagnosia, right–left disorientation, finger agnosia) occur more frequently with left- than right-hemisphere lesions, whereas the reverse is true for disorders affecting stimuli external to the body (constructional disability, disorders of topographical orientation, neglect of the side of space contralateral to the lesion, dressing apraxia). It is also generally agreed that the asymmetry of the hemispheres is less marked for spatial and perceptual functions than for language.

The Left Hemisphere and Personal Space

The spatial disorders characteristic of left-hemisphere damage have been said to affect personal as opposed to extrapersonal space (Butters, Soeldner, & Fedio, 1972; Semmes, Weinstein, Ghent, & Teuber, 1963) although the distinction is by no means absolute (cf. Corballis, Chapter 8, this volume). Left–right confusion—conventionally classified as a disorder of personal spatial orientation—may be demonstrated not only with respect to parts of one's own body, but also to the bodies of others and the designation of the two halves of space (Benton, 1959; Critchley, 1953). In one case of autotopagnosia the confusion extended to the identification of parts of inanimate objects as well as parts of the patient's body (De Renzi & Scotti, 1970). Although confusion is most readily demonstrated in verbal tasks (Benton, 1959; Head, 1926; Sauguet, Benton, & Hécaen, 1971) it is not merely an aphasic failure to use terms correctly (Benton, 1959; Gertsmann, 1958; Sauguet et al., 1971) and it may be demonstrated without the overt use of words in question and response (Sauget et al., 1971; Semmes et al., 1963). The conventional interpretation of these disorders implicates a disturbance of the "body schema" and the reader is referred to Benton (1959; Sauget et al., 1971) for a discussion of this concept and the role of disturbances of language and symbolic understanding in somatognosic disorders. These issues will not be discussed further here.

The Right Hemisphere and Visual Spatial Agnosia

The name *visual spatial agnosia* was given to the clinical picture of spatial disorder seen after right-hemisphere damage, presumably because the variety of performance deficits subsumed under this heading were "indicative of faulty appreciation of the spatial aspects of visual experience [Benton, 1979]." Beyond this rather general analysis no simple explanation is available that will account for all of these disorders.

They are not simply the consequence of subtle visual sensory impairment (Ettlinger, 1956) and although overt visual field defect is frequently associated with spatial disorder it is neither a necessary nor a sufficient condition for it. The distorted copy of the Rey Osterreith complex figure shown in Figure 13.5 was made by a patient with a full visual field, whereas the much better copy in Figure 13.6 is the work of a man with a complete homonymous hemianopia. Both had right-hemisphere missile wounds but in the former case the lesion

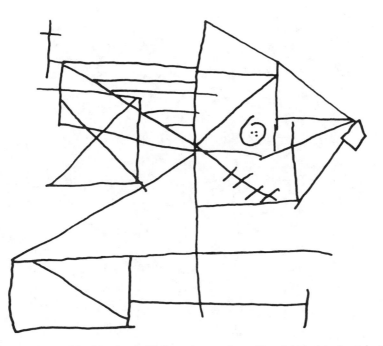

Figure 13.5.   Copy of the Rey Osterreith figure by a patient with a right parieto-frontal missile wound. (Reproduced from Ratcliff, 1980.)

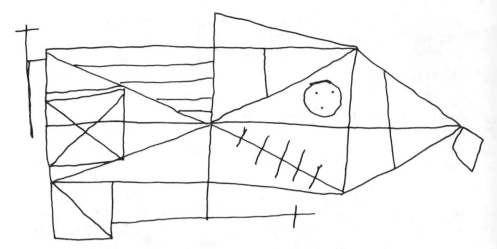

Figure 13.6. Copy of the Rey Osterreith figure by a patient with a right occipital missile wound. (Reproduced from Ratcliff, 1980.)

involved the parietal and frontal cortex, while in the latter it was restricted to the occipital lobe.

In other pathologies (particularly stroke) lesions that damage the parietal cortex will usually also damage the optic radiation and this, rather than a causal connection, is likely to explain the high incidence of visual-field defect in patients with visual–spatial disorders. The association of field defect with

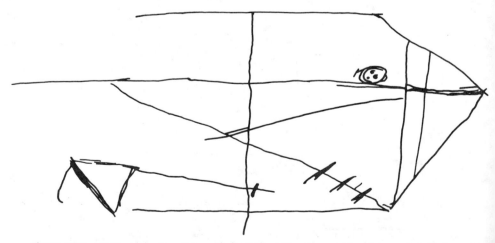

Figure 13.7. Copy of the Rey Osterreith figure by J.M. with a right parieto-occipital infarct. (Reproduced from Ratcliff, 1980.)

impairment on some tactile tasks (De Renzi & Scotti, 1969; De Renzi *et al.*, 1970) which can hardly be explained by visual failure, supports this hypothesis. Indeed the very existence of tactile spatial deficits indicates that spatial disorders cannot be explained solely on the basis of visual failure.

Neglect of the left side of space is an obvious feature of the constructional performance of some patients (see Figure 13.7 and Figure 13.8) and may even explain all of the constructional and route-finding difficulties of some individuals (Brain, 1941) but it does not account for all cases. The drawing in Figure 13.5, for example, shows considerable distortion on the right-hand side, as well as on the left, and typical spatial deficits can be demonstrated in patients who

Figure 13.8.   Writing and drawing by J.M. carried out in the following order: name (deleted), address, 42, 573, 1001, Many men went to market, clock, daisy, and copies of a triangle, star, and cube. Note neglect of left-hand side of page. (Reproduced from Ratcliff, 1980.)

show no measurable neglect (Newcombe, 1969; Newcombe & Russell, 1969; Ratcliff & Newcombe, 1973). The drawing of the floor plan of a patient's house in Figure 13.9 is a particularly good example that epitomizes visual spatial agnosia. It would be difficult to say that one side was more poorly executed than the other or even that any items have been omitted. Yet the patient's plan gives no information about the layout of the house. It is not a plan but a disorganized array of landmarks that appear to have been scattered piecemeal with no regard to their relative positions.

This drawing also illustrates another point. It was not a copy but an attempt to represent information that was presumably available in memory before the patient acquired his cerebral lesion. One cannot, therefore, explain his poor performance by invoking a failure at the level of perception. Rather one must say either that the spatial information that is so conspicuously lacking is no longer available to him or that he is unable to represent it in the form of a plan. The fact that patients' verbal descriptions of familiar environments can be deficient in a similar way (Benton, 1969; De Renzi & Faglioni, 1962; Lawson, 1969) and that they can become lost in them suggests that the difficulty is not simply an inability to draw plans. It is also important to note that the memory impairment in these cases is not a global amnesia but a specific loss of topographical information, although some information about places may be retained, presumably as a purely verbal association. A patient described by Benton (1969), for example, was able to state that Boston was in Massachusetts but not that it was north of New York.

This kind of selective memory impairment cannot account for the occasional patients who are well oriented in environments that were familiar to them before the onset of their cerebral disease and who seem to retain geographical

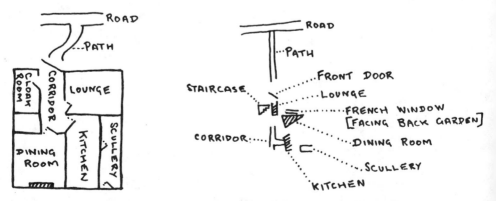

Figure 13.9. Drawing of the floor plan of his house by a patient with a right fronto-parietal tumor. The true plan is shown on the left. Labels have been added subsequently and are not in the patient's handwriting. (Redrawn from McFie et al., 1950.)

information but who lose their way in new surroundings (Scotti, 1968). A selective learning deficit or anterograde topographical memory impairment might be implicated in these cases but no kind of memory or learning defect can account for poor performance on tasks, like copying the Rey figure, where the stimulus is continuously present.

It seems, therefore, that none of the potential causes of spatial disorientation so far considered (visual sensory loss, left–sided neglect, constructional disability, poor spatial memory, a disorder of space perception) is by itself able to account for all forms of spatial disorder. However, each of the above factors might cause impairment on some spatial tasks such that between them they account for all such impairments. This possibility can be excluded only by reference to a task that controls for them all simultaneously and Ratcliff (1979) has reported an experiment in which he claimed that this requirement was fulfilled.

Ratcliff showed schematic drawings of a human figure, on which one of the hands was marked with a black disc, to patients with chronic penetrating missile wounds. The figure was exposed in one of four possible orientations (upright or inverted, front view or back view) which varied randomly from trial to trial. The patient's task was to say "right" or "left" depending on which hand was marked. The right posterior group were impaired relative to the left posterior group and control subjects on the inverted but not on the upright condition, where their performance was slightly better than that of men with left posterior lesions.

The stimuli in the upright and inverted condition were identical apart from their orientation, so any perceptual disorder, unilateral neglect, or right–left disorientation would have affected both conditions. They were continuously present throughout each trial and the response was verbal, so that memory impairment and constructional disability should not have affected the results in either condition. Ratcliff concluded that the patients with right posterior lesions had difficulty in mentally reorienting the inverted stimuli to bring them into a standard, familiar orientation before making the right–left judgment. Mental rotation (see Corballis, Chapter 8, this volume) is known to be involved in this task (Benson & Gedye, 1936) and Ratcliff linked his results with comments in the literature implicating a disorder of "spatial thought" (Benton, Levin, & Van Allen, 1974; Critchley, 1953) or "conceptual spatial performance" (Ettlinger, Warrington, & Zangwill, 1957; McFie et al., 1950; Paterson & Zangwill, 1944) in the spatial disorders associated with right-hemisphere pathology.

Is a disturbance of "spatial thought" the underlying impairment responsible for all the symptoms of spatial disorder discussed in this section, or is it simply to be added to the list of spatial deficits which cause some of these symptoms but are not always responsible for all of them? Clearly it would be difficult to exclude "disturbance of spatial thought" as a possible cause of any spatial performance deficit, partly because it is so poorly defined, but if there were a single underlying cause one might expect either that all the spatial deficits would occur together or that there would be a hierarchical relationship between them, the appearance or nonappearance of a particular symptom depending on the severity of the impairment. However, if they had different underlying causes one might expect to find double dissociations between them. Unfortunately, naturally occurring cerebral lesions do not necessarily destroy areas that constitute a functional unit while leaving other functional units intact, so intrahemispheric dissociations rarely appear, but there are some partial dissociations that provide clues to the processes that may be involved in the maintenance of spatial orientation.

Topographical orientation and locomotor route-finding tasks in which the patient has to maintain orientation while changing position with respect to the environment (see the discussions of "updating" by Pick & Reiser, Chapter 5, this volume; Potegal, Chapter 15, this volume) may be tapping skills different from those involved in spatial tasks that require the patient to organize the relative positions of items in a fixed array. Disorders of topographical orientation can follow unilateral right hemisphere lesions but the lesion is frequently bilateral when the disorder is not secondary to neglect (McFie et al., 1950). Ratcliff and Newcombe (1973) found that only patients with bilateral lesions were impaired on a locomotor maze task, although both right posterior and bilateral posterior groups were impaired on a stylus maze. Two other studies (Semmes, Weinstein, Ghent, & Teuber, 1955; Hécaen, Tzortzis, & Masure, 1972) have used a similar locomotor maze and found impairment in both left and right unilateral lesion groups, although the right-hemisphere group were more severely impaired in one of them (Hécaen, Tzortzis, & Masure, 1972). Locomotor and stylus maze performance has also been dissociated in the monkey (Orbach, 1955, 1959).

These human data are consistent with the suggestion that topographical orientation can be maintained, at least in part, by either of two strategies, one mediated by the right hemisphere and the other by the left. However, De Renzi et al. (1977) have interpreted the dissociation reported by Ratcliff and Newcombe in another way, pointing out that the stylus maze is a learning task while the locomotor maze is not, and have implicated the posterior part of the right hemisphere in spatial memory.

Constructional disorders are often treated under a separate heading in the clinical literature and Le Doux, Wilson, and Gazzaniga (1977) have argued, largely on the basis of split-brain work, that the right hemisphere is specialized for the "manipulo-spatial" aspects of responses such as those required in many of the spatial tasks that are poorly performed by patients with right-hemisphere lesions. However, it is rare to find constructional apraxia without associated impairment on nonconstructional spatial tasks in patients with right hemi-sphere lesions (Dee, 1970). When this does occur it probably reflects the greater sensitivity of constructional tasks to cerebral damage. Even in cases with left-hemisphere lesions, in which it has been argued that the construc-tional disability is caused by an executive rather than a perceptual deficit (see Warrington, 1969 for a review) it is usually accompanied by perceptual disorder (Dee, 1970).

There is some evidence to suggest that graphomotor constructional perfor-mance can be dissociated from performance on tasks using blocks or sticks (Benton, 1967; Dee, 1970) but the nature of the disability in constructional apraxia is far from clear. Its status as a separate category of deficit is justified on the grounds of its clinical importance rather than because it has been shown to reflect a unique or unitary dysfunction.

One further dissociation has occasionally been reported but it may be mis-leading. This is the preservation of good performance on tactually guided tasks in the presence of gross visual–spatial deficits. The majority of the literature indicates that tactile and visual spatial deficits occur together but this could either mean that a supramodal processor handles spatial information or that two modality specific processors lie so close to each other that they are usually damaged together—particularly by middle cerebral artery infarcts, the typical pathology in studies of spatial disorder. Dissociated visual and tactile deficits have not been found in the Oxford missile wound series and the occasional reports of such dissociation (e.g., Gilliatt & Pratt, 1952) do not necessarily disprove the supramodal processor view. In my experience disproportionately severe visual spatial impairments are associated with occipital lesions and hemianopia. Field defect alone is not sufficient to explain the deficits but it might do so indirectly if the lesion also involved splenial fibers of the corpus callosum, thus partially disconnecting the specialized right-hemisphere spatial processor from visual input, while still allowing access to the somatosensory and motor systems. This is the mirror image of the explanation sometimes advanced for dyslexia without dysgraphia.

It would be very surprising if there were not more potential subclasses of spatial disorder than have so far been discerned. Equally, it is not surprising that they have not yet been teased apart because naturally occurring cerebral lesions are rarely sufficiently focal to reveal them. At present it would be safest to limit the subdivision of spatial disorders to the distinction between disorders of sensory analysis and the more complex disorders discussed in this section with the proviso that the latter class can be subdivided into disorders affecting personal space or space external to the body depending on which hemisphere is damaged. With rather less confidence one may speculate that the disorders of orientation in extrapersonal space may be further subdivided into disorders that affect orientation in the larger environment through which a subject moves as opposed to those affecting orientation within the environment of a static subject. Either subclass may depend on a defect of spatial memory or impairment of the ability to integrate stimuli into a coherent spatial framework.

The term *disorders of perceptual integration* has been used in this context because these disorders are most readily demonstrable in tasks in which the subject must perform some operation on the stimulus (e.g., mentally reorienting it, analyzing the structure of a complex stimulus such that it can be faithfully reproduced or recalled, storing it in a form that allows the subject to take account of changes in perspective associated with his or her real or imagined movement through the environment) which would presumably be facilitated if it were recoded in some way. The right posterior cerebral cortex seems to be necessary for the performance of most such operations. By contrast, it is difficult to demonstrate right-hemisphere deficits in simple matching or discrimination tasks (Bisiach & Faglioni, 1974; Ratcliff, 1970). When such deficits are reported (Benton, Hannay, & Varney, 1975; Bisiach, Nichelli, & Spinnler, 1976; Warrington, & Rabin, 1970) they may reflect the superior ability of the right hemisphere to perform fine visual discriminations rather than its specialization for the processing of spatial information.

The Paradox of Left-Sided Neglect

Unilateral neglect stands apart from the other spatial disorders in that the neglecting patient is not simply unable to integrate stimuli falling into one half of space into a coherent spatial framework but appears to be unaware of their existence. It is more frequent and more severe after right-hemisphere lesions (Battersby, Bender, Pollack, & Khan, 1956; Brain, 1941; Colombo, De Renzi, & Faglioni, 1976; Costa, Vaughan, Horwitz, & Ritter, 1969; McFie *et al.*, 1950; Oxbury, Campbell, & Oxbury, 1974) although several authors have pointed out that the asymmetry in incidence may not be as great as it would appear because of the masking effect of severe aphasia in some patients with large left-hemisphere lesions.

Some authors (Albert, 1973; McFie *et al.*, 1950) have linked the preponderance of right-hemisphere lesions in visuospatial neglect with the known specialization of the right hemisphere for visuospatial processing. However if the specialized mechanisms of the right hemisphere were damaged one would not expect the resulting disorder to be limited to the left half of space. As Heilman (1979) has pointed out, right-hemisphere visuospatial dominance may explain why patients with right-hemisphere lesions and neglect have constructional apraxia but it does not explain why they have a higher incidence of neglect.

Yet the asymmetry in the incidence of neglect is an embarrassment to hypotheses that seek to explain it in terms of damage to unilateral mechanisms involved in the direction of attention to the contralateral half of space. Goldberg (Chapter 12, this volume) has demonstrated that neurons in area 7 play such a role but presumably this is true for these neurons in both hemispheres. A similar objection applies to explanations in terms of reduced sensory input (Battersby et al., 1956).

Heilman and his colleagues have suggested that neglect of one side of space results from hypoarousal of the contralateral hemisphere (Heilman, 1979; Heilman & Valenstein, 1972; Heilman & Watson, 1977; Watson, Heilman, Cauthen, & King, 1973; Watson, Heilman, Miller, & King, 1974). In Heilman's view (1979) there are two possible reasons for the higher incidence in right-hemisphere lesions. The left hemisphere might normally be preferentially activated because it is more involved in language processing than the right hemisphere and we tend to think in verbal terms. If this were so left-hemisphere lesions would only decrease a normal hemispheric imbalance in arousal, whereas right-hemisphere lesions would increase it—conceivable to such an extent that the right hemisphere could not be sufficiently activated, even by stimuli presented in the sensory fields that project directly to it. Alternatively, he suggests, the right hemisphere may actually be dominant for mediating an attentional response to stimuli in either half of space.

The former hypothesis has little to commend it. Left-sided neglect can readily be demonstrated on constructional and visual search tasks in which one would not expect preferential activation of the left hemisphere and although Heilman and Watson (1978) demonstrated reduced neglect in a cancellation task that encouraged visual–spatial processing (and presumably therefore right-hemisphere activation) they admit that they did not succeed in eliminating it. The latter hypothesis derives some support from E.E.G. evidence suggesting that the right parietal area can be focally activated by stimuli projected into either half field whereas the left hemisphere is only activated by right-field stimuli (Van Den Abell & Heilman, 1978). These findings may be related to the fact that right-hemisphere lesions affect simple reaction time more than equivalent left-hemisphere lesions (Howes & Boller, 1975).

If it is true that the right hemisphere is dominant for the mediation of attention to stimuli in the right as well as the left half of space and that its capacity to do so is sufficient to prevent neglect of the right in left hemisphere lesions, this would be an exception to the thesis advanced here and by Moscovitch (1979) that marked hemispheric asymmetries emerge only at later, more abstract stages of processing. Indeed it is difficult to find any explanation for unilateral neglect that is consistent with the hypothesis if neglect really is more frequently associated with right-hemisphere damage.

A possible way of doing so is to combine two of the explanations that do not seem to be satisfactory on their own. Suppose that each hemisphere contains a low-level mechanism, perhaps in area 7, that is capable of alerting the organism to the presence of a stimulus in the contralateral half of space. Suppose also that the right hemisphere's specialization for the processing of spatial information is such that it is "aware" that space extends all around the body and is capable of organizing a search that takes account of the possibility that stimuli may be present though not in the visual or sensory fields that project directly to it. If this were the case, damage to the left hemisphere would not result in severe unilateral neglect because the right hemisphere would be able to allow for the possibility that relevant stimuli were present, even though their presence had not been specifically signaled. Damage to the right hemisphere, by contrast, would leave the organism with only the low-level mechanism of the left hemisphere to guide it, and it would be alerted only to stimuli in the right half of space. It must be admitted, however, that this is no more than supposition.

Whatever the explanation for unilateral neglect, it must take account of the following facts: Lesions that cause it are usually large (Battersby, 1963; Hécaen, 1962) and either of sudden onset or rapidly progressive (Heilman, 1979); they usually include the parietal cortex (Critchley, 1953; Heilman & Watson, 1977) but may be limited to other areas (Heilman, 1979; Heilman & Valenstein, 1972; Oxbury et al., 1974); neglect of the left half of the body (Brain, 1941; Fredericks, 1969; Heilman, 1979) and even the left half of a scene that the patient is imagining (Bisiach & Luzzatti, 1978) may coexist with neglect of the left half of space; the left side of a central stimulus may be neglected in copying tasks although all or part of a stimulus placed further to the left may be reproduced (Gainotti, Messerli, & Tissot, 1972).

TENTATIVE CONCLUSIONS

It has been argued that some spatial disorders, those that reflect an impairment of basic sensory analysis of visual stimuli, result from damage to the human equivalent of structures that are known to exist in the prestriate and parietal cortex of infrahuman primates. These disorders are either seen only in association with bilateral lesions or occur equally frequently with lesions in either hemisphere.

More complex spatial disorders that reflect a failure to integrate stimuli into a coherent spatial framework result from interference with later stages (or deeper levels) of processing that are carried out predominantly by the specialized structures of the right hemisphere. Although there are some similarities between the more complex spatial disorders that follow damage to the human parietal lobe and those seen after parietal lesions in infrahuman primates (Milner, Ockleford, & Dewar, 1977; Sugishita, Ettlinger, & Ridley, 1978), the monkey does not provide as good a model for these disorders as it does for the lower level disturbances. This is not surprising. The mere fact that higher level spatial abilities are predominantly lateralized to the right hemisphere in human beings constitutes a difference in the functional organization of the brains of human beings and monkeys and the evolution of functionally specialized hemispheres is likely to have entailed some reorganization of those functions that have become lateralized.

A more fruitful approach to the analysis of these complex functions may be to follow some of the leads provided by cognitive psychology and psychometrics. Cognitive psychologists have been more concerned with language than with spatial orientation and have made valuable contributions to neuropsychology in that area (e.g., Coltheart, Patterson, & Marshall, 1980), but there is a wealth of psychometric data on spatial ability. McGee (1979; Chapter 9, this volume) in a recent review of the subject has pointed out that there is general agreement on the description of at least two spatial factors: visualization and orientation. The former consists of the ability to imagine the relative movement of parts of a stimulus with respect to each other, their folding and unfolding and their rotation in three dimensions. The latter refers to the ability to determine the spatial arrangement of elements in a stimulus pattern, to perceive such patterns as a whole, to compare them, and to remain unconfused by their rotation in the coronal plane. Most of the tests sensitive to right-hemisphere damage would seem to be tapping the latter ability[1] although a task requiring the subject to mentally fold a flat shape into a cube has been used with commissurotomized patients and linked with right-hemisphere processing (Levy, 1974).

This has been a selective and highly speculative review and its conclusions are certainly incomplete and quite probably wrong. However the conclusions themselves are less important than the way in which they were derived and the inadequacies they reveal in the clinical literature. Further progress in understanding the factors limiting the spatial performance of patients with cerebral lesions will only be achieved if the various deficits included in the general category of "disorders of spatial orientation" are more precisely defined and distinguished. This will require a more systematic approach than has characterized research in the area and it will be facilitated if those involved in clinical

---

[1]Note, however, that a function identified as "spacevisualization" was found to be affected by right parietal damage in a study in which factor analysis was used to investigate the perceptual deficits associated with cerebral lesions. "Perceptual integration" was affected by right hemisphere lesions, particularly those involving the posterior parietal and posterior temporal areas, and "perceptual differentiation" was affected by lesions in either parietal or temporal lobe (Greene, 1971).

research recognize the relevance of the clues that can be provided by recent advances in primate neurophysiology, cognitive psychology, and psychometrics.

REFERENCES

Albert, M. L. A simple test of visual neglect. *Neurology,* 1973, *23,* 658–664.

Bálint, R. Die Seelenlähmung des Schauens, optische Ataxie, räumliche Störung der Aufmerksamkeit. *Monatsschrift fur Psychiatrie und Neurologie,* 1909, *25,* 51–81.

Batterbsy, W. S. The cerebral basis of visual space perception. In L. Halpern (Ed.), *Problems of dynamic neurology.* Jerusalem: Hadassah Medical Organization, 1963.

Battersby, W. S., Bender, M. B., Pollack, M., & Kahn, R. L. Unilateral "spatial agnosia" ("inattention") in patients with cerebral lesions. *Brain,* 1956, *79,* 68–93.

Bayledier, C. Pulvinar–lateroposterior afferents to cortical area 7 in monkeys demonstrated by horseradish peroxidase tracing technique. *Experimental Brain Research,* 1977, *27,* 501–507.

Benson, A. J., & Gedye, J. L. Logical processes in the resolution of orientation conflict. *R. A. F. Institute of Aviation Medicine Report 259.* London: Ministry of Defense (Air) 1963.

Benton, A. L. *Right–left discrimination and finger localization.* New York: Hoeber, 1959.

Benton, A. L. Constructional apraxia and the minor hemisphere. *Confinia Neurologica,* 1967, *29,* 1–16.

Benton, A. L. Disorders of spatial orientation. In P. J. Vinken, & G. W. Bruyn (Eds.), *Handbook of clinical neurology* (Vol. 3). Amsterdam: North Holland, 1969.

Benton, A. L. Visuoconstructive disability in patients with cerebral disease: Its relationship to side of lesion and aphasic disorder. *Documenta Ophthalmologica,* 1973, *34,* 67–76.

Benton, A. L. Visuoperceptive, visuospatial, and visuoconstructive disorders. In K. M. Heilman & E. Valenstein (Eds.), *Clinical neuropsychology.* Oxford: Oxford University Press, 1979.

Benton, A. L., Hannay, J., & Varney, N. R. Visual perception of line direction in patients with unilateral brain disease. *Neurology,* 1975, *25,* 907–910.

Benton, A. L., & Hécaen, H. Stereoscopic vision in patients with unilateral cerebral disease. *Neurology,* 1970, *20,* 1084–1088.

Benton, A. L., Levin, H. S., & Van Allen, M. W. Geographic orientation in patients with unilateral cerebral disease. *Neuropsychologia,* 1974, *12,* 183–191.

Benton, A. L., & Van Allen, M. W. Impairment in facial recognition in patients with cerebral disease. *Cortex,* 1968, *4,* 344–358.

Bisiach, E., & Faglioni, P. Recognition of random shapes by patients with unilateral brain lesions as a function of complexity, association value and delay. *Cortex,* 1974, *10,* 101–110.

Bisiach, E., & Luzzatti, C. Unilateral neglect of representational space. *Cortex,* 1978, *14,* 129–133.

Bisiach, E., Nichelli, P., & Spinnler, H. Hemispheric functional asymmetry in visual discrimination between univariate stimuli: An analysis of sensitivity and response criterion. *Neuropsychologia,* 1976, *14,* 335–343.

Boller, F., Cole, M., Kim, Y., Mack, J. L., & Patawaran, C. Optic ataxia: Clinical–radiological correlations with the EMI scan. *Journal of Neurology, Neurosurgery and Psychiatry,* 1975, *38,* 954–958.

Brain, W. R. Visual disorientation with special reference to lesions of the right cerebral hemisphere. *Brain,* 1941, *64,* 244–272.

Butters, N., Soeldner, C., & Fedio, P. Comparison of parietal and frontal lobe spatial deficits in man: Extrapersonal vs. personal (egocentric) space. *Perceptual Motor Skills,* 1972, *34,* 27–34.

Carmon, A., & Bechtoldt, H. P. Dominance of the right cerebral hemisphere for stereopsis. *Neuropsychologia*, 1969, 7, 29–39.

Chedru, F. Space representation in unilateral spatial neglect. *Journal of Neurology, Neurosurgery and Psychiatry*, 1976, 39, 1057–1061.

Cole, M., Schutta, H. S., & Warrington, E. K. Visual disorientation in homonymous half-fields. *Neurology*, 1962, 12, 257–263.

Colombo, A., De Renzi, E., & Faglioni, P. The occurrence of visual neglect in patients with unilateral cerebral disease. *Cortex*, 1976, 12, 221–231.

Coltheart, M., Patterson, K., & Marshall, J. C. (Eds.). *Deep dyslexia*. London: Routledge and Kegan Paul, 1980.

Costa, L. D., Vaughan, H. G. Jr., Horwitz, M., & Ritter, W. Patterns of behavioural deficit associated with visual spatial neglect. *Cortex*, 1969, 5, 242–263.

Cowey, A. Cortical maps and visual perception. *Quarterly Journal of Experimental Psychology*, 1979, 31, 1–17.

Cowey, A., & Porter, J. Brain damage and global stereopsis. *Proceedings of the Royal Society of London, Series B*, 1979, 204, 339–407.

Critchley, M. *The parietal lobes*. London: Arnold, 1953.

Damasio, A. R., & Benton, A. L. Impairment of hand movements under visual guidance. *Neurology*, 1979, 29, 170–178.

Danta, G., Hilton, R. C., & O'Boyle, D. J. Hemisphere function and binocular depth perception. *Brain*, 1978, 101, 569–590.

Dee, H. L. Visuoconstructive and visuoperceptive deficits in patients with unilateral cerebral lesions. *Neuropsychologia*, 1970, 8, 305–314.

De Renzi, E., & Faglioni, P. Il disorientamento spaziale da lesione cerebrale. *Sistema Nervoso*, 1962, 14, 409–436.

De Renzi, E., Faglioni, P., & Scotti, G. Tactile spatial impairment and unilateral cerebral damage. *Journal of Nervous and Mental Disease*, 1968, 146, 468–475.

De Renzi, E., Faglioni, P., & Scotti, G. Hemispheric contribution to the exploration of space through the visual and tactile modality. *Cortex*, 1970, 6, 191–203.

De Renzi, E., Faglioni, P., & Spinnler, H. The performance of patients with unilateral brain damage on face recognition tasks. *Cortex*, 1968, 4, 17–34.

De Renzi, Faglioni, P., & Villa, P. Topographical amnesia. *Journal of Neurology, Neurosurgery and Psychiatry*, 1977, 40, 498–505.

De Renzi, E., & Scotti, G. The influence of spatial disorders in impairing tactual discrimination of shapes. *Cortex*, 1969, 5, 53–62.

De Renzi, E., & Scotti, G. Autotopagnosia: Fiction or reality? *Archives of Neurology*, 1970, 23, 221–227.

De Renzi, E., & Spinnler, H. Facial recognition in brain damaged patients. *Neurology*, 1966, 16, 145–152. (a)

De Renzi, E., & Spinnler, H. Visual recognition in patients with unilateral cerebral disease. *Journal of Nervous and Mental Disease*, 1966, 142, 515–525. (b)

Ettlinger, G. Sensory deficits in visual agnosia. *Journal of Neurology, Neurosurgery and Psychiatry*, 1956, 19, 297–307.

Ettlinger, G., Warrington, E. K., & Zangwill, O. L. A further study of visual–spatial agnosia. *Brain*, 1957, 80, 335–361.

Faglioni, P., Scotti, G., & Spinnler, H. Impaired recognition of written letters following unilateral hemisphere damage. *Cortex*, 1969, 5, 120–133.

Fredericks, J. A. M. Disorders of the body schema. In P. J. Vinken & G. W. Bruyn (Eds.), *Handbook of clinical neurology* (Vol. 4). Amsterdam: North-Holland, 1969.

Gainotti, G., Messerli, P., & Tissot, R. Qualitative analysis of unilateral spatial neglect in relation to laterality of cerebral lesions. *Journal of Neurology, Neurosurgery and Psychiatry,* 1972, *35,* 545–550.

Gerstmann, J. Psychological and phenomenological aspects of disorders of the body image. *Journal of Nervous and Mental Disease,* 1958, *126,* 499–512.

Geschwind, N. The anatomical basis of hemispheric differentiation. In S. J. Diamond & J. G. Beaumont (Eds.), *Hemisphere function in the human brain.* London: Elek Scientific Books, 1974.

Gilliatt, R. W., & Pratt, R. T. C. Disorders of perception and performance in a case of right-sided cerebral thrombosis. *Journal of Neurology, Neurosurgery and Psychiatry,* 1952, *15,* 264–271.

Gloning, I., Gloning, K., & Hoff, H. *Neuropsychological symptoms and syndromes in lesions of the occipital lobes and adjacent areas.* Paris: Gauthier–Villars, 1968.

Green, J. G. *A factoral study of perceptual function in patients with cerebral lesions.* Unpublished doctoral dissertation, University of Glasgow, 1971.

Hamsher, K. de S. Stereopsis and unilateral brain disease. *Investigative Ophthalmology,* 1978, *4,* 336–343.

Hamsher, K. de S., Levin, H. S., & Benton, A. L. Facial recognition in patients with focal brain lesions. *Archives of Neurology,* 1979, *36,* 837–839.

Hartje, W., & Ettlinger, G. Reaching in light and dark after unilateral posterior parietal ablations in monkey. *Cortex,* 1973, *9,* 346–354.

Head, H. *Aphasia and kindred disorders of speech.* Cambridge: Cambridge University Press, 1926.

Hécaen, H. Clinical symptomalogy in right and left hemispheric lesions. In V. B. Mountcastle (Ed.), *Interhemispheric relations and cerebral dominance.* Baltimore: Johns Hopkins Press, 1962.

Hécaen, H. Aphasic, apraxic and agnosic syndromes in right and left hemisphere lesions. In P. J. Vinken & G. W. Bruyn (Eds.), *Handbook of clinical neurology* (Vol. 4). Amsterdam: North–Holland, 1969.

Hécaen, H., Tzortzis, C., & Masure, M. C. Troubles de l'orientation spatiale dans une epreuve de recherche d'itineraire lors des lesions corticales unilaterales. *Perception,* 1972, *1,* 325–330.

Heilman, K. M. Neglect and related disorders. In K. M. Heilman & E. Valenstein (Eds.), *Clinical Neuropsychology.* Oxford: Oxford University Press, 1979.

Heilman, K. M., & Valenstein, E. Frontal lobe neglect in man. *Neurology,* 1972, *22,* 660–664.

Heilman, K. M., & Watson, R. T. The neglect syndrome—A unilateral defect in the orienting response. In S. Harnad, R. W. Doty, L. Goldstein, J. Jaynes & G. Krauthamer (Eds.), *Lateralization in the nervous system.* New York: Academic Press, 1977.

Heilman, K. M., & Watson, R. T. Changes in the symptoms of neglect induced by changes in task strategy. *Archives of Neurology,* 1978, *35,* 47–49.

Heywood, S. P. Asymmetries in returning the eyes to specified target positions in the dark. *Vision Research,* 1973, *13,* 81–94. (a)

Heywood, S. P. *Retinal and extra-retinal control of eye movements.* Unpublished doctoral dissertation, University of Oxford, 1973. (b)

Heywood, S. P., & Ratcliff, G. Long-term oculomotor consequences of unilateraal colliculectomy in man. In G. Lennerstrand & P. Bach-y-Rita (Eds.), *Basic mechanisms of ocular mobility and their clinical implications.* Oxford: Pergamon, 1975.

Holmes, G. Disturbances of visual orientation. *British Journal of Ophthalmology,* 1918, *2,* 449–468; 506–516.

Holmes, G. Disturbances of visual space perception. *British Medical Journal,* 1919, *2,* 230–233.

Holmes, G., & Horrax, G. Disturbances of spatial orientation and visual attention, with loss of stereoscopic vision. *Archives of Neurology and Psychiatry,* 1919, *1,* 385–407.

Howes, D., & Boller, F. Simple reaction time: Evidence for focal impairment from lesions of the right hemisphere. *Brain,* 1975, *98,* 317–332.

Jackson, J. H. Case of large cerebral tumour without optic neuritis and with left hemiplegia and imperception. *Royal London Ophthalmological Hospital Reports*, 1876, *8*, 434–444.

Humphrey, D. R. On the cortical control of visually directed reaching: Contributions by nonprecentral motor areas. In R. E. Talbot & D. R. Humphrey (Eds.), *Posture and movement.* New York: Raven Press, 1979.

Julesz, B. *Foundations of cyclopean perception.* Chicago: Chicago University Press, 1971.

Kimura, D., Barnett, H. J. M., & Burkhart, G. The psychological test pattern in progressive supranuclear palsy (Bulletin No. 477). London, Ontario: University of Western Ontario, Department of Psychology, 1979.

Kleist, K. *Gehirn-Pathologie vornehmlich auf Grund der Kriegserfahrungen.* Leipzig: Barth, 1934.

Lange, J. Agnosien and Apraxien. In O. Bumke & O. Foerster (Eds.), *Handbuch der Neurologie* (Vol. 6). Berlin: Springer, 1936.

Lawson, I. R. Confusion in the house: The assessment of disorientation for the familiar in the home. *Psychiatric Quarterly,* 1969, *43,* 225–239.

Le Doux, J. E., Wilson, D. H., & Gazzaniga, M. S. Manipulo-spatial apsects of cerebral lateralization: Clues to the origin of lateralization. *Neuropsychologia,* 1977, *15,* 743–750.

Lehmann, D. and Wälchli, P. Depth perception and location of brain lesions. *Journal of Neurology,* 1975, *209,* 157–164.

Le May, M. Morphological cerebral asymmetries of modern man, fossil man, and nonhuman primates. In S. R. Harnad, H. D. Steklis, & J. Lancaster (Eds.), *Origins and evolution of language and speech. Annals of the New York Academy of Sciences,* 1976, *280.*

Levy, J. Psychobiological implications of bilateral asymmetry. In S. J. Dimond & G. Beaumont (Eds.), *Hemisphere function in the human brain.* London: Elek Scientific Books, 1974.

McGee, M. G. Human spatial abilities: Psychometric studies and environmental, genetic, hormonal and neurological influences. *Psychological Bulletin,* 1979, *86,* 889–918.

McFie, J., Piercy, M. F., & Zangwill, O. L. Visual-spatial agnosia associated with lesions of the right cerebral hemisphere. *Brain,* 1950, *73,* 167–190.

McFie, J., & Zangwill, O. L. Visual–constructive disabilities associated with lesions of the left cerebral hemisphere. *Brain,* 1960, *83,* 243–260.

Meadows, J. C. The anatomical basis of prosopagnosia. *Journal of Neurology, Neurosurgery and Psychiatry,* 1974, *37,* 489–501.

Meier, M. J., & French, L. A. Lateralized deficits in complex visual discrimination and bilateral transfer of reminiscence following unilateral temporal lobectomy. *Neuropsychologia,* 1965, *3,* 261–272.

Milner, A. D., Ockleford, E. M., & Dewar, W. Visuo-spatial performance following posterior parietal and lateral frontal lesions in stumptail macaques. *Cortex,* 1977, *13,* 350–360.

Moscovitch, M. Information processing and the cerebral hemispheres. In *Handbook of behavioral neurobiology* (Vol. 2: Neuropsychology). New York: Plenum, 1979.

Mountcastle, V. B. The view from within: Pathways to the study of perception. *Johns Hopkins Medical Journal,* 1975, *136,* 109–135.

Mountcastle, V. B., Lynch, J. C., Georgopoulos, A., Sakata, H., & Acuna, C. Posterior parietal association cortex of the monkey: Command functions for operations within extra-personal space. *Journal of Neurophysiology,* 1975, *38,* 871–908.

Newcombe, F. *Missile wounds of the brain: A study of psychological deficits.* London: Oxford University Press, 1969.

Newcombe, F., & Russell, W. R. Dissociated visual perceptual and spatial deficits in focal lesions of the right hemisphere. *Journal of Neurology, Neurosurgery and Psychiatry,* 1969, *32,* 73–81.

Orbach, J. Non-visual functioning of occipital cortex in the monkey. *Proceedings of the National Academy of Science,* 1955, *41,* 264–267.

Orbach, J. Disturbances of the maze habit following occipital cortex removals in blind monkeys. *Archives of Neurology and Psychiatry,* 1959, *81,* 49–54.

Oxbury, J. M., Campbell, D. C., & Oxbury, S. M. Unilateral spatial neglect and impairments of spatial analysis and visual perception. *Brain,* 1974, *97,* 551–564.

Paterson, A., & Zangwill, O. L. Disorders of visual space perception associated with lesions of the right cerebral hemisphere. *Brain,* 1944, *67,* 331–358.

Paterson, A., & Zangwill, O. L. A case of topographical disorientation associated with a unilateral cerebral lesion. *Brain,* 1945, *68,* 118–212.

Perenin, M. T., & Jeannerod, M. Visual function within the hemianopic field following early cerebral hemidecortication in man—I. spatial localisation. *Neuropsychologia,* 1978, *16,* 1–14.

Piercy, M. F., Hécaen, H., & Ajuriaguerra, J. de. Constructional apraxia associated with unilateral cerebral lesions—Left and right sided cases compared. *Brain,* 1960, *83,* 225–242.

Ratcliff, G. *Aspects of disordered space perception.* Unpublished doctoral dissociation, University of Oxford, 1970.

Ratcliff, G. Spatial thought, mental rotation and the right cerebral hemisphere. *Neuropsychologia,* 1979, *17,* 49–54.

Ratcliff, G. The clinical significance of disorders of visuo-spatial perception. *Geriatric Medicine,* 1980, *10,* 71–74.

Ratcliff, G., & Cowey, A. Disturbances of visual perception following cerebral lesions. In D. J. Oborne, M. M. Gruneberg, & J. R. Eiser (Eds.), *Research in psychology and medicine.* London: Academic Press, 1979.

Ratcliff, G., & Davies–Jones, G. A. G. Defective visual localisation in focal brain wounds. *Brain,* 1972, *95,* 49–60.

Ratcliff, G., & Newcombe, F. Spatial orientation in man: Effects of left, right, and bilateral posterior cerebral lesions. *Journal of Neurology, Neurosurgery and Psychiatry,* 1973, *36,* 448–454.

Riddoch, G. Dissociation of visual perceptions due to occipital injuries, with especial reference to the appreciation of movement. *Brain,* 1917, *40,* 15–57.

Riddoch, G. Visual disorientation in homonymous half-fields. *Brain,* 1935, *58,* 376–382.

Robinson, D. L., Goldberg, M. E., & Stanton, G. B. Parietal association cortex in the primate: Sensory mechanisms and behavioral modulations. *Journal of Neurophysiology,* 1978, *41,* 910–932.

Rondot, P., Recondo, J. de, & Ribadeau Dumas, J. L. Visuomotor ataxia. *Brain,* 1977, *100,* 355–376.

Rothstein, T. B., & Sacks, J. G. Defective stereopsis in lesions of the parietal lobe. *American Journal of Ophthalmology,* 1972, *73,* 281–284.

Rubens, A. B. Anatomical asymmetries of human cerebral cortex. In S. R. Harnad, R. W. Doty, L. Goldstein, J. Jaynes, & G. Krauthamer (Eds.), *Lateralization in the nervous system.* New York, Academic Press, 1977.

Russell, W. R. Discussion on parietal lobe syndromes. *Proceedings of the Royal Society of Medicine,* 1951, *44,* 341–343.

Sauguet, J., Benton, A. L., and Hécaen, H. Disturbances of the body schema in relation to language impairment and hemispheric locus of lesion. *Journal of Neurology, Neurosurgery and Psychiatry,* 1971, *34,* 496–501.

Scotti, G. La perdita della memoria topografica: Descrizione di un caso. *Sistema Nervoso,* 1968, *20,* 352–361.

Semmes, J., Weinstein, S., Ghent, L., & Teuber, H. L. Spatial orientation in man after cerebral injury: I. Analysis by locus of lesion. *Journal of Psychology,* 1955, *39,* 227–244.

Semmes, J., Weinstein, S., Ghent, L., & Teuber, H. L. Correlates of impaired orientation in personal and extra-personal space. *Brain,* 1963, *86,* 747–772.

Smith, S., & Holmes, G. A case of bilateral motor apraxia with disturbance of visual orientation. *British Medical Journal*, 1916, *1*, 437–441.

Sugishita, M., Ettlinger, G., & Ridley, R. M. Disturbance of cage finding in the monkey. *Cortex*, 1978, *14*, 431–438.

Trojanowski, J. Q., & Jacobson, S. Areal and laminar distribution of some pulvinar cortical efferents in rhesus monkey. *Journal of Comparative Neurology*, 1976, *169*, 371–392.

Tzavaras, A., Hécaen, H., & Le Bras, H. Le probleme de la specificite du deficit de la reconnaissance du visage humain lors les lesions hemispheriques unilaterales. *Neuropsychologia*, 1970, *8*, 403–416.

Van den Abell, T., & Heilman, K. M. *Lateralized warning stimuli, phasic hemispheric arousal and reaction time.* Paper presented to the International Neuropsychological Society. Minneapolis, February, 1978.

Warrington, E. K. Constructional apraxia. In P. J. Vinken & G. W. Bruyn (Eds.), *Handbook of clinical neurology* Amsterdam: North–Holland 1969.

Warrington, E. K., & James, M. Disorders of visual perception in patients with unilateral cerebral lesions. *Neuropsychologia*, 1967, *5*, 253–266. (a)

Warrington, E. K., & James, M. An experimental investigation of facial recognition in patients with unilateral cerebral lesions. *Cortex*, 1967, *3*, 317–326. (b)

Warrington, E. K., James, M., & Kinsbourne, M. Drawing disability in relation to laterality of cerebral lesion. *Brain*, 1966, *89*, 53–82.

Warrington, E. K., & Rabin, P. Perceptual matching in patients with cerebral lesions. *Neuropsychologia*, 1970, *8*, 475–487.

Watson, R. T., Heilman, K. M., Cauthen, J. C., & King, F. A. Neglect after cingulectomy. *Neurology*, 1973, *23*, 1003–1007.

Watson, R. T., Heilman, K. M., Miller, B. D., & King, F. A. Neglect after mesencephalic reticular formation lesions. *Neurology*, 1974, *24*, 294–298.

Weiskrantz, O. L., Warrington, E. K. Sanders, M. D., & Marshall, J. Visual capacity in the menianopic field following a restricted occipital ablation. *Brain*, 1974, *97*, 709–728.

Witelson, S. F. Anatomic asymmetry in the temporal lobes: Its documentation, phylogenesis, and relationship to functional asymmetry. In S. J. Dimond & D. A. Blizard (Eds.), *Evolution and lateralization of the brain. Annals of the New York Academy of Sciences*, 1977, *299*.

Yealland, L. R. Case of gunshot wound involving visual centre, with visual disorientation. *Proceedings of the Royal Society of Medicine*, 1916, *9*, 97–99.

Yin, T. C. T., & Mountcastle, V. B. Visual input to the visuomotor mechanisms of the monkey's parietal lobe. *Science*, 1977, *197*, 1381–1383.

Zeki, S. M. Uniformity and diversity of structure and function in rhesus monkey prestriate visual cortex. *Journal of Physiology*, 1978, *277*, 273–290.

V

SUBCORTICAL CONTRIBUTIONS TO SPATIAL ABILITIES

Spatially Organized Behaviors of Animals: Behavioral and Neurological Studies[1]

In their natural habitats, animals face a variety of spatial discrimination problems, and the approach they take to solve them can have a substantial influence on their chance for survival. Not surprisingly, animals have developed many different and often highly sophisticated strategies to solve their problems. One of the challenges to researchers interested in spatially organized behavior is to understand these strategies, to create laboratory tests that will elicit them, and to design experiments to determine underlying psychological and neuroanatomical mechanisms.

The experiments in this chapter were designed in response to this challenge. Behaviorally, they demonstrate three different spatial abilities: (1) the use of working memory to recall a series of places that have been previously visited; (2) the determination of the topological relationship among a set of places so that a particular goal can be reached from a novel starting location; (3) the efficient organization of travel among a set of places so that the distance traveled to reach all of them can be minimized.

Neurologically, the experiments focus on the hippocampus. Previous re-

[1]This research was supported in part by research grants MH 24213 from the National Institutes of Health, NSF-BNS 7824160 from the National Science Foundation, and S07 RR07041 from the Biomedical Research Support Grant Program, Division of Research Resources, National Institutes of Health.

335

search has demonstrated that this brain structure is critical for some types of spatial memories. Still remaining are two questions. First, what types of spatial memories require the hippocampus? Second, to what extent should hippocampal function be described primarily in terms of spatial processes, or in terms of memorial processes? The present experiments varied both the spatial and the memorial requirements of a variety of different discrimination tasks. Animals with lesions of the hippocampal system were tested in these tasks, and the resulting dissociations suggest that the primary function of the hippocampus is more closely associated with a certain type of memory (working memory) than with a certain type of spatial process (cognitive mapping).

More generally, these experiments demonstrate the advantages of combining the behavioral and neurological approaches. The behavioral experiments not only describe the cognitive processes possessed by animals, they also introduce new behavioral testing procedures that can be used to examine the functional organization of the brain. Likewise, the neurological experiments not only describe the brain mechanisms underlying different types of spatial and memorial processes, they also indicate the extent to which these (and other) processes can function independently of each other. In this fashion, we can begin to obtain the information necessary to construct both cognitive and neuronal models of the strategies animals use to move through their environments and the ways in which they remember the information obtained from these movements.

SPATIAL MEMORY

Behavioral Studies

A fundamental problem for any animal is finding food, and a good memory can help solve this problem in an efficient manner. Not surprisingly, animals in their natural habitats often demonstrate a very good memory for the places they have been. The Hawaiian Honeycreeper, *Loxops virens,* is a relevant example. This bird obtains nectar from flowers of the mamane tree. Because the nectar supply of each flower is limited, the bird usually removes all of the available nectar on a single visit. The flower then takes several hours to produce a new supply of nectar. During this repletion period, the bird goes to other, unvisited flowers that still have their full nectar supply, and avoids previously visited flowers until sufficient time has passed for them to regenerate their nectar (Kamil, 1978). Other examples of foraging also indicate that predators searching for prey rely heavily on strategies requiring a very good spatial memory

(Kamil & Sargent, 1980; Peters, 1973, 1978; Pyke, 1979; Schoener, 1971; Whitham, 1977).

Results from laboratory experiments demonstrate that animals have a remarkable ability to process spatial information (see reviews in Olton, 1978a, 1979a; Sutherland & Mackintosh, 1971). They learn spatial discrimination tasks rapidly, perform then well, and often attempt to solve nonspatial discrimination tasks with a spatial strategy before finding the correct, nonspatial solution. Furthermore, performance in spatial discrimination tasks is often highly flexible so that profound and adaptive changes in the tendency to approach or avoid a place may follow just a single experience in that place (Dember & Fowler, 1958; Douglas, 1966; Honzik & Tolman, 1936; Maier, 1929; Olton, 1979a, 1979b; Olton, Collison, & Werz, 1977; Olton, Handelmann, & Walker, 1981; Olton & Samuelson, 1976; Olton, Walker, Gage, & Johnson, 1977; Stahl & Ellen, 1974; Tolman & Gleitman, 1949; Walker & Olton, 1979a).

To examine the characteristics of spatial memory, we developed a laboratory task that measured the ability of rats to remember a series of different places they had visited to avoid returning to them. The apparatus was an elevated radial arm maze. Each of the arms extended from a central platform like a spoke on a wheel. The maze was open so the rats could easily see the room in which it was placed. At the beginning of each trial a pellet of food was placed at the end of each arm. A food-deprived rat was placed in the middle of the maze and allowed to choose freely among the arms. The optimal strategy was to choose each arm once and not to return to it during that trial. In this way, the rat could get all the food in the minimum number of choices.

Rats have been tested with mazes that have 2, 4, 8, 17, and 32 arms (Beatty & Shavalia, 1980; Jarrard, 1978; Maki, Brokofsky, & Berg, 1979; Olton, 1978a, 1979; Olton & Collison, 1979; Olton et al., 1977; Olton & Samuelson, 1976; Olton & Scholsberg, 1978; Roberts, 1979; Suzuki, Augerinos, & Black, 1980; Walker & Olton, 1979a; Zoladek & Roberts, 1978). In all of these, they learned rapidly and performed well. On the simplest mazes, they took only a few test sessions to perform almost perfectly. On the larger mazes, they took about 30 test sessions to reach a level of performance that was highly accurate, but not perfect. For example, at the end of training on the maze with 17 arms, rats regularly went to an average of 15 different arms within the first 17 choices (Olton et al., 1977).

I have described this procedure as a test of spatial memory, but there are several strategies having little to do with space or memory that can be used to produce successful performance. For example, a rat might follow a response pattern, such as choosing adjacent arms. Alternatively, he might use intramaze cues, ones that are inherent in the test procedure (such as the smell of food), or ones that the rat adds (such as an "odor trail"). In all cases, the rat would have to remember only the general algorithm and apply it at the appropriate point in the test procedure.

The problem here is a classic one in psychology and has to do with the presence of "relevant, redundant" solutions to a discrimination. Because all of the strategies mentioned above can lead to correct performance in the radial arm maze, they are relevant, and because no changes were made in the apparatus during testing, they are also redundant. The solution to this problem, of course, is to eliminate the relevancy and/or redundancy of different strategies and determine which ones have the most important influence on choice behavior (see discussion in Olton, 1978a, 1979a, 1979b).

The possible importance of response strategies was evaluated by blocking the entrance to each arm and forcing rats to go to a particular subset of arms before allowing them to choose freely among all arms (Olton & Schlosberg, 1978; Roberts, 1978, 1979). In this procedure, the experimenter (rather than the rat) determined the first set of arms to be chosen, the spatial relationship of those arms, and the order in which they were to be chosen. When subsequently given a choice among all the arms, rats were just as accurate at responding to only the unchosen arms as they had been in the usual test procedure in which they determined all the choices themselves. Thus the rats did not have to use a response strategy to perform correctly.

The possible importance of intramaze cues was evaluated with two procedures. In one of them, the maze was rotated each time the rat came back to the center platform after getting a pellet of food at the end of an arm (Olton & Collison, 1979). For one group of rats, a small platform at the end of each arm contained the food pellet. When the maze was rotated, the food remained in the same location so that the rats were reinforced for choosing each location in the room irrespective of the arm that happened to be there. These rats performed as well as those in the usual procedure in which the maze remained stationary, demonstrating that they did not need the intramaze stimuli to choose accurately. For a second group of rats, a food cup attached to the end of each arm contained the food pellet. When the maze was rotated, the food moved with it so that these rats were reinforced for choosing each arm once irrespective of the location of that arm in the room. These rats performed no better than expected by chance, demonstrating that they could not use the intramaze stimuli to choose accurately.

In a second set of procedures, the maze remained in the same place in the testing room but the stimuli around it were moved (O'Keefe, & Conway, 1978; Suzuki et al., 1980). A black curtain was placed around the maze and specific stimuli (lights, cards with patterns on them, etc.) were attached to the curtain. When these stimuli were rotated around the maze, the rats chose to go toward the appropriate set of stimuli irrespective of the arm of the maze that happened to be there, again demonstrating that the rats used the stimuli surrounding the maze to perform appropriately, rather than stimuli in the maze itself.

All these data demonstrate that elevated maze procedures constitute a test of spatial memory. The tasks are spatial because the most important discriminative stimuli are those in the room surrounding the maze, and the tasks require memory for these stimuli because the rats do not use alternative strategies to determine which arms have been previously chosen.

Other experiments have described the characteristics of the memory used by the rats when performing on the radial arm maze. These characteristics are similar to those described for the memory processes of people who are remembering lists of items. A model, based on this analogy, is presented in Figure 14.1. In this model, the rat begins in the upper left corner at the box marked *Start*. It gets some sensory input by looking down an arm and stores this in a

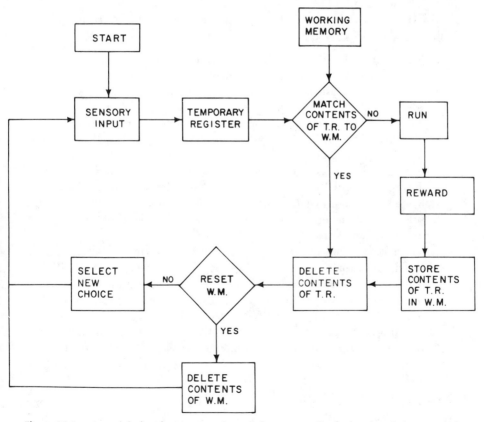

Figure 14.1. A model of performance on the radial arm maze. For further description, see text. (Reprinted with permission from Olton, 1978, p. 364.)

Temporary Register. The contents of this Temporary Register are compared to those of *Working Memory,* the process that stores information about previous repsonses. If the contents of the Temporary Register match any of the items in Working Memory, the arm indicated by the Temporary Register has been chosen before during that test session and should not be chosen again. The rat deletes the contents of the Temporary Register, makes a decision about whether to reset Working Memory, and moves to another arm, getting new sensory input. If the contents of the Temporary Register do not match any of the items in Working Memory, the arm indicated by the Temporary Register has not been chosen before during that test session. The rat runs down the arm, gets the food pellet, stores the contents of the Temporary Register in Working Memory so that it will not return to that arm during the remainder of that session, and moves to another arm as before. This procedure continues until the rat is removed from the apparatus. (Further discussion is presented in Honig, 1978; Olton, 1978a.)

Working memory processes the information stored in it as if it were a list of items. Some of the characteristics of this list include the following:

1. The amount of material is limited and the maximum amount of information that can be stored is probably about 15 items (Olton, 1979a; Olton, Becker, & Handelmann, 1980; Olton, Collison, & Werz, 1977; Roberts, 1979).
2. The memory does not decay for periods of up to several hours (Beatty & Shavalia, 1980; Olton, 1972; Olton and Samuelson, 1976; cf. "spontaneous alternation," Potegal, Chapter 15, this volume).
3. Substantial interference is caused by items stored in working memory during a particular test session, but relatively little interference is caused by items from a previous session or by events that are not relevant to that session (Maki *et al.,* 1979; Olton, Becker, & Handelmann, 1980; Olton & Samuelson, 1976).
4. There is a recency effect; later responses made during a trial are remembered better than earlier responses made in that trial (Roberts, 1978).
5. There is no primacy effect; responses made at the beginning of a trial are not remembered any better than responses made later in that trial (Olton *et al.,* 1977; Olton & Samuelson, 1976).
6. Working memory can be reset so that the interference produced from responses during one test session may have relatively little effect on the memory of responses from a subsequent test session (Olton & Samuelson, 1976).

In summary, then, animals in their natural habitats have an excellent memory for the places they have been, and laboratory experiments that elicit these abilities can provide us with a good description of the characteristics of this memory. There are three implications of these results. First, people interested in examining the characteristics of memory may find spatial procedures the most appropriate ones to use for their investigations (Olton, 1978a, 1979a). Second, animals have one of the prerequisities for efficiently organized spatial behaviors, namely, the ability to remember where they have been. Third, the processes underlying spatial memory in animals may be very similar to those already described for associative learning in people (c.f. Olton, Becker, & Handelmann, 1979a, 1979b, 1980).

Neurological Studies

The hippocampus has been implicated as a structure important for both spatially organized behaviors (see review in O'Keefe & Nadel, 1978) and working memory (see reviews in Hoehler & Thompson, 1979; Olton, 1980; Olton *et al.,* 1979a, 1979b, 1980; Rozin, 1976). As might be expected, disruption of normal hippocampal function in rats produced a severe impairment of choice accuracy when they were tested on the elevated radial arm maze. Before describing these results, however, I want to emphasize three features of the experimental design used for this research.

First, the rats were given extensive training, both preoperatively and postoperatively. The preoperative training ensures that the rats have the opportunity to learn the experimental procedures and to become familiar with the stimuli to be remembered. The postoperative training allows ample opportunity for any recovery of function to take place and reduces the influence of any temporary or generalized changes in arousal, activity, motivation, etc. Combined, these procedures emphasize the "associational" (memory) components of the task and minimize the "nonassociational" ones (Jarrard, 1975; Norman, 1973; Olton, 1978b; Olton & Papas, 1979).

Second, the lesions completely transected the extrinsic connections of the hippocampus at different places. The complete destruction is important because partial lesions are often followed by recovery of function. This recovery may occur even when only a very small portion of the relevant brain area is left intact (Acheson, Zigmond, & Striker, 1980; Geschwind, 1965; Lashley, 1939; Mishkin, 1966, 1972; Olton & Feustle, 1981; Olton & Papas, 1979; Settlage, 1939; Zigmond & Striker, 1972, 1973) and may be mediated by a variety of different processes. Thus, interpretation of the behavioral changes following brain lesions (especially in subcortical areas) is simplified when these lesions are complete and bilateral (irrespective of the amount of the tissue that is being considered). One of the difficulties with making a complete lesion, of course, is that when it is large enough to completely destroy the structure in question, it is usually large enough to produce extraneous damage to the surrounding brain areas as well. However, lesions in different portions of a neuroanatomical system produce different types of unintended damage. If the behavioral results of these lesions are similar, then they are most readily interpreted as being due to disruption of the system in question (the common element) rather than to disruption of other systems (a variable element).

Third, emphasis has been placed on enduring changes in behavior that do not return to normal even with continued testing (Olton, 1978b; Olton *et al.*, 1979a, 1979b, 1980). These enduring deficits should provide better information about structure–function relationships than relatively more transient ones because the latter might arise from any number of generalized neurophysiological or behavioral processes, as well as specific plastic changes in the morphology of the brain. Certainly any theory of brain function must eventually be able to explain all the behavioral changes following a brain lesion, but temporary changes are subject to so many interpretations that firm conclusions are relatively difficult to draw from them. In any case, permanent changes in behavior mean that the brain is unable to compensate for the loss of the structure that was destroyed, a point indicating that normal functioning of that structure is a prerequisite for normal behavior in that task.

Results of these lesion studies demonstrate that the hippocampal system is critical for normal performance in spatial working-memory tasks. Following bilateral destruction of the fimbria-fornix, entorhinal cortex, precommissural fornix as it passes through the septum, or postcommissural fornix, rats performed at chance levels and showed no signs of recovery of function even after as many as 50 tests, about five times as many as were needed to learn the task initially. These results were found in mazes with two arms (Walker & Olton, 1980), four arms (Walker & Olton, 1979b), eight arms (Jarrard, 1978; Olton, Walker, & Gage, 1978), and 17 arms (Olton & Werz, 1978), and persisted even when special procedures were introduced to give the rats more time to learn the stimuli to be remembered (Walker & Olton, 1979b) and to break up response sequences (Olton & Werz, 1978). In contrast, lesions of the posterolateral neocortex, amygdala, caudate nucleus, sulcal frontal cortex, or medial frontal cortex (Becker, Walker, & Olton, 1980; Craig & Jarrard, 1978) had at most a temporary effect on choice accuracy, and even in that case choice accuracy returned to normal after about 10 tests. Thus accurate performance on the radial arm maze is selectively sensitive to damage in the hippocampal system; sensitive, because complete bilateral damage to any of the extrinsic fiber connections affected performance, selective because damage in other brain systems had relatively little effect. As expected, partial destruction of the extrinsic fiber connections also produced a deficit in choice accuracy, but this was less severe and disappeared as postoperative testing continued (Olton & Feustle, 1981; Olton & Papas, 1979).

Other experiments have begun to determine the characteristics of the elevated radial arm maze task that require hippocampal involvement with particular emphasis on the relative importance of the memory characteristics of the task as compared to its spatial characteristics. The general strategy has been to manipulate one of these variables while holding the other constant (Black, 1975; Olton et al., 1979a, 1979b, 1980; Platt, 1964). In one experiment (Olton & Feustle, 1981) the working memory requirement was maintained but the stimuli to be remembered were in the arms of the maze (rather than in the room surrounding it) and had a changing toplogical relationship (rather than a constant one). The maze was enclosed so that the rats could not see out of it; each arm contained a specific set of discriminative stimuli, and the position of the arms around the central platform was changed after every choice while the rat was confined to the central platform. This type of procedure effectively prevented cognitive mapping. Here again, destruction of the fimbria-fornix produced a permanent impairment of choice accuracy. These results indicate that the hippocampal system is necessary for both nonspatial and spatial working memory, a finding that is consistent with theories emphasizing a memory function for the hippocampus, but not with theories emphasizing a spatial function.

In a second experiment, the working memory requirement was changed while the spatial characteristics were maintained (Olton & Papas, 1979). The apparatus was an open, elevated radial arm maze. One set of *baited* arms had one pellet of food at the end of each arm. A second set of *unbaited* arms never had food on them. The set of baited arms constituted the usual test of working memory and the optimal strategy for these arms was to choose each one once and not return to it for the remainder of that test session. The set of unbaited arms constituted a test of *reference* memory, and the optimal strategy for these arms was never to choose them. At the end of postoperative testing, those rats with complete lesions of the fimbria-fornix were performing no better than expected by chance in the working memory component of this task (i.e., they made many errors by returning to a baited arm from which they had already removed the food), but were performing as well as normal rats in the reference memory component (i.e., they almost never made an error by choosing an unbaited arm). These results demonstrate that the rats could discriminate between the baited and unbaited sets of arms, remember that the baited arms sometimes had food on them whereas the unbaited arms never had food on them, and inhibit responses to the unbaited arms. The only difficulty these rats exhibited was in remembering which of the baited arms they had already chosen during that test session. Thus their impairment was very specific and was associated with a particular type of memory requirement. Again these results are consistent with suggestions that the hippocampus is primarily involved in memory processes, and are inconsistent with suggestions that it is primarily involved in spatial processes.

Taken together, the results of these experiments demonstrate that the hippocampal system plays a critical role in spatial memory. This role is elicited by the memory components of the tasks, however, rather than their spatial characteristics. A more detailed elaboration of this point of view has been presented elsewhere (Olton *et al.*, 1979a, 1979b, 1980; Olton & Feustle, 1980; Olton & Papas, 1979) and is very similar to those descriptions of human amnesia emphasizing a failure of episodic memory in the presence of an intact semantic memory (Kinsbourne & Wood, 1975; Rozin, 1976; Tulving, 1972).

LOCATION LEARNING

Behavioral Studies

A spatial environment provides numerous opportunities to learn about the topological relationships among its different parts. Of particular interest in the following experiments is the question of whether or not rats can learn to go to a particular location irrespective of tbe direction (or route) they have to travel to get there. This ability is an integral component of most definitions of cognitive mapping (Kaplan, 1973, 1975; Menzel, 1973, 1978; O'Keefe, 1980; O'Keefe & Nadel, 1978, 1979; Peters, 1973, 1978; Tolman, 1948; Tolman, Ritchie, & Kalish, 1946) and will be called *location* learning to distinguish it from *place* learning as the term was used in the *place versus response* controversy (Becker, Walker, & Olton, 1980; Olton, 1979a, 1979b; Restle, 1957; Walker & Olton, in press).

The reason for this distinction, and the associated terminology, can be best illustrated by briefly returning to the experimental design most frequently used to demonstrate place learning. The maze had four arms in the shape of a +. For this example, the arms will be called north, east, south, and west, respectively, beginning with the top arm and moving in a clockwise direction. At the beginning of the experiment, the rat was placed in the south arm and given a choice between the east and west arms (the north arm was either removed or the entrance to it was blocked). Food was placed only in the east arm. Within a few trials, the rat learned to enter the east arm to get the food and to avoid the west arm. The rat was then given a *transfer test;* it was placed in the north arm and given a choice between the east and west arms (the south arm was removed or the entrance to it was blocked). If the rat went to the east arm, it was said to exhibit *place learning* because it went to the same place in the maze (i.e., the east arm) as it had when it was started from the south arm. If the rat went to the west arm, it was said to exhibit *response learning* because it made the same response (i.e., turned right at the choice point) as it had when it was started from the south arm (cf. Bremner, Chapter 4, this volume).

Although the results of this experiment can differentiate place learning from response learning, they cannot make an additional, and equally important distinction. Consider again the rat who exhibited place learning. This rat might have entered the east arm by orienting toward the stimuli at the end of that arm and moving toward them, or it might have entered the east arm to get to a particular spot in the maze, presumably the point where the food cup was located at the end of the east arm. We have called the first process *direction learning* and the second process *location learning* (Olton, 1979a, 1979b; Becker et al., 1980; Walker & Olton, in press).

This distinction is important for two reasons. First, direction learning and location learning represent different spatial processes. Direction learning provides information about orientation and is similar to specifying a relative direction. Location learning provides information about a particular point in space, which is specified by a position, including absolute magnitude and direction. Thus location learning requires more cognitive processing and specifies more details about the environment than does direction learning. Second, most discussions of cognitive mapping (Becker & Olton, 1980, 1982; Becker, Olton, Anderson, & Breitinger, 1981; Kaplan, 1973, 1975; Mech, 1970; Menzel, 1973; 1978; Becker, Walker, & Olton, 1980b; O'Keefe, 1980; O'Keefe & Nadel, 1978, 1979; Peters, 1973; Tolman, 1948; Tolman et al., 1946; Walker & Olton, in press; see also Pick & Reiser, Chapter 5, this volume) describe it in terms of location learning as defined here, rather than just direction learning. Thus the distinction between these two processes is important to provide operational definitions of cognitive mapping.

We chose the term *location learning* to describe this process because the term *place learning* has already been intimately associated with the experimental procedures in the place versus response controversy and we did not want to attempt to add a new meaning to an already well-defined word. Thus, although the terms *place* and *location* may be synonyms in casual conversation, in the present chapter they have technical definitions that set them apart, and *location learning* is much more specific than *place learning*.

Very few experiments have distinguisbed between location learning and place learning in rats (see reviews in Olton, 1979, in press a), and those that did concluded that rats were unable to demonstrate location learning (Blodgett, McCutchan, & Mathews, 1949; Ritchie, 1948). The procedure in those experiments was complicated, however, and the rats may have performed poorly because they failed to understand the experimental procedure (see Olton, 1979) rather than because they lacked the cognitive processes for location learning.

To resolve the issue of whether or not rats can demonstrate true location learning, we designed an experiment that had two parts (Becker, Walker, & Olton, 1980; Olton, 1979; Walker, 1978; Walker & Olton, in press). At first, each rat was trained to go to a particular goal from three different starting areas. Then it was given a transfer test in which it was placed in an entirely new starting area and asked to go to the same goal; a correct response required moving in a direction that had never before led to the correct goal. Successful performance on this transfer test indicates that the rat has learned the topological relationships among the stimuli in its environment and can correctly relate a new view of these stimuli to information gained from other views.

The maze used for one test is illustrated in Figure 14.2. It was composed of an elevated set of boxes and interconnected arms, with a guillotine door at the junction of each arm and box. The four black boxes were used as start boxes. Each of the other boxes was a goal box and had in the middle of it a small opaque dish that served as a food cup. At the start of each trial, a hungry rat was placed in one of the start boxes. A set of guillotine doors was raised to give the rat a choice between two of the immediately adjacent goal boxes. One of the goal boxes was correct and had food in the food cup, the other goal box was incorrect and had an empty food cup.

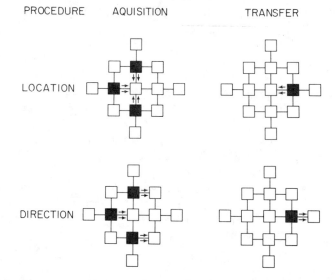

Figure 14.2. The maze and test procedures used to demonstrate true location learning. The maze was composed of a set of boxes connected by alleys. During acquisition, rats were trained to go from each of three start boxes (solid black squares) to the goal box indicated by the arrows in order to get food. Rats in the location procedure always went to the same goal box, which required them to move in different directions from each start box. Rats in the direction procedure always moved in the same direction, which required them to go to a different goal box from each start box. For the transfer tests, the rats were placed in a fourth start box, one in which they had never been before. For rats in the location procedure, the correct response was to go to the same goal box as before, which required them to move in a new direction. For rats in the direction procedure, the correct response was to go in the same direction as before, which required them to go to a new goal box. An important point in this design is that the transfer test itself (being placed in the new start box) was the same for both groups of rats, but the correct response (indicated by the arrows) was different.

Each rat was given five trials a day. For each trial, rats in the location procedure were given a choice between the center goal box, which was correct, and one of the other goal boxes, which was incorrect. During initial training, each rat was given trials from three different start boxes. After reaching a criterion of 20 consecutive errorless trials, each rat was given 10 transfer tests in which it was placed in the fourth start box and given a choice between the correct (center) goal box and one of the other goal boxes. An important point in this procedure is that the rat had never been in this particular start box prior to the transfer test. Thus the transfer test required the rat to make a choice from a perspective in the room that it had never seen before. As illustrated in Figure 14.2, successful performance on this transfer test required the rat to move in a new direction (toward the left side of the figure), one which had never before led to the correct goal box.

Results of this experiment showed that although the rats did not perform perfectly on the transfer tests, they did perform consistently better than expected by chance. The probability of a correct response on the very first transfer test was .75 and the probability of a correct response on all transfer tests was .86. Both figures were significantly greater than expected by chance.

A second group of rats was trained to go in a particular direction. Here the correct goal box was always in the same direction with respect to the start box so that correct performance required the rat to go to a different goal box from each start box. When given transfer tests, rats trained in this procedure also performed appropriately. The probability of a correct response was .75 for the first transfer test and .83 for all transfer tests. These results are important because they demonstrate that during the transfer tests the correct performance of rats given the location procedure was due to their previous training and not due to some structural characteristics of the transfer test itself. As indicated in Figure 14.2, the actual transfer test (being placed in a new start box) was the same for rats in both the location and direction procedures, but the correct goal was different. The fact that both groups of rats performed successfully (and made different responses) during the transfer tests indicates that during initial training, the rats in each group had learned the rule appropriate for successful performance in their procedure.

In a second experiment, which incorporated a more extensive training procedure prior to the critical transfer tests, rats performed almost perfectly (Becker, Walker, & Olton, 1980; see also Becker & Olton, 1980, 1982; Becker, Olton, Anderson, & Breitenger, 1981; Olton, Becker, & Handelmann, 1980). The procedure was designed to teach the rat to choose between two objects on the basis of their spatial locations and not on the basis of their position relative to the rat or their idiosyncratic stimulus characteristics. The test procedure gave the rat a choice between two objects placed in an arena. The correct object contained food in a food cup on top of it, the incorrect object did not. The location of the two objects remained the same from trial to trial, and the object with the food was always in the same location. The actual objects in these locations changed from trial to trial, as did their relative position with respect to the rat when it entered the arena. Thus only the spatial location of the objects determined the one that contained food.

The apparatus was an arena, approximately 1 m square. Around the outside edge of the arena was a wall, 35 cm high. Outside the wall was a runway, 10 cm wide. In the wall were 12 holes, 3 on each side of the arena. Each hole could be covered by a gulliotine door. Inside the arena were two objects, each 15 cm high. In the top of each object was a hole, 5 cm deep, that served as a food cup. A piece of cardboard, 10 cm square, was placed over each food cup to cover it.

At the beginning of each trial, food was placed in the food cup of one object but not in the other, and each food cup was covered by a piece of cardboard. The guillotine door covering one of the holes in the arena wall was removed allowing access between the arena and the runway. The rat was placed on the runway, ran through the opening into the arena, jumped on top of one of the objects and knocked the cardboard off the food cup. If the object was in the correct location, the rat obtained the food from the food cup. If the object was in the incorrect location, the food cup was empty and the rat was removed from the arena without being allowed to go to the object in the other location.

Throughout testing, the hole allowing the rat to enter the arena changed from trial to trial. Thus the position of the objects relative to the rat as it entered the arena varied, making this relationship an irrelevant stimulus. The two objects in the arena were chosen from a set of eight objects, differing in their pattern of black and white stripes, and also changed from trial to trial. Thus the stimulus characteristics of the objects varied, making these irrelevant stimuli. Only the location of the objects determined which one contained the food.

During initial training, each rat entered the arena through the center hole on each of three sides. After reaching a criterion of 18 correct responses in 20 consecutive trials, the rat was given transfer tests from each of the remaining six doors on those three sides, from the center door on the fourth side, and then from the remaining two doors on the fourth side. These transfer tests took place during several days and were accompanied by regular tests from the initial three doors.

The rats performed almost perfectly on all the transfer tests. Such accuracy might be expected during the first set of transfer tests which were conducted from the same sides of the arena that had been used during initial training. During the transfer tests from the fourth side, however, the rats were presented a view of the arena that they had never seen prior to those tests. Nonetheless, they still performed accurately, and many of them went directly from the opening in the wall to the correct location with no signs of hesitation.

Again we included a second group of rats, one trained to follow a different rule. For these rats, the two objects in the arena were the same for every trial, but the location of the objects changed from trial to trial, as did the hole that was uncovered in the wall between the runway and the arena. The same object always had food in it, irrespective of its spatial location. Thus the discrimination could be solved only on the basis of the stimulus characteristics of the objects themselves, and not their spatial location. The rats given this test procedure also chose very accurately during their transfer tests, again demonstrating that this performance was influenced by the past reinforcement history of the rats rather than by the physical properties of the transfer test itself.

These results confirm those of the previous experiment demonstrating that rats can solve tasks that require location learning, operationally defined as the ability to get to a particular goal using different routes and beginning from different starting locations. The performance of the rats in this procedure was substantially more accurate than that of the rats tested on the elevated maze. There are many possible reasons for this difference, but one of the most important may be the inclusion of transfer tests from the same sides of the arena that were used during initial training prior to the transfer tests from the fourth side. This additional training should have helped prepare the rats for the transfer procedure in general and also should have given them more opportunities to learn about the topological relationships of the stimuli in the environment. In any case, the parameters influencing the accuracy of location learning remain to be specified; such information will help to describe the conditions that enhance and/or suppress true location learning, and indicate the extent to which it is a common occurrence.

These data are important for two reasons. First, they demonstrate that rats are capable of true location learning, a process that has either been ignored (see Olton, 1979a, 1979b) or rejected (Blodgett, McCutchan, & Mathews, 1949; Ritchie, 1948) in previous experiments due to inappropriate experimental designs or test procedures. Second, they challenge us to develop models of cognitive processes that can explain this ability (Kaplan, 1973, 1975; Peters, 1973, 1978). We need a model that can compare new views of spatial stimuli to old ones to determine their interrelationships, to predict the spatial consequences of different movements even when the movements being considered have never been made before, and to make an appropriate decision based on the costs and benefits of each of the possible courses of action. This is a formidable list of requirements, and provides a substantial challenge to current theories of cognitive processes. (For studies of these processes in humans see Pick & Reiser, Chapter 5, this volume.)

Neurological Studies

Accurate performance on the transfer tests in the location procedure requires cognitive mapping but not working memory. Consequently, this test procedure can be used to compare the predictive validity of the cognitive mapping theory of hippocampal function (Black, 1975; Black, Nadel, & O'Keefe, 1977; Nadel, O'Keefe, & Black, 1975; O'Keefe & Black, 1975, 1978; O'Keefe & Nadel, 1978, 1979) with that of the working memory theory (Olton, Becker, & Handelmann, 1979a, 1979b, 1980; Olton & Feustle, 1981; Olton & Papas, 1979). The former predicts that rats with damage to the hippocampal system will be unable to perform correctly on the transfer tests (because they require cognitive mapping), whereas the latter predicts that they will perform normally on these transfer tests (because the tests do not require working memory).

We have tested rats with fimbria-fornix lesions on the elevated maze in the procedures used to demonstrate location learning (Walker & Olton, 1981). Of most importance is the performance of a group of rats trained to go from three start boxes to the center goal box, both preoperatively and postoperatively until they reached a criterion of stable performance. They were then given transfer tests from the fourth start box. During these tests, they performed as well as the normal rats described earlier in this section. These data demonstrate that even without an intact hippocampal system, rats can use a cognitive mapping strategy, operationally defined as successful performance on the transfer test. Rats with fimbria-fornix lesions were unable to choose correctly during a subsequent test of working memory using the same maze, indicating that the fimbria-fornix lesion was functionally effective in that testing situation.

Data from the experiments in the arena support these conclusions (Becker & Olton, 1980, 1982; Becker, Olton, Anderson, & Breitinger, 1981). Although the behavior of rats with fimbria-fornix lesions was not identical to that of normal rats, any differences were slight and the rats with lesions performed accurately on the critical transfer tests, even when all testing took place postoperatively. Thus rats with lesions were able to develop as well as use a cognitive mapping strategy after damage to the hippocampal system. Again these results are consistent with the working memory theory, but not with the cognitive mapping theory.

EFFICIENT ORGANIZATION OF MOVEMENT THROUGH SPACE: THE TRAVELING SALESMAN PROBLEM

Behavioral Studies

Animals not only remember the places they have been, they also organize trips among these places in an efficient manner. The question of the optimal path to take when traveling arises in many contexts, but is probably most familiar as the "traveling salesman problem." Consider a salesman who is in a city and has 10 different stores to visit. He wants to visit each one once to show his wares and talk to the purchasing department. In addition, he wants to go to these 10 stores in the most efficient manner, covering the least distance possible. Thus he must plan his itinerary to begin at the proper place and to go from store to store using the most direct route, one which will minimize the total distance covered. (Other factors, such as the speed and cost of various types of transportation, attractiveness of the route, etc., might also influence his choice, but for the present experiment these are not relevant and will be ignored.)

Chimpanzees given a traveling salesman problem to solve performed much better than expected by chance (Menzel, 1973, 1978). The experiment was conducted in an outdoor enclosure. The chimpanzees had been in the enclosure many times and were familiar with all parts of it. At the start of each trial, 18 pieces of food were placed out of sight in different places (under some sticks, behind a clump of grass, etc.). One person carried the chimpanzee who watched where another person placed the food. When placing the food, the experimenters moved about the enclosure inefficiently. Thus they might place the first piece of food on the left side, the second one on the right side, and then return to the left side to place the next one there. After the chimp had been shown where all the food was placed, it was carried to a cage at one side of the enclosure. The experimenters left the enclosure, climbed to a place where they could watch the chimp get the food, and then released the chimp from the cage.

The chimps obtained about 12 of the 18 peices of food, and did so in a relatively efficient manner, certainly better than expected by chance and better than the path the experienters had taken when placing the food. Most chimps made a large circle through the enclosure, picking up food on the way. Thus they took the information about the places in which food had been hidden, reorganized it, and then established their own route which was more efficient than the one that had been used to give them that information.

These experiments demonstrate that chimpanzees have the cognitive capacity to make the types of decisions required to solve the traveling salesman problem. However, they do not provide the quantitative data that are necessary to examine the various components of the successful performance and to develop an appropriate model of the cognitive processes underlying it. To obtain these data, we decided to test ferrets. Like chimpanzees, ferrets are very curious and interact well with people, two traits that considerably simplify the testing procedures. They are also small and adaptable, making laboratory tests a feasible undertaking.

The test environment was a room in which walls, 1.5 m high, enclosed an area 2 m × 6 m. In this enclosure were placed 10 objects, each with a food cup. These objects were built from scraps in the carpentry shop and pieces of junk lying around the lab. Attached to each object, about 20 cm above the floor, was a cup in which food could be placed. Because the food cup was elevated, the ferret had to make an obvious response to see if food was in it and this response was much different from the usual patterns of running throughout the enclosure. At one side of the enclosure was a cage that functioned as the starting point and a stool on which the experimenter sat while recording data.

On any given trial, four of the food cups were *baited* and had food in them. The other six food cups were *unbaited* and did not have food in them. The baited cups were changed randomly from trial to trial with the constraint that within 100 trials all food cups were used equally often. Each ferret was deprived of food as needed to produce sufficient motivation for it to search for food, a point that was reached after a loss of about 300 grams of body weight.

At the start of each trial, the experimenter picked up the ferret from its home cage and carried it around the room to each of the four cups that had food in them. At each of these baited cups the experimenter held the ferret next to the food cup and placed a piece of food in the cup. The ferret usually looked at the piece of food being placed in the cup and made some effort to get it, such as sticking out its tongue, moving toward the cup, etc. After showing the ferret all four baited cups, the experimenter carried it to the start cage and sat on the stool beside it. The experimenter released the ferret from the cage and recorded its movements through the enclosure, noting the path taken and the food cups that were checked. After the ferret had obtained the food from each of the four baited food cups, it returned to the experimenter and stood on its hind legs to get an additional 10 grams of food in a small bowl. This food was given only when the ferret had obtained the food from each of the baited food cups and had returned to the experimenter.

The optimal strategy on each trial was to visit only the four baited food cups, visit each of these only once, minimize the total distance traveled, and then return to the experimenter to get the bowl of food at the end. A failure to perform optimally resulted in errors, of course, and the types of errors were used to determine which aspects of the task were not solved appropriately.

During the first few trials, the ferrets were very inefficient in their performance and made many errors. They often visited unbaited food cups, returned to baited cups from which they had already removed the food, and went to the experimenter for the bowl of food before going to each of the baited food cups. The drawing of their path through the enclosure looked like a hopeless scribble drawn by a young child who was paying no attention to what he or she was doing.

After about 50 trials, performance stabilized and the ferrets chose very accurately. They usually made only one error of any kind during a trial, and often chose perfectly, first going only to the food cups that had food in them and then returning to ask for the bowl of food from the experimenter. Furthermore, they took a relatively direct route among the baited food cups, using a path that was different from, and more efficient than, the one used by the experimenter to show them the baited cups. This description indicates that the ferrets were able to meet all the requirements of the task relatively well, including the one of reorganizing the spatial information given to them at the start of each trial.

We have just begun the quantitative analyses of these data. The goal is to describe the choice behavior of the ferrets, and then model it with a computer program. Of particular interest is the pattern of errors exhibited by the ferrets, and the spatial distribution of their movements among the food cups. By determining these and comparing them to those produced by various computer programs, we will be able to determine how the ferrets perform relative to the levels expected by chance, and what types of restrictions we must place in the programs to have the computer perform as well as the live animals.

The first analysis has considered the pattern of errors, irrespective of their spatial relationships. Three different types of errors, summarized in Table 14.1, were recorded. Each trial was divided into five sequential periods, summarized in Table 14.2. We then calculated the number of errors of each type in each period. The behavioral data are taken from three ferrets. Each one had received approximately 100 trials prior to this analysis and had reached a stable level of performance. The computer data were taken from a PDP-8 using the random number generator of an OS-8 compiler. The program instructed the computer to draw numbers ranging from 1 through 11 from the random number generator. The numbers 1 through 10 were used to indicate objects 1 through 10, respectively, whereas the number 11 was used to indicate returning to the experimenter to ask for the bowl of food. Three runs, each of 10 trials, were used to obtain data from the computer.

The results of this analysis are presented in Table 14.3. There are two major points to be drawn from a comparison of the performance of the ferrets and the computer model. First, the ferrets were much more accurate than the

TABLE 14.1
Types of Errors Made by the Ferrets and the Memory Process That Failed[a]

Type of error	Memory process that failed
Choosing an unbaited cup	Remembering which cups were baited and which were unbaited
Returning to a baited cup	Remembering which baited cups had already been chosen in that test session
Asking for bowl of food before going to each of the four baited food cups	Remembering how many baited cups had already been chosen

[a]Each of the three types of errors considered in the analysis of the ferrets' performance is identified in the left column. The memory process that failed when that error was generated is indicated in the right column. All errors are errors of commission, that is, making a response when a response was inappropriate.

TABLE 14.2
Periods of the Test Session and Their Descriptions[a]

Period	Description
P0	Release from start cage
P1	Response to one of the baited food cups
P2	Response to a different baited food cup
P3	Response to a different baited food cup
P4	Response to a different baited food cup
	Asking for the bowl of food from the experimenter

[a]The left column indicates the five periods of each test session, the right column indicates the sequence of events during a test session based on the number of different food cups the ferret visited. The periods changed only when the ferret responded to a previously unchosen baited food cup, and the number of each period indicates the number of previously successful choices.

computer program, generally making only about 10% of the number of errors. These data indicate that the ferrets performed much better than expected by chance. Second, the pattern of errors made by the ferrets in the different periods of the test were different from the patterns of errors made by the computer program. The most striking example is the number of food cups chosen in P4, the interval between choosing the last baited food cup and returning to the experimenter. The ferrets almost always went directly from the last baited food cup to the experimenter so that during this period the mean number of unbaited cups chosen was .1 and the mean number of baited cups repeated was .2. In contrast, the computer program made many errors during this period; the mean number of unbaited cups chosen was 7.3 and the mean number of baited cups repeated was 4.0. These data indicate that the significance of the final choice was obviously much different to the ferrets (which received a substantial portion of their daily food intake at this time) than to the computer program (in which the number 11 had no more significance than any of the other numbers). Other patterns are also obvious. For example, the number of times the ferrets went to the experimenter before visiting all four baited food cups gradually increased during the test periods, a trend not seen in the output from the computer program.

The analyses presented here are, of course, rudimentary. We obviously want more data from both the ferrets and the computer, and want to transform the data so we can discuss the results in terms of probabilities rather than absolute numbers. Most importantly, we want to analyze the spatial distribution of the choices, a variable that has been ignored in the analyses presented earlier. Qualitative descriptions of the ferrets' behavior indicate that they

TABLE 14.3
The Mean Number of Errors Made by Three Ferrets and a Computer Program during Each Period of the Test Session[a]

Type of errors	Periods of test				
	P0	P1	P2	P3	P4
Responding to an unbaited food cup					
Ferrets	.2	.3	.3	.3	.1
Computer program	1.7	2.3	2.1	8.3	7.3
Returning to a previously chosen baited food cup					
Ferrets	—	.0	.2	.6	.2
Computer program	—	.3	1.0	3.7	4.0
Asking for the bowl of food from the experimenter before getting the food from all four baited food cups					
Ferrets	.0	.0	.1	.4	—
Computer program	.6	.3	.4	1.5	—

[a]Each number is the mean number of errors of that type (see Table 14.1) during that period of the test session (see Table 14.2). Some errors cannot occur in some test periods, and consequently there are no data for these.

moved among tbe baited food cups in a relatively efficient manner, but only a more quantitative analysis will tell us how well they actually chose, the ways in which their performance deviated from the optimal, and the factors responsible for this deviation. Still, these initial descriptions are valuable for two reasons. First, they provide some specific information about the ways in which ferrets approach the traveling salesman problem. Second, they indicate the general type of approach that can be taken to obtain quantitative data about efficient movement through space.

CONCLUSIONS

Animals demonstrate remarkably accurate and intelligent behavior in spatial testing procedures. This facility could arise for many reasons, but two seem most important. First, a spatial procedure often elicits an animal's natural search strategies, ones which they are already predisposed to apply to any given discrimination problem. Thus animals are well prepared to deal with the information that is given to them and can attend to the specific demands of the task at hand. Second, a spatial procedure represents the end point of a continuum of stimulus redundancy. Consider, for example, a comparison of two sets of discriminative stimuli. In one case, these might be two abstract stimuli (tones or lights) presented in an operant box; only these two stimuli are relevant to the discrimination and all the other stimuli in the environment must be ignored. In another case, the discriminative stimuli might be two different places on a maze in a room with many stimuli. In this case, the stimulus complex associated with one place is markedly different from the stimulus complex associated with the other place, and virtually all the stimuli are both relevant and redundant. Indeed, the basic concepts of "place" and "cognitive mapping" almost always assume that there are many stimuli with a constant topological relationship. According to this view, the distinction between spatial and nonspatial tasks is a quantitative one rather than a qualitative one, and the important dimension is the number of relevant–redundant cues and/or the ratio of relevant–redundant cues to irrelevant–nonredundant cues (Olton, 1978a). In essence, the greater redundancy provides more stimulus associations which in turn allows a greater depth of processing, a variable than can have a profound influence on performance in a variety of different settings.

From this perspective, the unimportance of the hippocampus in cognitive mapping tasks such as true location learning is not surprising. Animals with damage to the hippocampal system can usually learn any task in which the stimulus relationships remain constant. As described above, environments that support cognitive mapping represent the ultimate in constant relationships, and the fact that the hippocampus is unnecessary to perform these correctly is consistent with the rest of the literature describing the behavioral changes following hippocampal damage (see Olton, Becker, & Handelmann, 1979a, 1979b).

The critical importance of the hippocampus for working memory procedures, whether these involve spatial or nonspatial stimuli, is also consistent with previous findings from both animals and people (Olton, Becker, & Handelmann, 1979a, 1979b, 1980; Olton & Feustle, 1981). When stimulus relationships change and require the temporal coding of working–episodic memory, damage to the hippocampal system usually produces a profound amnesia. These results leave us in the somewhat unsatisfactory state of concluding that the primary functions of the hippocampus are not uniquely tied to spatial behaviors, but they do tell us something about the neural mechanisms of memory that animals use in spatial (as well as nonspatial) tasks, and allow us to look elsewhere for the brain mechanisms involved in cognitive mapping.

ACKNOWLEDGMENTS

The author thanks J. Becker and L. Squire for discussions about the human amnesic syndrome, M. O'Reagan, B. Taylor, and M. Shera for testing the ferrets, and M. Richardson for typing the manuscript.

REFERENCES

Acheson, A. L., Zigmond, M. J., & Stricker, E. M. Compensatory increases in tyrosine hydroxylase activity in rat brain after intraventricular injections of 6-hydroxydopamine. *Science*, 1980, *207*, 537–540.

Beatty, W. W., & Shavalia, D. S. Spatial memory in rats: Time course of working memory and effect of anesthetics. *Behavioral and Neural Biology*, 1980, *28*, 454–462.

Becker, J. T., & Olton, D. S. Object discrimination by rats: The role of frontal and hippocampal systems in retention and reversal. *Physiology and Behavior*, 1980, *24*, 33–38.

Becker, J. T., & Olton, D. S. Cognitive mapping and hippocampal system function. *Neuropsychologia*, 1982, *19*, 733–741.

Becker, J. T., Olton, D. S., Anderson, C. A., & Breitinger, E. R. P. Cognitive mapping in rats: The effects of hippocampal and frontal system lesions on retention and reversal. *Behavioral Brain Research*, 1981, *3*, 1–22.

Becker, J. T., Walker, J. A., & Olton, D. S. Neuroanatomical bases of spatial memory. *Brain Research*, 1980, *200*, 307–320. (a)

Becker, J. T., Walker, J. A., & Olton, D. S. *Cognitive mapping by rats: True location learning.* Unpublished manuscript, 1980. (b)

Black, A. H. Hippocampal electrical activity and behavior. In R. L. Isaacson & K. H. Pribam (Eds.), *The hippocampus* (Vol. 2: Neurophysiology and behavior). New York: Plenum Press, 1975. Pp. 129–167.

Black, A. H., Nadel, L., & O'Keefe, J. Hippocampal function in avoidance learning and punishment. *Psychological Bulletin*, 1977, *84*, 1107–1129.

Blodgett, H. C., McCutchan, K., & Mathews, R. Spatial learning in the T-maze: The influence of direction, turn, and food location. *Journal of Experimental Psychology*, 1949, *39*, 800–809.

Craig, S. L., & Jarrard, L. E. Unpublished data described in L. E. Jarrard, Selective hippocampal lesions: Differential effects on performance by rats of a spatial task with preoperative versus postoperative training. *Journal of Comparative and Physiological Psychology*, 1978, *92*, 1119–1127.

Dember, W. N., & Fowler, H. Spontaneous alternation behavior. *Psychological Bulletin*, 1958, *55*, 412–428.

Douglas, R. J. Cues for spontaneous alternation. *Journal of Comparative and Physiological Psychology*, 1966, *62*, 171–183.

Geschwind, N. Disconnexion syndromes in animals and man. Part I. *Brain*, 1965, *88*, 237–294.

Hoehler, F. K., & Thompson, R. F. The effect of temporal single alternation on learned increases in hippocampal unit activity in classical conditioning of the rabbit nictilating membrane response. *Physiological Psychology*, 1979, *7*, 345–351.

Honig, W. K. Studies of working memory in the pigeon. In S. H. Hulse, H. Fowler, & W. K. Honig (Eds.), *Cognitive processes in animal behavior*. Hillsdale, New Jersey: Lawrence Erlbaum, 1978, 211–248.

Honzik, C. H., & Tolman, E. C. The perception of spatial relations by the rat: A type of response not easily explained by conditioning. *Journal of Comparative Psychology*, 1936, *22*, 287–318.

Jarrard, L. E. Role of interference in retention by rats with hippocapal lesions. *Journal of Comparative and Physiological Psychology*, 1975, *89*, 400–408.

Jarrard, L. E. Selective hippocampal lesions: Differential effects on performance by rats of a spatial task with preoperative versus postoperative training. *Journal of Comparative and Physiological Psychology*, 1978, *92*, 1119–1127.

Kamil, A. C. Systematic foraging by a nectar-feeding bird (*Loxops virens*). *Journal of Comparative and Physiological Psychology*, 1978, *92*, 388–396.

Kamil, A. C., & Sargent, T. (Eds.). *Optimal foraging.* New York: Garland/STPM Press, 1980.

Kaplan, S. Cognitive maps in perception and thought. In R. Downs & D. Stea (Eds.), *Image and environment.* Chicago: Aldine, 1973. Pp. 63–78.

Kaplan, S. Adaptation structure, and knowledge. In R. G. Golledge & G. T. Moore (Eds.), *Perspectives in environmental cognition.* Stroudsburg: Dowden, Hutchinson, and Ross, 1975.

Kinsbourne, M., & Wood, F. Short-term memory processes and the amnesic syndrome. In D. Deutsch & J. A. Deutsch, (Eds.), *Short-term memory processes.* New York: Academic Press, 1975. Pp. 257–291.

Lashley, K. S. The mechanism of vision: XVI. The function of small remnants of the visual cortex. *Journal of Comparative Neurology,* 1939, *70,* 45–67.

Maier, N. R. F. Reasoning in white rats. *Comparative Psychology Monographs,* 1929, *6,* pp. 1–93.

Maki, W. S., Brokofsky, S., & Berg, B. Spatial memory in rats: Resistance to retoractive interference. *Animal Learning and Behavior,* 1979, *7,* 25–30.

Mech, D. *The wolf: The ecology and behavior of an endangered species.* Garden City, New York: Natural History Press, 1970.

Menzel, E. Chimpanzee spatial memory organization. *Science,* 1973, *182,* 943–945.

Menzel, E. Cognitive mapping in chimpanzees. In S. H. Hulse, H. Fowler, & W. K. Honig, (Eds.), *Cognitive processes in animal behavior,* Hillsdale, New Jersey: Lawrence Erlbaum, 1978. Pp. 375–422.

Mishkin, M. Visual mechanisms beyond the striate cortex. In R. Russel, (Ed.), *Frontiers in physiological psychology.* New York: Academic Press, 1966. Pp. 93–119.

Mishkin, M. Cortical visual areas and their interactions. In A. G. Karcsmar & J. C. Eccles, (Eds.), *Brain and human behavior.* Heidelberg: Springer–Verlag, 1972. Pp. 93–119.

Nadel, L., O'Keefe, J., & Black, A. H. Slam on the brakes: A critique of Altman, Brunner, and Bayer's response-inhibition model of hippocampal function. *Behavioral Biology,* 1975, *14,* 151–162.

Norman, D. A. What have animal experiments taught us about human memory? In J. A. Deutsch, (Ed.), *The physiological basis of memory.* New York: Academic Press, 1973. Pp. 185–213.

O'Keefe, J. A review of hippocampal place cells. *Progress in Neurobiology,* 1980, *13,* 419–439.

O'Keefe, J., & Black, A. H. Single unit and lesion experients on the sensory inputs to the hippocampal cognitive map. In *Functions of the septo-hippocampal system.* Ciba Foundation Symposium 58. New York: Elsevier. Pp. 179–198.

O'Keefe, J., & Conway, D. H. Hippocampal place units in the freely moving rat: Why they fire where they fire. *Experimental Brain Research,* 1978, *31,* 573–901.

O'Keefe, J., & Nadel, L. *The hippocampus as a cognitiye map.* Oxford: Oxford University Press, 1978.

O'Keefe, J., & Nadel, L. Precis of the hippocampus as a cognitive map. *The Behavioral and Brain Sciences,* 1979, *3,* 487–533.

Olton, D. S. Discrimination behavior in tbe rat: Differential effects of reinforcement and nonreinforcement. *Journal of Comparative and Physiological Psychology,* 1972, *79,* 284–290.

Olton, D. S. Characteristics of spatial memory. In S. H. Hulse, H. Fowler, & W. K. Honig (Eds.), *Cognitive processes in animal behavior.* Hillsdale, New Jersey: Lawrence Erlbaum, 1978. Pp. 341–373. (a)

Olton, D. S. The function of septo-hippocampal connections in spatially organized behavior. In *Functions of the septo-hippocampal connections.* Ciba Foundation Symposium 58. New York: Elsevier, 1978. Pp. 327–349. (b)

Olton, D. S. Mazes, maps, and memory. *American Psychologist,* 1979, *34,* 583–596. (a)

Olton, D. S. Inner and outer space: The neuroanatomical bases of spatially organized behaviors. *The Behavioral and Brain Sciences,* 1979, *32,* 311–312. (b)

Olton, D. S. Behavioral neuroscience. In *McGraw–Hill yearbook of science and technology.* New York: McGraw–Hill, 1980. Pp. 99–101.

Olton, D. S. Staying and shifting: Their effects on discrimination learning. In M. L. Commons (Ed.), *Quantitative analyses of behavior: Matching and maximizing accounts*. New York: Pergamon Press, in press.

Olton, D. S., Becker, J. T., & Handelmann, G. E. Hippocampus, space and memory. *The Behavioral and Brain Sciences*, 1979, *2*, 313–322. (a)

Olton, D. S., Becker, J. T., & Handelmann, G. E. A re-examination of the role of hippocampus in working memory. *The Behavioral and Brain Sciences*, 1979, *2*, 352–365. (b)

Olton, D. S., Becker, J. T., & Handelmann, G. E. Hippocampal function: Working memory or cognitive mapping? *Physiological Psychology*, 1980, *8*, 239–246.

Olton, D. S., & Collison, C. Intramaze cues and "odor trials" fail to direct choice behavior on an elevated maze. *Animal Learning and Behavior*, 1979, *7*, 221–223.

Olton, D. S., Collison, C., & Werz, M. A. Spatial memory and radial arm maze performance in rats. *Learning and Motivation*, 1977, *8*, 289–314.

Olton, D. S., & Feustle, W. A. Hippocampal function required for nonspatial working memory. *Experimental Brain Research*, 1981, *41*, 380–389.

Olton, D. S., Handelmann, G. E., & Walker, J. A. Characteristics of spatial memory and foraging behavior in rodents. In A. Kamil & T. Sergeant (Eds.), *Optimal foraging*. New York: Garland STPM Publications, 1981. Pp. 333–354.

Olton, D. S. & Papas, B. C. Spatial memory and hippocampal function. *Neuropsychologia*, 1979, *17*, 669–682.

Olton, D. S., & Samuelson, R. J. Remembrance of places passed: Spatial memory in rats. *Journal of Experimental Psychology: Animaal Behavioral Processes*, 1976, *2*, 97–116.

Olton, D. S., & Scholsberg, P. Food searching strategies in young rats: Win-shift predominates over win-stay. *Journal of Comparative and Physiological Psychology*, 1978, *92*, 609–618.

Olton, D. S., Walker, J. A., & Gage, F. H. Hippocampal connections and spatial discrimination. *Brain Research*, 1978, *139*, 295–308.

Olton, D. S., Walker, J. A., Gage, F. H., & Johnson, C. T. Choice behavior of rats search for food. *Learning and Motivation*, 1977, *8*, 315–331.

Olton, D. S., & Werz, M. A. Hippocampal function and behavior: Spatial discrimination and response inhibition. *Physiology and Behavior*, 1978, *20*, 597–605.

research. Stroudsburg: Dowden, Hutchinson, and Ross, 1973. Pp. 119–153.

Peters, R. Communication, cognitive mapping, and strategy in wolves and hominids. In R. L. Hall and H. S. Sharp (Eds.), *Wolf and man*. New York: Academic Press, 1978. Pp. 95–107.

Platt, J. R. Strong inference. *Science*, 1964, *146*, 347–353.

Pyke, G. H. Optimal foraging in bumblebees: Rule of movement between flowers within inflorescences. *Animal Behaviour*, 1979, *27*, 1167–1181.

Restle, F. Discrimination of cues in mazes: A resolution of the "place vs. response" question. *Psychological Review*, 1957, *64*, 217–228.

Ritchie, B. F. Studies in spatial learning: VI. Place orientation and direction orientation. *Journal of Experimental Psychology*, 1948, *38*, 659–669.

Roberts, W. A. *Some studies of spatial memory in the rat*. Paper presented at the meeting of the Canadian Psychological Association, Ottawa, June 9, 1978.

Roberts, W. A. Spatial memory in the rat on a hierarchical maze. *Learning and Motivation*, 1979, *10*, 117–140.

Rozin, P. The psychobiological approach to human memory. In M. R. Rosenzweig & E. L. Bennett (Eds.), *Neural mechanisms of learning and memory*. Cambridge: MIT Press, 1976. Pp. 3–46.

Schoener, T. W. Theory of feeding strategies. In R. F. Johnston (Ed.), *Annual review of ecology and systematics* Vol. 11. Palo Alto: Annual Reviews, 1971.

Settlage, P. H. The effect of occipital lesions on visually-guided behavior in the monkey: 1. Influence of the lesions on final capacities in a variety of problem situations. *Journal of Comparative Psychology*, 1939, *29*, 91–131.

Stahl, J. A., & Ellen, P. Factors in the reasoning performance of the rat. *Journal of Comparative and Physiological Psychology*, 1974, 87, 598–604.

Sutherland, N. S., & Mackintosh, N. H. *Mechanisms of animal discrimination learning*. New York: Academic Press, 1971.

Suzuki, S., Augerinso, G. & Black, A. H. Stimulus control of spatial behavior on the eight arm maze in rats. *Learning and Motivation*, 1980, 11, 1–18.

Tolman, E. Cognitive maps in rats and men. *Psychological Review*, 1948, 55, 189–208.

Tolman, E. C., & Gleitman, H. Studies in learning and motivation: I. Equal reinforcements in both end-boxes followed by shock in one end box. *Journal of Experimental Psychology*, 1949, 39, 810–819.

Tolman, E. C., Ritchie, B. F., & Kalish, D. Studies in spatial learning: I. Orientation and the short-cut. *Journal of Experimental Psychology*, 1946, 36, 13–24.

Tulving, E. Episodic and semantic memory. In E. Tulving & W. D. Donaldson (Eds.), *Organization of memory*. New York: Academic Press, 1972.

Walker, J. A., & Olton, D. S. The role of response and reward in spatial memory. *Learning and Motivation*, 1979, 10, 73–84. (a)

Walker, J. A., & Olton, D. S. Spatial memory deficit follwoing fimbria-fornix lesions: Independent of time for stimulus processing. *Physiology and Behavior*, 1979, 23, 11–15. (b)

Walker, J. A., & Olton, D. S. Fimbria-fornix lesions impair spatial working memory but not cognitive mapping. *Journal of Comparative and Physiological Psychology*, in press.

Whitham, T. G. Coevolution of foraging in Bombus and nectar dispensing in Chilopsis: A last dreg theory. *Science*, 1977, 197, 593–595.

Zigmond, M. J., & Stricker, E. M. Deficits in feeding bebavior after intraventricular injection of 6-hydroxydopamine in rats. *Science*, 1972, 177, 1211–1214.

Zigmond, M. J., & Stricker, E. M. Recovery of feeding and drinking by rats with intraventricular 6-hydroxydopamine or lateral hypothalamic lesions. *Science*, 1973, 182, 717–720.

Zoladek, L., & Roberts, W. The sensory basis of spatial memory in the rat. *Animal and Learning and Behavior*, 1978, 6, 77–81.

Note added in proof. A recent series of experiments by Morris have also made the distinction between place learning and location learning. Using a water maze, Morris and his associates have been able to demonstrate that rats can learn to go to a particular location, getting there in a variety of different directions, even without local cues (Morris, R. G. M. Spatial localisation does not require the presence of local cues. *Learning and Motivation*, 1981, 12, 239–260).

MICHAEL POTEGAL

Vestibular and Neostriatal Contributions to Spatial Orientation

> The possibility suggests itself that the semi-circular canals are in some way respon-
> sible for [spatial orientation]—but how they are responsible for it—if they do figure
> in it at all—is at present beyond our knowledge [Watson, 1907, p. 89].

Two themes are developed in this chapter. The first, historically older, theme is that the vestibular system, a classic "lower sensory system," may also contribute to higher level functions in the domain of spatial orientation. The second, much more recent, theme is that the basal ganglia may be involved in the processing of this vestibular information as part of their contribution to a system for egocentric spatial orientation. The presentation of the first theme is, in turn, divided into two parts. First, indirect evidence for the plausibility of a vestibular contribution to spatial orientation is discussed, then direct be-havioral demonstrations are reviewed.

THE VESTIBULAR-NAVIGATIONAL GUIDANCE HYPOTHESIS: DEAD-RECKONING BASED ON VESTIBULAR INPUT

In a letter to *Nature* in 1873 Darwin suggested that animals might be able to navigate by a system of "dead reckoning" similar to that used by mariners at sea. In the simplest form of dead reckoning it is necessary to know only: (*a*) the

361

SPATIAL ABILITIES
Development and Physiological Foundations

direction in which one is going; (*b*) the steady velocity at which one is traveling; and (*c*) the period of time during which one has been traveling. By then multiplying the velocity by the time, one can determine how far one has gone in the known direction. In a more complex journey, involving multiple changes in speed, the simple process of multiplying velocity by time to find distance must be replaced by the mathematical operation of *integration,* a computation that is like a series of successive, instantaneous multiplications and additions. In the most complex journey, in which both speed and direction may be varied, a sophisticated operation involving a combined integration of linear and angular velocities is required to keep track of one's position.

More than 30 years prior to Darwin's letter, Flourens (1830) had pioneered in demonstrating that the vestibular organs were indicators of movement. Some 60 years after Flourens's efforts and 20 years after Darwin's letter, Exner (1893) united the work of Flourens and his successors to Darwin's hypothesis in his proposal that the vestibular system could provide the velocity signals for the organism's dead reckoning of distance and direction. The utricle, and perhaps the saccule, are sensitive to linear acceleration whereas, in the velocity range of normal movements, the semicircular canals have been shown to integrate angular acceleration into angular velocity (e.g., Melville Jones & Milsum, 1970). With appropriate time keeping these signals can be combined and integrated by the central nervous system (CNS) to determine the distance and direction traveled (Barlow, 1964; Mayne, 1974). Since the vestibular end organs detect acceleration by virtue of their displacement from their inertial position, such vestibular guidance can be classified as a form of inertial navigation (Barlow, 1964). Mayne (1974) has pointed out that tilt and linear acceleration have similar effects on the sensory cells of the utricle. To discriminate linear acceleration from tilt, which must be done before the acceleration signal can be integrated, the CNS may use combinations of static and dynamic utricular receptors as well as input from the semicircular canals. It should be clear even from this sketchy description that considerable processing must be done on the vestibular input signals. For fuller exposition of the mathematical details see Mayne (1974).

The analogy between maritime "dead reckoning" and the vestibularly guided navigation of organisms can be extended in several respects. Mariners using dead reckoning at sea must rely on compass direction, time, and estimated velocity for navigation. Their navigational accuracy is improved, of course, when they can also use sun position and the relative movement of nearby land masses (see Pick & Reiser, Chapter 5, this volume). Since, as will be described, vestibular navigation is prone to cumulative errors, supplementary corrective input from other sensory modalities will be as useful to this system as external guides are to mariners at sea. For example, the motor outflow and somatosensory feedback information generated during normal locomotion could be used to supplement vestibular feedback. In general, the mariner's calculated position is plotted on a navigational chart and "updated" periodically with new calculations. Like navigation at sea, updated, vestibularly derived position information is probably most useful to the organism when "plotted" on some kind of internal spatial map (cf. Böök & Gärling, 1981; Hart & Berzok, Chapter 7 this volume). Some characteristics of this putative map are discussed in the second section of this chapter.

INDIRECT EVIDENCE BEARING ON THE HYPOTHESIS OF VESTIBULAR-NAVIGATIONAL GUIDANCE

To make the vestibular-navigational hypothesis more plausible the useful operating range of the system must be established. This involves answering the following questions: Within what velocity and acceleration ranges does vestibular transduction occur? How much drift or noise is there in the vestibular input? How does the nervous system use this information and, in particular, what is the evidence that it can actually perform the required computations? What might be the physiological and anatomical substrates of this computation? Evidence bearing on these questions is presented in the following sections.

The Sensitivity and Reliability of Vestibular Acceleration Transduction

The semicircular canals detect rotational accelerations; psychophysical estimates of the threshold for horizontal angular acceleration detection in human beings vary from .035 to 4.0 degrees/sec^2 with most estimates being in the range of 0.1–2.0 degrees/sec^2 (Geudry, 1974, Table 1). Angular accelerations normally occurring during active movement are well above these threshold values (Hallpike & Hood, 1953; Van Egmond, Groen, & Jongkees, 1949).

Some indications of the accuracy of the vestibular system in transducing above-threshold acceleration input is suggested by the psychophysical measures of vestibular sensation obtained from subjects being passively rotated (Geudry, 1974). The most relevant measures for our purpose were obtained using triangular velocity wave form inputs that are a quasi-natural form of rotational stimulation. Even with this, most appropriate, form of stimulation, subjects showed varying degrees of stimulus underestimation as a function of the technique used to evaluate their sensation. However, with the technique yielding the best estimate, it is interesting that the subjects' judgments of the total angle through which they had been displaced were far more accurate (to within a few percentage points) than their velocity judgments (Geudry, 1974, Figure 15). It appears as if the subjects were particularly good at just the sort of velocity-over-time integration demanded by the vestibular-navigational hypothesis. It should be noted that each of Geudry's subjects was quite consistent in judgments over the range of stimulus values but that there were large differences among subjects. This may imply that either there is considerable intersubject variability in the adequacy of vestibular navigation and/or that conscious, vestibularly derived sensations of movement may be related to the integrated velocity signal by some kind of "constant" that varies from individual to individual. In the latter context it should be pointed out the detection of conscious sensation in psychophysical studies may not reflect much about the possibly unconscious processing of vestibular input occurring during normal locomotion. For example, note the remark made by Garling, Böök, and Lindberg:"During actual walking the process underlying maintenance of orientation is not very open to inspection [1979, p. 8]."

The reliability of vestibular acceleration transduction is limited by the noise (random variation) and drift (systematic shifts) occurring at the various stages of signal processing. Rough estimates of noise can be derived from neurophysiological observations. The coefficient of variation (standard deviation/mean) of firing rates of first order semicircular canal neurons during accelerations varies widely. However, in both cats (Blanks, Estes, & Markham, 1975) and monkeys (Goldberg & Fernandez, 1971) a substantial subpopulation of neurons exists whose coefficient of variation is less than .1. This means that fluctuations in acceleration transduction will be less 10% on the average. Furthermore, Melville Jones and Milsum (1970) suggest that "ensemble averaging" across the whole population of responding neurons can improve the signal-to-noise ratio. Young has suggested that the overall noise in this system may be approximated by "white Gaussian noise with standard deviation equal to the threshold for the oculogyral illusion [1973, p. 210]." If so, then noise would not be a major problem for vestibular acceleration transduction reliability in the range of normal movement.

The problem posed by rotational drift may be somewhat greater. Estimates of rotational drift depend upon whether the vestibular system is viewed in a sensory detection mode or a motor control mode. In the former case, rotational drift during locomotion can be viewed as externally imposed (e.g., by left–right asymmetries in body parts which produce differential stride lengths of the two legs [Lund, 1930]). Each step would produce a small net angular deviation in one direction (i.e., a slight turn). Each such turn involves an angular acceleration. The maximum acceleration that will remain undetected by the vestibular sensory apparatus can be estimated from the psychophysical measurements of rotational threshold. Although many such measurements exist (see Geudry, 1974) the most relevant appear to be Doty's (1969) plot of detection latency versus angular acceleration. Assuming the extreme case of .75 degree/sec² angular acceleration during half of a 1 sec step, the resulting angular deviation is approximately .2 degree/step. Such a deviation would produce a trajectory with a limit circle radius of several hundred meters. This calculation is not inconsistent with a number of observations that blindfolded human beings instructed to walk in a straight line tend to veer in one direction (Lund, 1930; for review see Howard & Templeton, 1966, pp. 257–258). If the vestibular system is considered in a motor control mode, that is, as a generator of motor commands influencing stride length, asymmetries arising *within* the vestibular system must be considered. In this case psychophysical measurements are irrelevant and any amount of net angular deviation is possible. The extreme example of this is the strong turning tendency that follows unilateral vestibular damage. It therefore seems that rotational drift poses a major problem for a vestibular navigational system. As Barlow (1964) has pointed out, this problem is exacerbated by the process of integration, which tends to accumulate errors. Thus, as the performance of blindfolded subjects suggests, vestibular navigation uncorrected by external landmarks would only be useful for short-term (perhaps on the order of a few minutes), short-distance journeys.

With regard to the transduction of horizontal acceleration the 1–3 m/sec^2 maximal accelerations of normal human locomotion (Eberhardt, 1976; Herman, Wirta, Bampton, & Finley, 1976) would seem to be well above the psychophysically derived thresholds for detection of experimentally imposed linear horizontal accelerations, which are on the order of .1 m/sec^2 (see Geudry, 1974). However, peak value alone does not determine the detectability of locomotor accelerations. These accelerations occur within a brief (approximately .2 sec) portion of the stride cycle. The transduction of this acceleration impulse is a function of the "time constant" of the utricle (and possibly the saccule). If the utricular response is not rapid enough relative to the duration of the acceleration impulse, the transduction will be reduced. Unfortunately, psychophysical estimates of the time constant of utricular response range from well below .2 sec (DeVries, 1950) to well above it (Young & Meiry, 1968). This theoretical discussion may well be moot however in view of the fact that "during horizontal linear oscillation, observers have an immediate sensation of linear velocity. . . . Subjects can integrate this velocity signal and make estimates of linear displacement which appear to be systematic [Geudry, 1974, p. 74]." However, once again, detection of conscious sensation may not reflect much about the "unconscious" processing of vestibular input that may occur during normal locomotion. There are no relevant data known to this author regarding noise or drift in the detection of linear acceleration.

Integration of Vestibular Input in the CNS

Melville Jones and Milsum (1970) and others have demonstrated that angular acceleration is integrated into angular velocity by mechanical action of the vestibular end organs. Does the CNS have the capacity to further integrate this velocity signal into displacement information? The previously cited observations from vestibular psychophysics suggest that the answer is yes. Given that it is angular velocity information that is transmitted up the vestibular branch of the VIII nerve it is noteworthy that subjects are better at estimating angular displacement than angular velocity. This suggests that integration is an optimal mode of operation of this system. As noted earlier, it also appears that subjects who are exposed to sinusoidal linear oscillation develop immediate sensations of changing linear velocities and are able to integrate these into estimates of displacement.

Neurophysiological observations also tend to support the existence of CNS "integrators." One of these is involved in the vestibulo-ocular reflex; it provides an eye position-related drive to abducens motor neurons (Skavenski & Robinson, 1973). Robinson (1974) suggests that this same integration circuitry is involved in generating both slow and quick phases of rotary nystagmus. It has been argued that it also plays a role in the maintenance of fixation (Henn & Cohen, 1976; Keller, 1974; Robinson, 1974). Robinson (1974) initially proposed that this integrator lies within the pontine paramedian reticular formation with some additional neural circuitry in the cerebellum. More recently, Raphan, Matsuo, and Cohen (1979) have tentatively restricted this localization to the periabducens region. Another integrator has been incorporated into a model that accounts for the phenomenon of "velocity storage" in vestibular and optokinetic nystagmus with reasonable success (Raphan, Matsuo, & Cohen, 1979). These authors tentatively suggest that the neurons responsible for this computation may be located in the prepositus nucleus.

Raphan et al. (1979) point out that when the input of an ideal integrator falls to zero, the output is maintained at its final value indefinitely. Such a device would obviously be useful in, for example, maintenance of gaze; the pulse of input to abducens motorneurons that is responsible for a transient saccade would, when simultaneously fed into the integrator, produce a steady-state output maintaining fixation at the end of the saccade (cf. Robinson, 1974, Fig. 1). When the input to a nonideal ("leaky") integrator, falls to zero, the output also decays to zero with a characteristic time constant. The time constant of the leaky integrator for eye position can be estimated from the drift of eye position in the dark; in the cat it is greater than 20 sec (Robinson, 1974). Similar estimates of this time constant in human beings have been on the order of 20 sec (Becker & Klein, 1973). Raphan et al. (1979) estimated that the time constant for the integrator involved in velocity storage is about 10 sec.

It would seem that integrators for vestibular navigation would have to have time constants at least an order of magnitude greater to even reach the drift-imposed limit of accuracy suggested by Barlow (1964). However, there are no directly relevant experimental data known to the author that could be used to estimate these time constants. Under the somewhat dubious assumption that the phenomenon of spontaneous alternation (see below) taps vestibular navigational integrators, they would appear to have time constants on the order of hours. An alternative scheme would be to use the output of the integrator to update the organism's internal map representing its position in space (cf. Lindberg & Gärling, 1981). Storage of location information then becomes the more general (but no less mysterious) problem of memory.

Central Nervous System Structures That Might Be Involved in Vestibular-Navigational Computation

It has been established that vestibular information finds its way to all the major divisions of the central nervous system including the forebrain. Which of these loci might be involved in vestibular-navigational computation? The first locus for processing vestibular velocity signals is at their site of entry into the CNS, the vestibular nuclei. It is certainly true that information arriving from the vestibular branch of the VIII nerve undergoes a variety of transformations at this site (e.g., see section entitled "Convergence between Vestibular and Somatosensory Modalities," p. 370). The next most obvious site is the cerebellum. The flocculo-nodular lobe and fastigial nuclei, in particular, receive not only secondary fibers from the vestibular nuclei but primary fibers from the vestibular end organs as well (see Pompeiano, 1974 for a review of vestibulo–cerebellar relations). Furthermore, the cerebellum has been thought to contribute to vestibulo–ocular integration (Robinson, 1974).

In addition to these well-known sites for vestibular information processing, there are some less well-known vestibular projections to the forebrain that may be involved in navigational computations. The recent history of these particular ascending vestibular projections begins with the finding of vestibular projection areas in cortex. In 1932 Spiegel reported on an area near the posterior ectosylvian gyrus of the cat in which topical application of normally subconvulsive doses of strychnine caused convulsions when the animal was rotated. EEG studies showed increased high frequency activity in this area in response to angular acceleration (see Copack, Dafny, & Gilman, 1972 for review). Razumeyev and Shipov (1970, pp. 260–266) cite four Russian studies published between 1946 and 1963 in which the most pronounced EEG changes following caloric and D.C. stimulation or destruction of the labyrinth were found in "parietal" areas of cats and humans. In a review paper Spiegel, Szekely, Moffet and Egyed (1970) refer to several more recent studies in which EEG changes or long latency potentials accompanying rotation were confined to, or especially marked in, posterior cortical regions of cats rendered cataleptic with bulbocapnine.

The rather diffuse posterior localization obtained with relatively prolonged rotational, caloric, or D.C. electrical stimulation of the labyrinth can be contrasted to the distribution of responses to punctuate mechanical or electrical stimulation of vestibular end organ, nerve, or nuclei. In the first experiment of this kind (Walzl & Mountcastle, 1949) potentials evoked by pulsed electrical stimulation of the vestibular branch of the VIII nerve were recorded along a narrow strip of cortex on the anterior bank of the anterior suprasylvian fissure in the cat. A number of subsequent investigators have confirmed vestibular projections to the same or immediately adjacent areas of cat cortex (for review see Potegal, Copack, de Jong, Krauthamer, & Gilman, 1971). Microelectrode recordings from single cells in various areas of cat cortex seem consistent with the differential responses of the anterior and posterior cortex to punctate and prolonged vestibular stimulation, respectively (e.g., Kornhuber & deFonseca, 1964). The postcruciate dimple (PCD) of the posterior sigmoid gyrus is a second, small anterior cortical area in the cat from which vestibular responses have been recorded with gross electrodes and microelectrodes (e.g., Sans, Raymond, & Marty, 1970; Ödkvist, Liedgren, Larsby, & Jerlvall, 1975). Thus, in the cat, short latency evoked responses to punctuate vestibular stimulation are definitely confined to two anterior cortical areas; EEG changes and long latency potentials to prolonged stimulation may be preferentially localized in posterior regions.

Vestibular cortical projection areas have also been defined relative to the intraparietal sulcus in the rhesus monkey (Fredrickson, Figge, Scheid, & Kornhuber, 1966; Schwartz & Fredrickson, 1971) and the central sulcus in the squirrel monkey (Ödkvist, Schwarz, Fredrickson, & Hassler, 1974). Localization relative to functional (or, at least, electrophysiologically defined) regions of the cortex across species is more variable, appearing within the forelimb fields of the postcentral somatosensory area (S1) in the guinea pig (Ödkvist, Rubin, Schwarz, & Fredrickson, 1973a), the rabbit (Ödkvist, Rubin, Schwarz, & Fredrickson, 1973b), and the squirrel monkey (Ödkvist *et al.*, 1974) but not in the cat and the rhesus monkey. The vestibular and auditory cortical projections are contiguous in the cat (and perhaps in human beings, see further on) but not in other species. One point of agreement is that these projection areas are not classical primary sensory areas in that they do not respond just to vestibular input (see section entitled "Convergence between Vestibular and Somatosensory Modalities," p. 370). There also has been speculation about more general principles that might unify these observations, for example, that the more anterior areas, which may be cytoarchitectonically identifiable as Brodmann's area 3a, receive mainly muscle spindle afferents and subserve motor coordination whereas the more posterior areas, corresponding to Brodmann's area 2v, receive joint afferents and subserve conscious orientation in space (Deecke, Schwartz, & Fredrickson, 1977). However, there are currently no valid cross-species generalizations about the location of vestibular projection areas relative to functional or cytoarchitectonic divisions of cortex.

There are two sources of data relevant to the rostral transmission of vestibular information in the human brain. A number of attempts to make scalp measurements of brain activity (evoked potentials or EEG changes) induced in human subjects by rotation have produced inconsistent results probably because of the difficulties in controlling stimulus parameters and eliminating confounding effects. Salamy, Potvin, Jones, and Landreth (1975) tried to reduce these sources of artifact and argued that a negative–positive wave complex with peaks at 190 and 350 msec after rotation onset represented a response to vestibular stimulation. If so, it is rather surprising that the complex could be recorded on all leads even given the somewhat diffuse response to prolonged rotational stimulation found in animal studies. At any rate, their claim would have been stronger if this complex were shown to be absent in labyrinthine defective, but otherwise normal, subjects.

The second source of data are the reports of vestibular sensations induced by electrical stimulation of the exposed cortex of patients undergoing neurosurgical procedures. In a review of 108 cases Penfield (1957, Fig. 3) found that seven points from which more or less specific labyrinthine responses could be elicited were located within the superior temporal convolution near the auditory area. Hécean also reviewed Penfield's surgical notes and identified the superior temporal convolution as the vestibular locus in seven cases (Hécean, Penfield, Bertrand, & Malmo, 1956, Fig. 22). It is unclear if the same seven cases are being reported in both reviews. Both papers briefly mention that similar sensations can occur during direct electrical stimulation of the parietal lobe or as auras accompanying seizure activity there (cf. Foerster, 1936).

Considerable experimental effort has been expended to establish the route by which vestibular information ascends to the cortex and other structures. In contrast to earlier negative reports (see Copack et al., 1972; Deecke, Schwarz, & Fredrickson, 1974; Raymond, Sans, & Marty, 1974; for review) more recent studies have indicated that direct vestibulothalamic projections exist. These axons lie within (Lang, Büttner–Ennever, & Büttner, 1979) or dorsomedial to (Abraham, Copack, & Gilman, 1977; Deecke et al., 1974); the classical auditory pathway of the lateral lemniscus. Vestibulofugal axons in the medial longitudinal fasciculus, the route for vestibulo-occular control, do not contribute substantially to the thalamic input whereas the role of vestibular information ascending directly or indirectly through the reticular formation is uncertain. The earliest electrophysiological studies suggested that the magnocellular part of the medial geniculate body (MGmc), a polymodal area, was a major area of vestibulothalamic activity (e.g., Potegal et al., 1971; for review see Copack et al., 1972). Although subsequent studies in the cat (Blum, Abraham, & Gilman, 1979) and the squirrel monkey (Liedgren, Milne, Rubin, Schwarz, & Tomlinson, 1976) have identified neurons in MGmc that do, in fact, receive vestibular input, the majority of recent studies (e.g., Day, Blum, Carpenter, & Gilman, 1976; Raymond, Demêmes, & Marty, 1976) have focused attention upon the ventral posterolateral nucleus (VPL, a classical somatosensory relay nucleus) and the ventrolateral nucleus (VL). In monkeys the pars oralis of the VPL (VPLo) and the VPI lying ventral to it in particular, have been found to contain neurons responding to vestibular stimulation (Deecke *et al.*, 1977; Liedgren, Milne, Rubin, Schwarz, & Tomlinson, 1976). In general, vestibular sensitive neurons are rather sparsely scattered in small groups within the thalamic nuclei (cf. Lang *et al.*, 1979). It is of some interest that Tasker and Organ (1971) elicited sensations of movement and vertigo by stimulation of an area adjacent to the mouth field within the ventrocaudal somatosensory nuclei of the thalamus of a neurosurgical patient.

The finding that a particular thalamic nucleus receives vestibular input does not necessarily indicate that it transmits vestibular information to the cortex. The consensus of two neuroanatomical studies of the cat using horseradish peroxidase is that: (*a*) there is a projection of MGmc to the anterior ectosylvian vestibular area; and (*b*) there is a projection from the VPL to the PCD that is partially or completely separate from any projection of the VPL to the anterior ectosylvian area (Day *et al.*, 1976; Liedgren *et al.*, 1976). A species difference seems to exist between the cat and the monkey in that there does not appear to be a cortical vestibular projection from the MGmc in the monkey. A difference may also exist between monkey species in that data from lesion and/or antidromic stimulation experiments indicate that VPI is the major thalamocortical relay nucleus for vestibular information in the rhesus monkey (Deecke *et al.*, 1974) whereas VPLo serves this function in the squirrel monkey (Liedgren *et al.*, 1976).

These electrophysiological and neuroanatomical studies establish a potential neuronal substrate for vestibular-navigational computations. In this context it is of interest that Büttner, Henn, and Oswald (1977) found that vestibular velocity information is preserved from the vestibular nuclei to the thalamus, implying that the integration of the velocity signal to displacement required by the guidance system occurs more rostrally (e.g., in cortex or basal ganglia). The vestibular input to the basal ganglia will be described in the second part of this chapter. As a final point, it must be said that it is remarkable that in the more than three decades since Walzl and Mountcastle's (1949) original demonstration of ascending vestibular projections, their findings have been replicated and extended in numerous electrophysiological and neuroanatomical studies, yet there has not been a single behavioral study of the function of this system.

Convergence between Vestibular and Somatosensory Modalities

Investigators of the thalamus and the cortex have emphasized that neurons in these structures that respond to vestibular stimulation often also respond to somatosensory stimulation. At the thalamic level, 40–82% of vestibularly sensitive neurons of monkey VPI/VPLo are also responsive to somatosensory input (Buttner & Henn, 1976; Deecke *et al.*, 1977; Liedgren *et al.*, 1976). At the cortical level, the incidence of convergence ranges from 56% to 67% to 100% for deep and superficial peripheral nerve stimulation in cat PCD (Odkvist *et al.*, 1975) and in rhesus monkey cortex (Schwarz & Fredrickson, 1971) respectively. Such sensory convergence at more rostral levels is not surprising since neurons of the vestibular nuclei have been shown to respond to kinesthetic input (Fredrickson, Schwarz, & Kornhuber, 1965). These observations have suggested to a number of investigators that ascending vestibular projections are part of a system for monitoring posture and movement of body parts. Indeed the activity of a few of the units recorded in monkey cortex (Schwarz & Fredrickson, 1971) and thalamus (Deecke *et al.*, 1977) seem to reflect various specific body postures with limb and vertebral column position transmitted to the neurons via joint and muscle sense and head attitude relative to gravity transduced by the otolith organs. These signals are compatible in that they are static representations of position. Such neurons would presumably play a role in the internal representation of posture on a moment to moment basis.

What are the implications of vestibular–somatosensory convergence for the vestibular navigational hypotheses? It is conceivable that kinesthetic information could be used to "gate out" irrelevant vestibular input. For example, if a subject who was vestibularly monitoring his progress while walking a straight path were to turn his head toward an unexpected noise the kinesthetic input from the "irrelevant" head turn could be used to "gate out" the irrelevant vestibular feedback from the ongoing vestibular-navigational computation. It is less likely that kinesthetic input is used to directly augment vestibular-guided navigation, since kinesthetic information is conventionally described as a static position sense whereas the vestibular signal is velocity related. It is computationally more difficult to combine a position and velocity signal than it is to combine two velocity signals, as in the case of visual–vestibular interaction (see further on). Conceivably, the CNS could integrate vestibular velocity input before combining it with kinesthetic input. However, Buttner *et al.*'s (1977) observation on the similarity of the velocity-representing characteristics of thalamic and vestibular-nuclear neurons suggests that this does not happen, at least at the thalamic level. The velocity signal from the muscle afferents may be a better candidate for augmenting vestibularly guided navigation; perhaps proprioceptive information generated by rhythmic head movements during locomotion might facilitate vestibular input from the same source.

At this point it should be recalled that 18–60% of the vestibular responding units did not respond to somatosensory stimulation. Furthermore, it may be that the two studies showing the highest rates of convergence (Deecke *et al.*, 1977; Schwarz & Fredrickson, 1971) are overestimates because the labyrinthine polarization they used for stimulation can produce nonspecific effects. Conversely, the study using the most natural form of vestibular stimulation (Büttner & Henn, 1976) found the smallest number of somatosensory convergent vestibular units. The remaining nonconvergent thalamic units, in which the vestibular velocity signal is preserved (Buttner *et al.,* 1977) and whose activity is not linked to eye movements (Magnin & Fuchs, 1977), would appear to be the most currently promising candidates for the transmission of information involved in vestibular-navigational computation.

Convergence between Vestibular and Visual Modalities

The study of vestibular–visual interaction, an active area of research, currently focuses on the modulation by visual stimulation of subjects' perceptual and/or oculomotor responses to vestibular input or vice versa (see Sedgwick, Chapter 1, this volume). These phenomena reflect, at least in part, the operation of a system that functions to monitor the consequences of the organism's movement through the environment. The interactions begin at the very earliest stages of processing in the CNS. In the vestibular nuclei of the goldfish (Allum, Graf, Dichgans, & Schmidt, 1976) and the monkey (Waespe & Henn, 1979) all neurons responsive to rotation of the animal in the dark are also responsive to rotation of the visual surround when the animal is stationary. Movement of the visual surround is more effective than vestibular input in driving the neurons when low accelerations and frequencies are applied for prolonged periods. Vestibular stimulation is more effective for high acceleration, high frequency transients. When both visual and vestibular stimulation are present, as would normally be the case when the organism actively locomotes through the environment with its eyes open, vestibular nucleus neurons accurately follow the movement over a great range of rotational velocities, independently of the acceleration and frequencies involved. It is important to note that the basis of this combination of inputs is that they are compatible; both visual and vestibular input to vestibular nuclear neurons are velocity signals.

One very interesting implication of vestibular–visual convergence is the possibility that the vestibular-navigational system can accept and operate upon one aspect of visual input (movement of surround) directly in addition to using other aspects of visual input (e.g., visual landmarks) as post-hoc corrections for its computations. The use of visual surround movement could be expected to increase the range and accuracy of "vestibular"-navigational guidance. Against this interpretation, however, is the finding that not all thalamic vestibular-sensitive units have well-defined responses to visual stimulation (Büttner & Henn, 1976; Magnin & Fuchs, 1977).

DIRECT EVIDENCE FOR THE VESTIBULAR-NAVIGATION HYPOTHESIS

Producing direct, unequivocal evidence that organisms can and do use vestibular-navigational guidance is not a trivial problem. First, it must be determined if, under well-controlled experimental conditions, vestibular input is sufficient for (or even contributes significantly to) the organism's tracking of its location in space. Second, it must be determined if there are any naturally occurring spatial tasks or conditions for which organisms choose or are forced to use vestibular navigation. The answer to both these questions depends, of course, on the species being considered as well as on a host of other factors.

The experimental fulfillment of the demand for sufficiency has two parts. It must first be demonstrated that the organism in question can monitor its spatial location in the absence of task-relevant cues other than vestibular input. For the obvious exteroceptive cues of vision and audition as well as various somatosensory and motor output cues this demand can be met by passively transporting blind or blindfolded organisms along some trajectory in the presence of masking noise and then demanding a response that is a function of the trajectory (e.g., pointing to or returning to the starting point). The experimental challenge in designing this task lies in anticipating and excluding or making irrelevant subtle, esoteric and species-specific sensory cues such as echolocation (Riley & Rosenzweig, 1957) or magnetic senses (Baker, 1980; Walcott, Gould, & Kirschvink, 1979). The complementary part of the sufficiency demonstration is that disruption or elimination of vestibular input should eliminate, or at least significantly impair, spatial orientation ability in the experimental situation. The difficulty here is to demonstrate the specificity of the resulting orientation deficit. In rats, for example, elimination or disruption of vestibular input by surgery or forced rotation can produce a generalized emotionality. A further difficulty is that vestibular disruption can create motor impairments so that even if the organism "knows" its spatial location it may have difficulty in executing the response that displays its knowledge. It is therefore necessary to show that the vestibular disruption leaves intact performance on tasks matched as closely as possible to the navigational task but lacking the demand for vestibular guidance. That is, the control task should be of equal difficulty, have the same type of reward, demand a similar form of response, etc.

The evidence to date is sparse and only partially meets all of these demands. Some studies have been done on human subjects actively locomoting through their environment. Cohen, Cooper, and Ono (1963), like Geudry (1974), found that estimates of distance traversed vary substantially between subjects. Garling and his colleagues (1979) have inferred the existence of a "central processing" of information that is ongoing during active locomotion. If subjects perform a distracting mental arithmetic task while walking along a path to its endpoint they are substantially less accurate than nondistracted subjects in verbally indicating their distance and direction from the starting point (Lindberg & Gärling, 1981). The effect is greater in blindfolded than in sighted subjects. Distance estimates are also affected by the velocity of locomotion (Cohen, Cooper, & Ono, 1963) suggesting another limitation on the system. It is possible that Garling's "central processing" involves vestibular-navigational computation. Certainly, it is a commonplace clinical observation that labyrinthine defective humans become disoriented in the dark. The Russian investigator Beritoff (1965) has provided some experimental extensions of this clinical observation using a task similar to Garling's except that his blind-folded, labyrinthine defective subjects actually had to return physically to their starting point rather than just verbally indicate its location. However, Worchel (1952) reported that deaf, vestibularly deficient subjects performed as well as, or even better than, normal subject on various versions of this task. Worchel explained this rather surprising result as the consequences of compensatory development of their kinesthetic information processing.

As indicated, a more stringent test of the vestibular navigational hypothesis involves the passive transport of subjects who are blindfolded or transported in the dark. A number of investigators have found that human subjects transported short distances in carts (Beritoff,1965; Juurmaa & Suonio, 1975, Liebig, 1933; Mayne, 1974; Worchel, 1951) or driven longer distances in cars (Garling, Mantyla, & Säisä, 1978) can still maintain their orientation, although not as well as actively locomoting subjects. Under conditions of blindfolding and passive transport human subjects may well be forced to rely on vestibular-navigational guidance although somatosensory information may play a supporting role. Beritoff (1965) has put this hypothesis to the test; he found that labyrinthine defective subjects were severely impaired under these conditions. Beritov's observations, made years ago, remain the most direct evidence for vestibular-navigational guidance in human beings. However, the recent demonstrations of a magnetic sense in human beings (Baker, 1980, but see Gould & Able, 1981) indicate that Beritoff's results should be reconfirmed with appropriate controls.

The study of the vestibular-navigational hypothesis in nonhuman subjects began with Exner's (1893) investigation of the possible role of the vestibular system in the homing behavior of pigeons. He attempted to interfere with the function of the vestibular systems of pigeons while they were being transported from their roosts by either rotating them, anesthetizing them, or electrically stimulating their vestibular end organs. He found no differences between any of the treated birds and the controls in the duration or accuracy of their homeward flights. Barlow (1964) reviewed eight subsequent studies performed between 1911 and 1951 using variants of these techniques all of which had inconclusive or negative results. However, these studies are subject to several criticisms, including the possibility that the birds were compensating for the loss of vestibular information by using their magnetic orientation ability (e.g., Walcott et al., 1979). From a theoretical point of view (Barlow, 1964) any inertial guidance system operating without correction from such navigational aids will accumulate errors. It is therefore inappropriate to look for pure vestibular guidance in long distance homing; vestibular guidance must be sought in shorter journeys.

Short-term, short-distance maze performance and route finding by laboratory animals would appear to be the appropriate experimental domain for tests of the vestibular-navigation hypothesis. For the greater part of this century investigators have been suggesting that the vestibular system contributes to this sort of performance (e.g., Watson, 1907; Zoladek & Roberts, 1978). This conclusion is usually based on the finding that above-chance maze performance persists after surgical and/or experimental elimination of other sensory cues, singly or in combination. For example, Dashiell (1959) proposed a form of navigational guidance called "direction orientation" which is essentially the ability of a rat to maintain its original direction of locomotion while negotiating a series of obstacles. Dashiell proposed a vestibular involvement in this behavior on the grounds that it was unaffected by maze rotation (which made intramaze cues irrelevant) and that it persisted after enucleation. When evidence of this relative imperviousness to sensory disruption is coupled to evidence that animals can find "short cuts," the possibility that they are using a fully developed vestibular navigational guidance system becomes apparent. However, it should be noted that such classic demonstrations of short-cut ability as Tolman, Ritchie, and Kalish's 1946 study may have been artifactual (Young, Greenberg, Paton, & Jane, 1967). Furthermore, this sort of evidence is, at best, indirect. More direct tests have been inconclusive because of poor experimental design (Ruger, 1918) or ambiguous results (Lindenlaub, 1960).

Until 1978 the most direct evidence for the vestibular-navigational hypothesis in nonhuman animals was Beritov's (1965) demonstrations that vestibularly impaired cats and dogs were unable to perform on passive transport tasks when blindfolded. Two surprising aspects of his report are the apparent rapidity with which his animals learned the task and the lack of disturbance produced by acute blindfolding. His statement that labyrinthine defective animals also showed substantial impairments on these tasks even without blindfolds is also somewhat surprising. Unfortunately, the English translation of Beritov's work does not offer detailed experimental protocols of his training and testing procedures.

To rectify this situation, Miller, Barnett, and Potegal (1978) developed a passive transport task for the animal most widely used in American behavioral studies, the laboratory rat. Thirsty rats were allowed to drink briefly from a water bottle and were then transported away from the bottle in a small, enclosed vehicle along a path containing a right angle. To get more water after being released from the vehicle the rats had to return to the bottle from which they had just drunk and avoid the seven other identical bottles placed around the room. After three consecutive successes at a given path length the path length was increased. We found that the rats could successfully perform this task at path lengths of at least 1.3 m, at which distance they had to bypass nearby incorrect bottles to get to the more distant correct bottle. We also found that rats could successfully relearn this task after enucleation. The inference that this postenucleation passive transport performance was based on vestibular guidance was supported in Miller, Potegal, and Abraham's (1980) study in which the earlier work was replicated and it was additionally found that lesions of the rostral poles of the vestibular nuclei produced a severe impairment of passive transport performance. In a recently completed study Miller, Potegal, and Abraham (1981) have found that these lesions also significantly impair acquisition of this task in sighted rats. Identical lesions did not impair an equally difficult olfactory trail-finding task. This series of studies comprises the best evidence known to this author for vestibular-navigational guidance in rats. The second section of this chapter discusses the evidence that the basal ganglia are involved in the corresponding computations.

Evidence for vestibular-navigational guidance in certain specific experimental situations is suggestive but relatively sparse. Data concerning the range of naturally occurring tasks and conditions that call on this system are nonexistent. In the absence of data one can speculate on situations that would seem particularly appropriate for vestibular navigation. For example, this form of navigation might be used by fish and other water dwelling animals that may be passively transported by water currents in dark or visually featureless environments. Another possibility is the use of this system by the young of species that are passively carried by their parents. More generally, vestibular navigation might be used during active locomotion by animals like rats living in dark burrows. Finally, the evidence for vestibular–visual interactions suggests that this system might contribute to spatial orientation even in organisms locomoting under visual guidance in the absence of distinctive landmarks.

SUMMARY OF THE EVIDENCE FOR THE
VESTIBULAR-NAVIGATION HYPOTHESIS

The sensitivity of the vestibular end organs is sufficient to transduce the angular and linear accelerations occurring in normal locomotion. Although noise does not appear to be a major problem for vestibularly guided navigation, the problem of cumulative drift limits the usefulness of this system to short range journeys. The vestibular-navigation hypothesis demands that the vestibular input be integrated by the CNS. Studies of vestibular psychophysics suggest that subjects do indeed respond as if they were integrating their own vestibular input and, under some conditions, that this is an optimal mode of operation. Neurophysiological studies of vestibularly modulated eye movements suggest that several vestibular integrators exist in the CNS subserving different functional systems. In addition to vestibular nuclei and cerebellum, possible locations for the vestibular navigational integrators are suggested by the finding of vestibular projections ascending to the forebrain. Vestibulo–thalamic fibers arise in the vestibular nuclei, run in or medial to the lateral lemniscus, and terminate in various thalamic nuclei. The thalamic nuclei that transmit the majority of vestibular information to the cortex vary according to species: MGmc and VPL in the cat, VPLo in the squirrel monkey, and VPI in the rhesus monkey. The most direct evidence comes from human and animal studies in which navigation has been shown to be possible in the absence of active locomotion and visual input (i.e., during passive transport). Navigation in most, but not all, of these studies has then been shown to be impaired under these conditions by peripheral or central vestibular dysfunction.

THE PUZZLE OF SPONTANEOUS ALTERNATION

It has long been known that a normal laboratory rat will alternate its choices of the arms of a T maze on successive trials with a probability of approximately 80%. This behavior, called spontaneous alternation (SA) because it appears without training or reward, has been variously used as an experimental model of memory (the rat must remember which arm it entered on the last trial to choose the other on the next trial), curiosity (the alternation of arms is supposedly a systematic exploration of the most novel stimulus on any trial), and behavioral inhibition (the rat is supposedly refraining from entering the same arm twice). Since the publication of Douglas's (1966a) controversial work SA has also been regarded as an example of vestibularly guided navigation. That is, Douglas claimed that rats alternated "direction in space" rather than left–right body turns. The bases for this claim were his observations that: (a) If the maze were rotated 90 degrees between trials rats failed to alternate; and (b) if the maze were rotated 180 degrees between trials rats would actually repeat body turns apparently "in order to" alternate direction in space. Since these observations seemed to rule out the possibility that kinesthetic body turn information alone was guiding alternation behavior, Douglas suggested that vestibular guidance might be involved. His evidence for this hypothesis was that: (a) Horizontal (but not vertical) spinning of rats between trials disrupted SA and; (b) middle ear infections producing vestibular signs reduced SA levels (Douglas, 1966b). Douglas also found that the use of different mazes for the first and second trial did not affect SA. From these and additional data Douglas argued that, with the exception of the rat's "odor trail," intramaze cues did not affect SA.

If Douglas's analysis were correct and complete SA would seem to be the ideal behavior for the investigation of the vestibular navigation hypothesis. It is simple, requires no training or reward, and is supposedly uninfluenced by cues in most other sensory modalities. Observations already made on SA could provide information about the operating characteristics of vestibular navigation. For example, Douglas, Mitchell, and Del Valle's (1974) finding that SA fails to occur if the angle between the maze arms is less than 60 degrees might indicate the limits of accuracy of the system. Estimation of the time constants of the integrators might be drawn from studies indicating that rats will alternate with an hour or more elapsing between successive trials (e.g., O'Connell, 1971).

Unfortunately, only some of Douglas's original observations have been systematically supported by subsequent work. The least consistent evidence concerns the lack of influence of other sensory cues. For example, Eisenberger, Myers, Sanders, and Shanab (1970) reported that intramaze visual cues can affect SA. Potegal, Day, and Abraham (1977) found above-chance SA rates with two mazes oriented at 90 degrees, indicating that kinesthetic–motor information may be sufficient for SA in at least some rats. In a more recent unpublished study Miller and Potegal compared SA in a situation with two unadorned mazes oriented in parallel (i.e., only vestibular navigation should have been operative) to a situation in which a single maze with arms strongly differentiated by visual, tactile, and olfactory cues was rotated 90 degrees between trials (i.e., although the rats' vestibular input would have indicated that it was not possible to alternate "direction in space" it would have been possible to alternate to the multiple intramaze cues). The rates of alternation in the "vestibular navigation" and "multi-cue but not vestibular" conditions were 80.0% and 91.4%, respectively. These rates are significantly above chance and the multicue, 90 degree rotation rate was significantly higher than the two maze, parallel "vestibular navigation" rate [$t(6) = 4.44$, $p < .01$].

The most consistently replicated of Douglas's findings is that vestibular dysfunction induced by labyrinthine damage (Douglas, 1966b; Franken & Baker, 1969; Potegal et al., 1977) or vestibular nuclear lesions (Miller & Potegal, unpublished) reduces SA rates. In their most recent study Douglas, Clark, Erway, Hubbard, and Wright (1979) report that mice with a genetic deficit which eliminates utricular function but does not have "any demonstrable effect on the form or function of the semicircular canals [Douglas, et al. 1979, p. 467]" completely fail to show SA. This is a curious finding since one would expect that canal rather than utricular function would be important in vestibular navigational control of SA which is essentially a series of turns. A more serious objection is presented by the repeated failure of Bronstein, Dworkin, Bilder, and Wolkoff (1974) to replicate the disruptive effects of intertrial spinning on SA. A similar failure to replicate was experienced by Abraham, Kornhauser, and Potegal (unpublished). It therefore may be the "side effects" of vestibular dysfunction, for example, changes in emotionality and/or activity, that give rise to the SA deficit rather than impairments in vestibularly guided navigation. That there is practically exclusive control of SA by vestibular guidance is untrue; whether such guidance plays any specific role remains unproven.

ARE THE BASAL GANGLIA PART OF THE ANATOMICAL SUBSTRATE FOR VESTIBULAR-NAVIGATIONAL COMPUTATION?

Historically, the basal ganglia have been thought to function as part of the motor systems of the CNS. This idea is rooted in the experimental evidence that stimulation of the caudate nucleus can affect movement and the clinicopathological evidence for an association between degeneration of the basal ganglia and several types of movement disorders (e.g., Parkinson's disease and Huntington's chorea).

Another development in the analysis of basal ganglia function began with the discovery by Battig, Rosvold, and Mishkin (1960) and subsequent confirmation by others that lesions of the caudate nucleus in monkeys were sufficient to cause an impairment on a delayed response task (for reviews see Öberg & Divac, 1979, Potegal, 1972). In delayed response the animal's task on each trial is to remember *where* the food has just been hidden. A number of other studies extended these results by showing that caudate nucleus lesions interfered with several other types of spatial performance (for reviews see Oberg & Divac, 1979; Potegal, 1972; Thompson, Guilford, & Hicks, 1980).

In 1969 this writer proposed that caudate spatial and motor function might be related, that is, "the caudate nucelus contains or is contained in a system in which potential orientation movements of the head and eyes are the code for spatial location. Specifically, the internal representation of any point, *P,* in space is the set of motor programs which would turn the head and eyes to bring *P* into focus, given the position of the torso at the moment [Potegal, 1972, p. 481]." (See for comparison Steinbach & Smith, 1976; Winterkorn, 1980.) In a test of this hypothesis, the writer employed an elevated maze with 12 radial arms extending out from a center disk; goal boxes were located at the end of each arm. In the test situation, a rat always started from one goal box; the food reward was located in another goal whose position was determined relative to the rat's starting box, that is, the location of the reward was egocentrically defined by the rat's startpoint. In this 12-choice situation, rats with lesions of the caudoputamen were impaired compared to sham operated animals. In a second experiment involving only two choices, rats searched for a food reward that was hidden according to one of two equally difficult spatial schemas. For some rats the reward was always located egocentrically. For other rats, it was located in the leftmost (or rightmost) of the two choices regardless of the left–right relation of those choices to the rat's position [cf. the relative position tasks Niki (1974) used with monkeys and Butterworth (cited in Bremner, Chapter 4, this volume) used with human infants]. It was found that partial lesions of the caudoputamen impaired performance on the egocentric task but not on the relative task, supporting the hypothesis that the basal ganglia may be necessary for egocentric localization.

A most interesting feature of this (or any) egocentric system is that an updating of the central representation of a target position must follow any movement of the observer. For example, suppose the position of a particular object directly in front of an observer were represented by her as "straight ahead." If she were to move 1 m to her right, the target would then have to be represented as "1 m to the left." Similar conclusions about the logical necessity for an updating function have been reached by Pick and Reiser (Chapter 5, this volume). To test the possibility that the caudate is also involved in the compensatory updating function an experiment was performed with patients afflicted with Huntington's Chorea (Potegal, 1971). Each patient, wearing a pair of goggles with a movable opaque shield, stood in front of a horizontal white table top that had a single black target dot at its center. After the patient had viewed the dot, his vision was occluded by lowering the opaque shield of the goggles. He then took a single step to his side and, from his new position, marked the remembered position of the dot with a stylus. It was found that the patients' accuracy in this situation was significantly worse than that of normal subjects. The patients were unimpaired on other spatial–motor tasks of equal difficulty with this apparatus which did not involve compensation for movement.

A possible role of vestibular input in such an egocentric orientation system is suggested by Hörnsten's (1979) report that unilateral vestibular damage in humans affects egocentric localization through the displacement of the "zero position" of the eyes. However, it should be obvious that the main connection between the vestibular navigational hypothesis and the role of the basal ganglia in egocentric localization is the possibility that vestibular navigational computations performed by the basal ganglia may be used for the compensatory updating of the internal representation of target location as the subject moves around. Of course, vestibular feedback is not the only possible source of information that can be used; kinesthetic feedback and motor output are probably utilized as well (Potegal, 1972).

For this conjoint hypothesis to have face validity it is necessary that the basal ganglia, in fact, receive vestibular input. Since Muskens's (1922) neuroanatomical speculations a number of experimental and clinical reports have linked vestibular and basal ganglia function (see Copack et al., 1972; Potegal, 1972). Of more direct relevance are a series of electrophysiological investigations demonstrating ascending vestibular projections to basal ganglia (Copack et al., 1972; Liedgren & Schwarz, 1976, Potegal et al., 1971; Spiegel, Szekely, & Gildenberg, 1965).

Although the electrophysiological data indicate that vestibular input is available to the basal ganglia, they offer no clue about how this input is transformed in these structures. Prior to the emergence of the major problems in Douglas's interpretation of SA, the study of the effects of lesions of the caudoputamen on SA appeared to be a valid approach to the question. Although earlier studies (Baettig, 1963; Borst, Delacour, & Libouban, 1970) had failed to find an SA deficit following lesions of the caudoputamen, later ones did (Divac, Wikmark, & Gade, 1975; Karpiak, Rapport, & Bowen, 1974; Kirkby, 1969; Potegal & Squire, 1974). However, the subsequent doubt cast on the vestibular navigational basis of SA by the conflicting experimental results renders these results of uncertain relevance. More relevant evidence can be found in the recent work of Abraham and Potegal (1979). In this study rats were trained on a passive transport task as already described, given caudoputamen or hippocampal lesions, and then retrained. We found that lesions of the posterior caudoputamen had a significant effect on passive transport saving scores. In contrast, the hippocampal lesions had no effect on passive transport retraining scores. Neither lesion in other groups of rats had any effect on an olfactory trail finding task. Thus the effects of caudatoputamen lesions on the passive transport task satisfy at least some of the criteria for specificity.

At these lines of research stand there seems to be some evidence that the basal ganglia are involved in egocentric orientation. Vestibularly guided navigation is one form of egocentric orientation. Electrophysiological and neuroanatomical data on the ascending vestibular projections indicate that the basal ganglia must certainly be considered among the candidate brain structures that may carry out vestibular-navigational computations. Vestibular information may be only one of the forms of input that the basal ganglia process in this regard.

Some speculative connections can also be made to other ideas about spatial orientation presented in this book. Abraham and Potegal (1979) did not find a deficit on the passive transport task following hippocampal lesions. Yet an abundance of evidence suggests that the hippocampus in rats is involved in their spatial orientation performance in some way (O'Keefe & Nadel, 1978; Olton, Chapter 14, this volume). Perhaps the hippocampus and basal ganglia in rats function in a complementary fashion in which the hippocampus mediates nonegocentric (possibly largely visual) orientation (Sedgwick, Chapter 1, this volume; Hermelin & O'Connor, Chapter 2, this volume), whereas control switches to the basal ganglia when egocentric (largely nonvisual) strategies are involved. A fascinating conjecture which is, at the time of this writing, completely unsullied by any data is the possibility that the early predominance of egocentric strategies in human development (e.g., Bremner, Chapter 4, this volume; Pick & Reiser, Chapter 5, this volume) reflect the influence of the basal ganglia. During the course of development other, possibly cortical, structures become active with a corresponding switch to alternative spatial strategies. The work of Goldman and her colleagues (e.g., Goldman, 1974) on the developmentally successive roles of caudate and frontal cortex in delayed response performance in monkeys is not inconsistent with this conjecture.

REFERENCES

Abraham, L., Copack, P., & Gilman, S. Brainstem pathways for vestibular projections to cerebral cortex in the cat. *Experimental Neurology,* 1977, *55,* 436–448.

Abraham, L., & Potegal, M. Caudate nucleus involvement in spatial orientation. *Society for Neuroscience Abstracts,* 1979, *5,* 212.

Allum, J., Graf, W., Dichgans, J., & Schmidt, C. L. Visual vestibular interactions in the vestibular nuclei of the goldfish. *Experimental Brain Research,* 1976, *26,* 463–485.

Baettig, K. Effect of lesions on spontaneous alternation and choice behavior. *Psychological Reports,* 1963, *13,* 493–494.

Baker, R. R. Goal orientation by blindfolded humans after long-distance displacements: Possible involvement of a magnetic sense. *Science,* 1980, *210,* 555–557.

Barlow, J. S. Inertial navigation as a basis for animal navigation. *Journal of Theoretical Biology,* 1964, *6,* 76–117.

Battig, K., Rosvold, H., & Mishkin, M. Comparison of frontal and caudate lesions on delayed response and delayed alternation in monkeys. *Journal of Comparative and Physiological Psychology,* 1960, *53,* 400–404.

Becker, W., & Klein, H. M. Accuracy of saccadic eye movements and maintenance of eccentric eye position in the dark. *Vision Research,* 1973, *13,* 1021–1034.

Beritoff, J. S. Neural mechanisms of higher vertebrates. Trans. W. T. Liberson. Boston: Little, Brown & Co., 1965.

Blanks, R. H. I., Estes, M. S., & Markham, C. H. Physiologic characteristics of vestibular first-order canal neurons in the cat. II. Response to constant angular acceleration. *Journal of Neurophysiology,* 1975, *38,* 1250–1268.

Blum, P. S., Abraham, L. D., & Gilman, S. Vestibular, auditory and somatic input to the posterior thalamus of the cat. *Experimental Brain Research,* 1979, *34,* 1–9.

Böök, A., & Gärling, T. Maintenance of orientation during locomotion in unfamiliar environments. *Journal of Experimental Psychology: Human Perception and Performance,* 1981, *7,* 995–1006.

Borst, A., Delacour, J., & Libouban, S. Effets, chez le rat, de lesions du noyau caude sur le conditionment de réponse alternée. *Neuropsychologia,* 1970, *8,* 89–101.

Bronstein, P. M., Dworkin, T., Bilder, B. H., & Wolkoff, D. F. Repeated failures in reducing rat's spontaneous alternation through the intertrial disruption of spatial orientation. *Animal Learning and Behavior,* 1974, *2,* 207–209.

Büttner, U., & Henn, V. Thalamic unit activity in the alert monkey during natural vestibular stimulation. *Brain Research,* 1976, *103,* 127–132.

Büttner, U., Henn, V., & Oswald, H. P. Vestibular related neuronal activity in the thalamus of the alert monkey during sinusoidal rotation in the dark. *Experimental Brain Research,* 1977, *30,* 435–444.

Cohen, J., Cooper, P., & Ono, A. The hare and the tortoise: A study of the tau-effect in walking and running. *Acta Psychologica,* 1963, *21,* 387–393.

Copack, P., Dafny, N., & Gilman, S. Neurophysiological evidence of vestibular projections to thalamus, basal ganglia, and cerebral cortex. In T. C. Frigyesi, E. Rinvik, & M. Yahr (Eds.), *Corticothalamic projections and sensorimotor activities.* New York: Raven Press, 1972. Pp. 309–331.

Darwin, C. Origin of certain instincts. *Nature*, 1873, *7*, 417–418.

Dashiell, J. F. The role of vision in the spatial orientation of the white rat. *Journal of Comparative and Physiological Psychology*, 1959, *52*, 522–526.

Day, M., Blum, P., Carpenter, M. B., & Gilman, S. Thalamic components of ascending vestibular projections. *Society for Neuroscience Abstracts*, 1976, *2*, 1058.

Deecke, L., Schwarz, D. W. F., & Fredrickson, J. M. Nucleus ventroposterior inferior (VPI) as the vestibular thalamic relay in the rhesus monkey. I. Field potential investigation. *Experimental Brain Research*, 1974, *20*, 88–100.

Deecke, L., Schwarz, D. W. F., & Fredrickson, J. M. Vestibular responses in the rhesus monkey ventroposterior thalamus. II. Vestibulo-proprioceptive convergence at thalamic neurons. *Experimental Brain Research*, 1977, *30*, 219–232.

DeVries, H. L. The mechanics of the labyrinth otoliths. *Acta oto-laryngologica*, (Stockholm) 1950, *38*, 263–273.

Divac, I., Wikmark, G. E., & Gade, A. Spontaneous alternation in rats with lesions in the frontal lobes: An extension of the frontal lobe system. *Physiological Psychology*, 1975, *3*, 39–42.

Doty, R. L. Effect of duration of stimulus presentation on the angular acceleration threshold. *Journal of Experimental Psychology*, 1969, *80*, 317–321.

Douglas, R. J. Cues for spontaneous alternation. *Journal of Comparative and Physiological Psychology*, 1966, *62*, 171–183. (a)

Douglas, R. J. Spontaneous alternation and middle-ear disease. *Psychonomic Science*, 1966, *4*, 243–244. (b)

Douglas, R. J., Clark, G. M., Erway, L. C., Hubbard, D. G., & Wright, C. G. Effects of genetic vestibular defects on behavior related to spatial orientation and emotionality. *Journal of Comparative and Physiological Psychology*, 1979, *93*, 467–480.

Douglas, R. J., Mitchell, D., & Del Valle, R. Angle between choice alleys as a critical factor in spontaneous alternation. *Animal Learning & Behavior*, 1974, *2*, 218–220.

Eberhardt, H. D. Physical principles of locomotion. In R. M. Herman, S. Grillner, P. S. G. Stein, & D. G. Stuart (Eds.), *Neural control of locomotion*. New York: Plenum Press, 1976.

Eisenberger, R., Myers, A. K., Sanders, R., & Shanab, M. Stimulus control of spontaneous alternation in the rat. *Journal of Comparative and Physiology Psychology*, 1970, *70*, 136–140.

Exner, S. Negative Versuchsergebnisse uber das Orientierungsvermugen der Brieftauben. *Sitzung-Berichte der Akademie der Wissenschaften in Wien*, 1893, *102*, 318–331.

Flourens, M. J. P. Experiences sur les canaux semi-circulaires de l'oreille, dans les oiseaux. *Mémoires de l'Academie des Sciences* (Paris), 1830, *9*, 455–466.

Foerster, O. Motorische Felder und Bahnen. In B. Blunke & O. Foerster (Eds.), *Handbuch der neurologie*, 1936 VI: IV S., 1–357.

Franken, R. E., & Baker, J. G. The effects of drive level on cues utilized in spontaneous alternation. *Psychonomic Science*, 1969, *16*, 239–240.

Fredrickson, J. M., Figge, U., Scheid, P., & Kornhuber, H. H. Vestibular nerve projection to the cerebral cortex of the rhesus monkey. *Experimental Brain Research*, 1966, *2*, 318–327.

Fredrickson, J. M., Schwarz, D. W. F., & Kornhuber, H. H. Convergence and interaction of vestibular and deep somatic afferents upon neurons in the vestibular nuclei of the cat. *Acta oto-laryngologica* (Stockholm), 1965, *61*, 168–188.

Gärling, T., Böök, A., & Lindberg, E. *Maintenance of orientation in the horizontal plane during locomotion: Empirical findings and implications for cognitive mapping*. Paper presented at the International Conference on Environmental Psychology, University of Surrey, Guildford, England, July 1979.

Gärling, T., Mäntylä, T., & Säisä, J. The importance of vision during locomotion for the acquisition of an internal representation of the spatial layout of large scale environments: Blindfolded and sighted car passengers with and without a distracting tack learning to localize invisible targets during a town route. *Ümeå Psychological Reports*, 1978, *147*, 1–25.

Geudry, F. E. Psychophysics of vestibular sensation. In H. H. Kornhuber (Ed.), *Handbook of sensory physiology* (Vol. VI/2). Berlin: Springer–Verlag, 1974.

Goldberg, J. M., & Fernandez, C. Physiology of peripheral neurons innervating semicircular canals of the squirrel monkey. III. Variations among units in their discharge properties. *Journal of Neurophysiology*, 1971, *34*, 676–684.

Goldman, P. An alternative to developmental plasticity: Heterology of CNS structures in infants and adults. In D. G. Stein, J. J. Rosen, & N. Butters (Eds.), *Plasticity and recovery of function in the central nervous system*. New York: Academic Press, 1974.

Gould, J. L., & Able, K. P. Human homing: An elusive phenomenon. *Science*, 1981, *212*, 1061–1063.

Hallpike, C. S., & Hood, J. D. The speed of the slow component of ocular nystagmus induced by angular acceleration of the head: Its experimental determination and application to the physical theory of the cupular mechanism. *Proceedings of the Royal Society London, Series B*, 1953, *141*, 216–230.

Hécaen, H., Penfield, W., Bertrand, C., & Malmo, R. The syndrome of apractognosia due to lesions of the minor cerebral hemisphere. *A. M. A. Archives of Neurology and Psychiatry*, 1956, *75*, 400–434.

Henn, V., & Cohen, B. Coding of information about rapid eye movements in the pontine reticular formation. *Brain Research*, 1976, *108*, 307–325.

Herman, R., Wirta, R., Bampton, S., & Finley, F. R. Human solutions for locomotion: Single limb analysis. In R. M. Herman, S. Grillner, P. S. G. Stein, & D. G. Stuart (Eds.), *Neural control of locomotion*. New York: Plenum Press, 1976.

Hörnsten, G. Constant error of visual egocentric orientation in patients with acute vestibular disorder. *Brain*, 1979, *102*, 685–700.

Howard, I. P., & Templeton, W. B. *Human spatial orientation*. London: John Wiley & Sons, 1966.

Juurmaa, J., & Suonio, K. The role of audition and motion in the spatial orientation of the blind and the sighted. *Scandinavian Journal of Psychology*, 1975, *16*, 209–216.

Karpiak, S., Rapport, M., & Bowen, F. Immunologically induced behavioral and electrophysiological changes in the rat. *Neuropsychologia*, 1974, *12*, 313–322.

Keller, E. L. Participation of medial pontine reticular formation in eye movement generation in the monkey. *Journal of Neurophysiology*, 1974, *37*, 316–332.

Kirkby, R. Caudate nucleus lesions impair spontaneous alternation. *Perceptual and Motor Skills*, 1969, *29*, 550.

Kornhuber, H. H., & Da Fonseca, J. S. Optovestibular integration in the cats cortex: A study of sensory convergence on cortical neurons. In M. B. Bender (Ed.), *The oculomotor system*. New York: Hoeber, 1964.

Lang, W., Büttner-Ennever, J. A. & Büttner, U. Vestibular projections to the monkey thalamus: An autoradiographic study. *Brain Research*, 1979, *177*, 3–17.

Liebig, F. G. Über unsere Orientierung im Raume bei Ausschluss der Augen. *Zietschrift für Sinnesphysiologie*, 1933, *64*, 251–282.

Liedgren, S. R. C., Milne, A. C., Rubin, A. M., Schwarz, D. W. F., & Tomlinson, R. D. Representation of vestibular afferents in somatosensory thalamic nuclei of the squirrel monkey (*Saimiri sciurius*). *Journal of Neurophysiology*, 1976, *39*, 601–612.

Liedgren, S. R. C., & Schwarz, D. W. F. Vestibular evoked potentials in thalamus and basal ganglia of the squirrel monkey (*Saimiri sciurius*). *Acta oto-laryngologica* (Stockholm), 1976, *81*, 73–82.

Lindberg, E., & Gärling, T. Acquisition of locational information about reference points during blindfolded and sighted locomotion: Effects of a concurrent task and locomotion paths. *Scandanavian Journal of Psychology*, 1981, *22*, 101–108.

Lindenlaub, E. Neue Befunde uber die Antangsorientierung von Mausen. *Zietschrift für Tierpsychologie*, 1960, *17*, 555–578.

Lund, F. H. Physical asymmetries and disorientation. *American Journal of Psychology*, 1930, *42*, 51–62.

Magnin, M., & Fuchs, A. F. Discharge properties of neurons in the monkey thalamus tested with angular acceleration, eye movement and visual stimuli. *Experimental Brain Research*, 1977, *28*, 293–299.

Mayne, R. A systems concept of the vestibular organs. In H. H. Kornhuber (Ed.), *Handbook of sensory physiology* (Vol. VI/2). Berlin: Springer–Verlag, 1974.

Melville Jones, G., & Milsum, J. H. Characteristics of neural transmission from the semi-circular canal to the vestibular nuclei of cats. *Journal of Physiology*, 1970, *209*, 295–316.

Miller, S., Barnett, B., & Potegal, M. *Cues for path-finding with passive movement exposure to the path*. Paper presented at the Eastern Psychological Association meeting, Washington, D.C., April 1978.

Miller, S., Potegal, M., & Abraham, L. *Vestibular involvement in spatial orientation*. Paper presented at the Eastern Psychological Association meeting, Hartford, Connecticut, April 1980.

Miller, S., Potegal, M. & Abraham, L. Vestibular involvement in spatial orientation. *Society for Neuroscience Abstracts*, 1981, *7*, 484.

Milsum, J. H., & Melville Jones, G. Dynamic asymmetry in neural components of the vestibular system. *Annals of the New York Academy of Sciences*, 1969, *156*, 851–871.

Muskens, L. The central connections of the vestibular nuclei with the corpus striatum and their significance for ocular movements and for locomotion. *Brain*, 1922, *45*, 454–478.

Niki, H. Prefrontal unit activity during delayed alternation in the monkey. II. Relation to absolute versus relative direction of response. *Brain Research*, 1974, *68*, 197–204.

Öberg, R. G. E., & Divac, I. "Cognitive" functions of the neostriatum. In I. Divac & R. G. E. Öberg (Eds.), *The neostriatum*. Oxford: Pergamon Press, 1979.

O'Connell, R. Delayed alternation in rats with position preferences. *Psychonomic Science* 1971, *22*, 137–139.

Ödkvist, L. M., Liedgren, S. R. C., Larsby, B., & Jerlvall, L. Vestibular and somatosensory inflow to the vestibular projection area in the post cruciate dimple region of the cat cerebral cortex. *Experimental Brain Research*, 1975, *22*, 185–196.

Ödkvist, L. M., Rubin, A. M., Schwarz, D. W. F. & Fredrickson, J. M. Vestibular cortical projection in the guinea pig. *Experimental Brain Research*, 1973, *18*, 279–286. (a)

Ödkvist, L. M., Rubin, A. M., Schwarz, D. W. F., & Fredrickson, J. M. Vestibular cortical projection in the rabbit. *Journal of Comparative Neurology*, 1973, *149*, 117–120. (b)

Ödkvist, L. M., Schwarz, D. W. F., Fredrickson, J. M., & Hassler, R. Projection of the vestibular nerve to the area 3a arm field in the squirrel monkey (*Saimiri sciurius*). *Experimental Brain Research*, 1974, *21*, 1–19.

O'Keefe, J., & Nadel, L. The hippocampus as a cognitive map. Oxford: Oxford University Press, 1978.

Penfield, W. Vestibular sensation and the cerebral cortex. *Annals of Otology* (St. Louis), 1957, *66*, 691–698.

Pompeiano, O. Cerebello–vestibular interrelations. In H. H. Kornhuber (Ed.), *Handbook of sensory physiology* (Vol. VI/2). Berlin: Springer–Verlag, 1974.

Potegal, M. The role of the caudate nucleus in spatial orientation of rats. *Journal of Comparative and Physiological Psychology*, 1969, *69*, 756–764.

Potegal, M. A note on spatial-motor deficits in patients with Huntington's disease: A test of a hypothesis. *Neuropsychologia*, 1971, *9*, 233–235.

Potegal, M. The caudate nucleus egocentric localization system. *Acta Neurobiologia Experimentalis,* 1972, *32,* 479–494.

Potegal, M., Copack, P., deJong, J. M. B. V., Krauthamer, G., & Gilman, S. Vestibular input to the caudate nucleus. *Experimental Neurology,* 1971, *32,* 448–465.

Potegal, M., Day, M. J., & Abraham, L. Maze orientation, visual and vestibular cues in two-maze spontaneous alternation of rats. *Physiological Psychology,* 1977, *5,* 414–420.

Potegal, M., & Squire, L. Contribution of the caudo-putamen to spontaneous alternation in rats. *Society for Neuroscience Abstracts,* 1974, 378.

Raphan, Th., Matsuo, V., & Cohen, B. Velocity storage in the vestibular ocular reflex arc (VOR). *Experimental Brain Research,* 1979, *35,* 229–248.

Raymond, J., Demêmes, D., & Marty, R. Voies et projections vestibulaires ascendentes emanant des noyaux primaires: Étude radioautographique. *Brain Research,* 1976, *111,* 1–12.

Raymond, J., Sans, A., & Marty, R. Projections thalamiques des noyaux vestibulaires: Étude histologique chez le chat. *Experimental Brain Research,* 1974, *20,* 273–284.

Razumeyev, A. N., & Shipov, A. A. Problems of space biology (Vol. 10: Nerve mechanisms of vestibular reactions). (NASA Technical Translation TT F-605). Washington, D.C.: National Aeronautics and Space Administration, 1970.

Riley, D. A., & Rosenzweig, M. R. Echolocation in rats. *Journal of Comparative and Physiological Psychology,* 1957, *50,* 323–328.

Robinson, D. A. The effect of cerebellectomy on the cat's vestibulo-ocular integration. *Brain Research,* 1974, *71,* 195–207.

Ruger, H. A. Some experiments in the transfer of habits in the white rat. *Psychological Bulletin,* 1918, *15,* 42.

Salamy, J. Potvin, A., Jones, K., & Landreth, J. Cortical evoked responses to labyrinthine stimulation in man. *Psychophysiology,* 1975, *12,* 55–61.

Sans, A., Raymond, J., & Marty, R. Réponse thalamique et corticales á la stimulation electrique du nerf vestibulaire chez le chat. *Experimental Brain Research,* 1970, *10,* 265–275.

Schwarz, D. W. F., & Fredrickson, J. M. Rhesus monkey vestibular cortex: A bimodal primary projection field. *Science,* 1971, *172,* 289–291.

Skavenski, A. A., & Robinson, D. A. Role of the abducens nucleus in the vestibulo-ocular reflex. *Journal of Neurophysiology,* 1973, *36,* 724–738.

Spiegel, E. The cortical centers of labyrinth. *Journal of Nervous and Mental Diseases,* 1932, *75,* 504–512.

Spiegel, E. A., Szekely, E. G., & Gildenberg, P. L. Vestibular responses in midbrain, thalamus, and basal ganglia. *Archives of Neurology* (Chicago), 1965, *12,* 258–269.

Spiegel, E. A., Szekely, E. G., Moffet, H., & Egyed, J. Cortical projection of labyrinthine impulses: Study of averaged evoked responses. In *Fourth symposium on the role of the vestibular organs in space exploration* (NASA SP-187), Washington, D.C.: National Aeronautics and Space Administration, 1970.

Steinbach, M., & Smith, D. R. Effects of strabismus surgery on spatial localization. *Society for Neuroscience Abstract,* 1976, *2,* 282.

Tasker, R. R., & Organ, L. W. Mapping of the somatosensory and auditory pathways in the upper midbrain and thalamus of man. (International Congress Series 253, Neurophysiology studied in man). *Excerpta Medica* (Amsterdam), 1971, 169–187.

Thompson, W. G., Guilford, M. O., & Hicks, L. H. Effects of caudate and cortical lesions on place and response learning in rats. *Physiological Psychology,* 1980, *8,* 473–479.

Tolman, E. C., Ritchie, B. F., & Kalish, D. Studies in spatial learning. I. Orientation and the short cut. *Journal of Experimental Psychology,* 1946, *36,* 13–24.

van Egmond, A. A. J., Groen, J. J., & Jongkees, L. B. W. The mechanisms of the semicircular canals. *Journal of Physiology* (London), 1949, *110,* 1–17.

Waespe, W., & Henn, V. Motion information in the vestibular nuclei of alert monkeys: Visual and vestibular input versus optomotor output. *Progress in Brain Research*, 1979, *50*, 683–693.

Walcott, C., Gould, J. L., & Kirschvink, J. L. Pigeons have magnets. *Science*, 1979, *205*, 1027–1028.

Walzl, E., & Mountcastle, V. Projection of vestibular nerve to cerebral cortex of the cat. *American Journal of Physiology*, 1949, *159*, 595.

Watson, J. B. Kinaesthetic and organic sensations: Their role in the reaction of the white rat to the maze. *Psychological Review Monograph Supplements*, 1907, *8*, (33).

Winterkorn, J. M. S. Deficits in spatial localization following monocular paralysis in adult cats. *Society for Neuroscience Abstracts*, 1980, *6*, 474.

Worchel, P. Space perception and orientation in the blind. *Psychological Monographs*, 1951, *65*, (332).

Worchel, P. The role of vestibular organs in spatial orientation. *Journal of Experimental Psychology*, 1952, *44*, 4–10.

Young, H. F., Greenberg, E. R., Paton, W., & Jane, J. A. A reinvestigation of cognitive maps. *Psychonomic Science*, 1967, *9*, 589–590.

Young, L. R. On visual–vestibular interaction. In *Fifth symposium on the role of the vestibular organs in space exploration* (NASA SP-314). Washington, D.C.: National Aeronautics and Space Administration, 1973.

Young, L., & Meiry, J. A revised dynamic otolith model. NASA SP-152, Washington, D.C.: National Aeronautics and Space Administration, 1968.

Zoladek, L., & Roberts, W. A. The sensory basis of spatial memory in the rat. *Animal Learning and Behavior*, 1978, *6*, 77–81.

Author Index

Subject Index

DEVELOPMENTAL PSYCHOLOGY SERIES